D1015610

MATERIALS
FOR
ELECTRICAL INSULATING
AND DIELECTRIC
FUNCTIONS

Hayden Series in Materials for Electrical and Electronic Design

ALEX. E. JAVITZ

Technical Consultant
Senior Member, IEEE, ACS, SPE

Editor-in-Chief

MATERIALS
FOR
ELECTRICAL INSULATING
AND DIELECTRIC
FUNCTIONS

HARRY L. SAUMS

Consultant, Magnet Wire

WESLEY W. PENDLETON

Manager of Electrical Section
Magnet Wire Engineering Center
Anaconda Wire & Cable Co.

HAYDEN BOOK COMPANY, INC.
Rochelle Park, New Jersey

ISBN 0-8104-5634-6
Library of Congress Catalog Card Number 73-11926

Printed in the United States of America

1   2   3   4   5   6   7   8   9   PRINTING

73 74 75 76 77 78   YEAR

FOREWORD TO SERIES

This series of integrated engineering books has been specifically planned to fill the practical needs of all those engaged in the application of materials in electrical and electronic design, regardless of their formal discipline. The purpose of this series, therefore, encompasses not only the design activities of electrical/electronics engineers but also those of chemists, physicists, metallurgists, and others associated with them. Materials are treated according to *functional* classes, rather than according to generic categories. Obviously, this approach is the more logical one, since it follows sound principles of design practice. Throughout this series, there is a continuing emphasis on the key role of materials and on the concept that all basic design functions —research, analysis, and application—draw their support and strengths from the capabilities that materials furnish.

A complete volume is devoted to each of the following major functional areas:

> Materials for Conductive and Resistive Functions
> Materials for Semiconductor Functions
> Materials for Structural and Mechanical Functions
> Materials for Magnetic Functions
> Materials for Electrical Insulating and Dielectric Functions
> Composite Materials for Combined Functions

Consistent with the special nature of each functional class, a three-dimensional method of exposition is employed in each of these volumes:

1. An examination of the particular design functions with which a given volume is concerned.
2. Appropriate tutorial discussion covering the fundamental structure and behavior of the materials needed to accomplish such functions, and the relation of these fundamental properties to ultimate performance characteristics.
3. A summary review of these materials, both commercially available and experimental, stressing typical property data; examples and analyses of applications, both conventional and advanced; and objectives and results of current research work.

In essence, each volume demonstrates how materials fulfill the design functions required of them in accordance with the interactions of their fundamental properties, processing and fabrication variables, performance and operational parameters, and environmental conditions.

Extensive references are provided to facilitate further study. In a field marked by rapid developments, such as that of materials, comprehensive reference sources are essential to help the reader keep up to date.

A seventh volume in this series, *Materials Science and Technology for Design Engineers*, is an interpretation of both classical and contemporary aspects of the field and is also planned to serve as a practical working tool. It deals with fundamental theory of materials, effects of structure on macroscopic properties, new areas of materials research and development, new techniques in reliability investigation, broad areas of critical environments, and analytical methods for determining design cost effectiveness.

The particular scope and purpose of the present volume, *Materials for Electrical Insulating and Dielectric Functions*, is described in the author's Preface that follows.

Alex. E. Javitz
Editor-in-Chief

PREFACE

This book is a concise summary of the theory and practice of electrical insulation and dielectrics, a subject obviously of paramount importance in *every* category of electrical and electronic design. The approach is functional and presents information on both fundamental theory and practical application technology. The book is intended as a practical reference not only for electrical and electronics engineers, but also for *all* disciplines concerned with insulating and dielectric applications — chemists, physicists, ceramicists, and others.

The subject matter is developed in *engineering terms* on an engineering level. Extensive use is made of graphs and data tables in order to make pertinent information available to the reader with a minimum of effort and time on his part. Of the twelve chapters in the book, the first three are respectively devoted to (1) an exposition of electrical insulating and dielectric functions and the forms in which materials for these functions are made and used; (2) an analysis of the fundamental structure of insulation and of basic dielectric phenomena; and (3) a discussion of the application significance of such phenomena. The next two chapters move into the design-vital area of functional application problems. One chapter deals with problems arising out of environmental factors; the other chapter is concerned with design problems in selected product categories.

The first five chapters summarized above lead logically into the last seven chapters that describe and discuss the various types and forms of insulation available for applications. The various categories of materials discussed are grouped by common denominators of form, composition, or other criteria, as may be most appropriate. The emphasis is on providing useful data applicable to design problems. Every effort has been made to cover the majority of commercially available materials and to include the most promising of the experimental products. Sufficient background is included on processing to indicate the relationship between processing and performance properties.

In combining theory and practice this work does not assume a prior knowledge on the part of the reader, and so he can enter at any point commensurate with his level of background and experience. References are provided to help him obtain additional information if so desired. Moreover, an Appendix provides a discussion on the influence of professional societies and other bodies on electrical insulation test procedures and classifications, checklists of design guidelines, and other useful information.

The kindness of many individuals and industrial organizations, as well as of government agencies, in providing helpful data is greatly appreciated. Specific acknowledgments appear in the text, figures, tables, and references.

This book is dedicated to Alex. E. Javitz, who has functioned not only as its editor but has also made a substantial contribution to the subject matter. Chapter 5 was prepared by him in accordance with information and data from authorities in their respective fields of practice. Moreover, throughout our writing of the manuscript and its final preparation he has been a "tower of strength" to us. We are grateful for his help and guidance.

HARRY L. SAUMS

W. W. PENDLETON

CONTENTS

CHAPTER 1 | INTRODUCTION

1.1 Definitions

The term *electrical insulation* is used to denote materials that have very high resistivity and can thus be used as isolators, that is, as separators between conductors of different potential. In essence, insulation has the basic function of confining current flow within the conductive circuit of a given device or piece of equipment, thus protecting the latter from short-circuits, current leakage, and similar malfunctions.

Insulation may be a solid, liquid, or gaseous substance. It may be a discrete (single) material, a physically cohesive composite structure (like a plastic laminate), or a suitable combination of materials (like a cable insulation system comprising several different types of materials). In a general sense, the term electrical insulation encompasses the terms *dielectrics* and *insulators*.

The latter terms, however, have their own shades of meaning. The appellation *dielectric* is frequently applied to a discrete material or a discrete class of material of very high resistivity, such as liquid dielectrics, gaseous dielectrics, film dielectrics, and plastic dielectrics. The term, it will be noted, is used with any of the three states of matter—solid, liquid, or gaseous.

In a still broader sense, the term *dielectric* may be applied to nonmetals that can be used for isolating electrodes, for storage of electric fields, and as media for control of elecron flow. An example is capacitor dielectrics.

Unfortunately, there is an overlap in the use of the terms, *electrical insulation* and *dielectrics*. Many authors use them interchangeably. Most physicists and physical chemists usually favor the term *dielectric* (as in *dielectric phenomena*). Engineers tend to use the term *electrical insulation* in reference to materials employed for insulation design functions.

The term *insulator* is actually a generic description of dual-function insulating

1

components made of solid materials—such as ceramics, porcelain, or molded plastics —that provide both isolation of conducting elements and mechanical support for the structure containing these elements. These forms of insulation are described further in Par. 1.6.

1.2   Specific Functions of Insulation

The actual functions of insulation are diverse. The following sections of this chapter will enumerate the major functions by type of component or part and by major parameters of concern. Representative materials used in each case will be briefly described.

1.3   Separation-Type Insulation

The isolation of conductors from ground and from each other is the classical example of separation type of insulation. Where low voltages exist and space is not a serious problem, cotton, which is the oldest form of wire insulation, can still be used, either in single or double servings. A composite insulation consisting of cotton over enamel also became popular after the enamels were introduced.

Paper, glass fiber, and other fibrous materials, can be used as spacing insulations. All types can be made with resin bonds. Glass fiber with an organic resin bond (phenolics, polyesters, epoxies, among others) is used in Class 130° C equipment, and with a silicone resin bond for 180° C operation. An example of the latter is glass-fiber insulation applied in traction motors where excessive overloads are encountered. Class 180° C glass-fiber insulation is also used in dry-type transformers where high-temperature operation is expected.

Other materials used for separation or spacing insulation include mixtures of glass fibers and polyester fibers, high-temperature polyamide paper, synthetic mica, asbestos, cyanurated paper, highly compressed magnesium oxide powder, and combinations of fused glass and ceramic coatings, the last named being applied for aerospace and military ultrahigh temperature use.

1.4   Barrier Insulation

As its name implies, barrier insulation is required to develop higher dielectric strength than that of the spacing type. Examples of typical materials include enameled magnet wire, film tapes, large mica flakes, varnished materials, solid insulators, and oil-impregnated pressboard. Barrier insulation may be rapidly destroyed by corona if it is organic in nature. Consequently it is desirable to inhibit corona by such protective techniques as impregnation, potting, encapsulating, varnishing, and heatsealing. Ordinary gaseous media should be avoided, but the use of electronegative gases can provide the desired protection against corona. (See Chapter 2 for a further description of these gases.)

Some inorganic barrier insulations—such as glass, mica, various ceramics, and combinations of such materials—can withstand corona (as well as radiation effects). Because of their stability, these materials are used in high-voltage applications.

Spacing and barrier types of insulation are often used in combination to achieve dielectric strength as well as mechanical strength.

1.5   Creepage Insulation

Creepage is really a special case of spacing insulation. In air a definite law exists for the interrelation of creepage distance and flashover voltage as seen in Fig. 1.1.[1] The approximate empirical equation is

$$d = V^{1.8} / 100 \qquad (1.1)$$

in which:

d = creepage distance, in inches
V = rms flashover voltage, in kV

Contamination—such as moisture, dirt, carbonized surface, and the presence of certain poor dielectric liquids—can reduce the level of flashover voltage needed for a given surface gap. The dotted line (A) in Fig. 1.1 represents such a condition for a partially carbonized path (tracking) in a phenolic-cellulose laminate used to test the thermal life of enameled magnet wire.

Creepage resistance can be improved by selecting good insulation surfaces not affected by moisture, by removing air or replacing it with other gases or by liquids, by careful maintenance, and by imposing longer distances such as by means

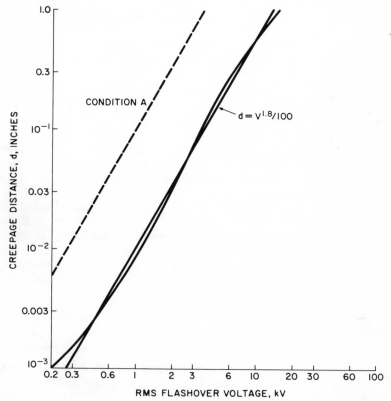

Fig. 1.1   Variation of creepage distance with a-c voltage in air[1]

of ribbing, grooving, and the like. Surfaces exposed to weather should be designed for maximum resistance to contamination. Insulator strings on overhead transmission lines are specially constructed to shed moisture, protect undersurfaces, and provide long paths for creepage. Transformer bushings are grooved for the same reason.

1.6   Support Function

Supporting insulation has the dual function of serving as a mechanical structure and a dielectric barrier. In simple cases, the mechanical support is required to provide adequate tensile or compressive strength; in some cases it is required to resist centrifugal forces.

Simple examples are stand off-insulators for supporting bus bars, and wedges to support coils in a high-voltage generator. More complicated supporting insulation may take on several functions, such as ground insulation for motor slots where both support and voltage-to-ground insulation are needed.

Naturally, materials used for such support-type insulators must exhibit high electrical resistivities as well as high mechanical strength and stability. Both electrical and mechanical integrity have to be maintained regardless of environmental conditions or the kind of electrical energy imposed during operation.

Early materials used for support functions, such as paper, cardboard, wood and hemp, have gradually given way to pressboard, resin/paper laminations, and resin/cotton cloth and glass cloth laminations. Mica in laminated and tape form has long been used in high-voltage structures. Some inorganic boards of mica and glass are available for high heat uses. Glass and ceramics are useful as insulators supporting high-voltage conductors.

Dielectrically, the supporting insulation must function as a barrier to high-voltage leakage. Rules of creepage distances must be followed, and the surface should be resistant to arcing and tracking.

On large rotating machines the insulation is built up by use of mica tape and impregnated with heat-resistant resin. After curing, the resulting ground insulation is void-free and cross-linked and offers high resistance to mechanical distortion, high-voltage corona attack, and arcing. It is also sufficiently flexible to be wound into machine stator slots.

The properties most needed in laminated insulation are high flexural strength, high impact strength, high bonding strength, low water absorption, high dielectric strength, low dielectric constant, and low dissipation factor. Other properties may also be important, such as density, hardness, modulus of elasticity, tensile strength, compressive strength, shear strength, coefficient of expansion, resistivity, and arc resistance. Special attributes of certain resin laminates are as follows:

Silicones—high-temperature stability of $200°$ C
Epoxies—low water absorption; high chemical resistance
Polyesters—good balance of electrical and mechanical properties
Melamine-formaldehyde—high resistance to arcing and tracking.

Breakdown voltage for most laminates is 60 to 80 kV on a ⅛-in. thickness in the perpendicular direction, but only 25 to 35 kV in the parallel direction. (More detailed data will appear in a later chapter.)

1.7   Enclosures

Insulation materials are used to provide protective enclosures for electrical and electronic components and devices to a constantly increasing extent. Control of corona, resistance to radiation and heat, miniaturization of high-impulse transformers, improvement in the rigidity of components, and ability to obtain reliable operation in the vacuum of space are among the many examples of the successful use of insulating materials for this function.

The techniques of encapsulation, potting, and embedment are outstanding in implementing the enclosure function for many components and small devices. A number of resin compounds are used, primarily the epoxies, but also the urethanes and silicones to an important degree. However, all elements within encapsulated and similar systems must be screened to assure that no chemical incompatibility exists. Such a condition might result in the formation of gaseous or other decomposition products, leading to contamination or degradation within the tightly sealed unit.

Another example of incompatibility would be a disparity in the coefficients of thermal expansion of the various elements of the system and the resin compound, leading to cracking and possible failure.

Even where the enclosure is a metallic container, the insulation plays a role in protecting the entry of connecting leads and as a support for the electrical structure within the enclosure. Examples of such systems are hermetically sealed motors used with refrigeration devices. Here, too, compatibility within a given system is essential.

1.8   The Protective Function

Beyond its use in protective enclosures, as described in the preceding section, insulation has a broad protective function against environmental and mechanical forces. This function is exercised through the use of such materials as varnishes, tapes, jacketing, and the like. In a dual-coated magnet wire, for example, the overcoat plays a protective role. The use of powdered resin coatings (by fluidized-bed or other techniques) in motor slot insulation can be considered as a protective process for motor windings.

The varnishing of windings has long been a means for their protection from contamination, rough handling, abrasion, and chemical attack. Taping of coils provides an abrasion shield against possible injury to the wires through outside contact. Even the bobbins on which coils are wound can be considered as protective elements as well as wire supports.

Cable jacketing provides abrasion protection needed when cables are positioned during installation. Jacketing is used in various types of cable systems, such as coaxials, to provide protective shielding for various conditions.

1.9   Thermal Dissipation

The increasing employment of elevated-temperature electrical and electronic devices has imposed on insulating materials the need to help in the thermal dissipation of heat developed within such units. Although many organic polymers are poor in thermal conductivity, certain materials, or combination of materials, are able to provide the necessary combination of high thermal conductivity and low electrical conductivity. Mica is a good example; so are glass fibers.

Organic materials can usually be improved in thermal conductivity by heavy loading with inorganic fillers, such as silica, powdered glass, powdered mica, chopped glass fibers, asbestos bits, and other additives.

Figure 1.2[2, 3] shows the properties of some of the glass- and asbestos-filled laminates with improved thermal conductivity. The best performance is shown by a phenolic-asbestos type, although a laminate with a higher-temperature resin might be expected to make the best showing. Silicone-glass laminates are also very good, having the additional thermal stability necessary to maintain integrity at temperatures in excess of 200° C.

Heat is also dissipated by means of liquid-filled apparatus. Here oil or chlorinated hydrocarbon compounds, such as Askarels, have been used for years in cooling transformers, reactors, switchgear, and the like. The mechanism is convection — the oil picks up heat which it circulates to cooler sections of the equipment.

Silicone fluids have been used for heat dissipation in installations operating at temperatures above 150° C. The thermal stability of the silicone molecule prevents the oxidation that occurs with oil. Fluorocarbon liquids vaporize with heat, thus giving a form of vapor cooling to provide thermal dissipation.

Other organic esters and ethers in liquid form, some capable of high-temperature cooling, have been formulated for cooling electrical devices. Even gases such as hydrogen are used to cool large electrical generators.

1.10   High-Voltage Insulation

Isolation of high-voltage conductors and insulation of high-voltage devices and apparatus in general are important insulation functions. In cables this function is fulfilled by a number of insulation types: oil-impregnated paper, rubber, plastics, and the like. In high voltage motors and generators this function is completed by molded mica tape, molded plastic laminates, asphaltic media, and the like. In oil transformers, oil-impregnated fiber board has been used as barrier tubes. Oil itself is an excellent high-voltage medium surpassed only by high-pressure electronegative gases and extremely high vacuum.

High-voltage media must either be little affected by corona or used in corona-free devices. A prime requisite for solids is dense internal structure not harboring internal voids. Impregnation methods that insure complete fill are also needed. Complete removal of all water is of major importance. Liquid media must be free of decomposition products, water, dirt, etc. Gases such as SF_6 should not be operated in corona conditions.

In certain application areas, such as underground power vaults for transformers, liquids must be fire-resistant. To meet this requirement, chlorinated hydrocarbons (Askarels) have found extensive application. Occasionally both gases and liquids in combinations will find important uses. Electronegative gases plus oil-fill are used in some transformers.

The application of high vacuum to particle accelerators and X-ray transformers has offered excellent high-voltage dielectric protection. The vacuum is good only at extremely low-torr values, however, and attainable only with great attention to structural sealing.

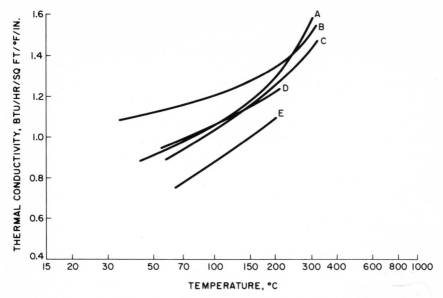

Fig. 1.2 Thermal conductivity related to temperature for several laminates (A—Phenolic resin/asbestos; B—triaryl cyanurate polyester/ glass; C—silicone resin/glass; D—polyester resin/glass; E— phenolic resin/glass)[2,3]

1.11   Feed-Through Function

By the flux-plotting techniques it can be shown that severe voltage gradients exist at the intersection of a metal tank enclosure in a piece of electrical equipment and any kind of feed-through bushing containing a lead with high voltage on it.

Design can be used to alter this condition by (a) condenser bushing techniques, (b) shaped bushings, (c) corona guard rings, (d) semiconducting stress grading, and (e) rod and torous insulators.

In condenser bushing practice a number of layers of insulation are separated by metallic foils that introduce capacitance values increasing in capacity from case to conductor. These capacitances are arranged to stress each layer equally and also provide a more uniform flux-plot diagram from case to foil edges.

With shaped bushings the ring of maximum stress occurs at the section of greatest thickness. Also some advantage can be gained by designing the bushing so that it follows the flux lines of the plot with a "series" arrangement of dielectrics, alternating a high-density insulator in areas of greater field intensity with air in the less intense areas.

Corona rings attached to high-voltage leads tend to convert the field shape to one of lower intensity at the case. Metal guard rings either grounded or at an intermediate voltage have been used to modify the field to one of less severity. Application of a semiconducting surface treatment to bushings may be effective in grading the stress and changing the field pattern. This has been found effective on ceramic bushings of small size.

The rod and torous pattern is similar to the guard ring technique except that the ring is made a part of the case adjacent to the bushing. Contact with the bushing should be by embedment to prevent overstressing the air in the gap adjacent to the bushing.

1.12   Forms of Insulation

For any insulating application, there is a form or configuration or commercially available materials able to accomplish any given task. Some materials are available in tape form, which can be used to build up insulation layer by layer. Still other materials are in the form of liquid resins or resin systems used for potting or embedment of components or circuit elements. In other cases, insulation is available in prefabricated form, ready for use without alteration. Certain types of wire and cable, in fact, represent a form of a prefabricated insulation system.

Insulation materials, as already noted, are available in solid, liquid, and gaseous form. They are available in rigid and semirigid laminates, flexible form (as in film, fabric and tape), transformer oils, encapsulating compounds, and composites of various types, among other materials.

Insulation systems often utilize several states of matter such as liquid-solid combinations (in oil-filled transformers), gas-solid (in pressurized high-voltage devices), and gas-liquid-solid (in high-voltage apparatus). One prime requirement for such systems is that all elements must remain compatible even after long-time aging. It is also necessary that no one element be overstressed.

Insulations in composite form have assumed rapidly growing importance. In particular, they serve multifunctional purposes if insulations are required to fill two or more requirements simultaneously.

The combination of mechanical and electrical stress is often seen in such devices as motor rotors, electrical contactor coils, heavy-duty traction motors, and overhead transmission-line insulators. In increasing numbers, applications call on the use of materials able to fulfill exacting thermal and electrical functions. Dry-type transformers furnish an example.

High-voltage cable represents a combination of electrical and mechanical stress, especially in riser sections. Potted coils represent a combination of functional requirements calling for environmental protection, support, and heat dissipation. High-voltage stress is still another parameter in potted coil insulation requirements. In general, and in more instances than is commonly realized, electrical insulation has to contend with radiation effects, shock, vibration, abrasion, and many other conditions.

The actual forms of composite insulating materials are manifold. Magnet wire, as well as other wire and cable insulating systems, are excellent examples of composite forms. Many different types of films, enamels, tapes, fibers, and other materials are combined.

In high-voltage motor and generator ground insulation, mica is often used with other materials, including paper, glass, and resins, to provide a composite insulation admirably suited for protection against high gradients.

All oil-filled capacitors, transformers, and cables are examples of composites functioning as separation insulation, barrier insulation, and liquid coolants.

Laminates, as already noted, are also examples of composite insulation designed to serve as support and barrier insulations.

1.13   Flexibility

As indicated above, flexibility can be an attribute of form in certain insulating materials. For example, it is necessary that insulation be flexible to perform properly when applied in leads, cables, magnet wire, building wire, connectors, and the like. Wire and cables in these categories have to be pulled, wound, shaped, pressed, and, in some cases, even hammered into place. Cables of all types are not only wound on reels and dereeled but must also be pulled through ducts in underground installations and pulled over pulleys in overhead systems without suffering rupture and without forming internal voids.

In the specific case of magnet wire, extremely good adhesion as well as flexibility are required. The insulation must not crack when rapidly elongated to the breaking point and must then survive further bending tests by being bent on a mandrel of small diameter. Some types of magnet wire coils, such as the yoke coils in TV tubes, must be wound and bent to an extremely severe angle, then heated to obtain bonding of an adhesive applied to the wire, and finally must be molded under considerable pressure until set.

Harness wire must maintain complete isolation of its parts despite the strains of being bent and shaped in many directions around sharp corners.

1.14   Porosity

Porosity is an important consideration when liquid or gaseous media are used with separation insulation. Sufficient porosity must exist to assure complete impregnation and free flow of these media. However, design problems arise. For example, the dielectric strength of oil-impregnated paper rises linearly with the density of the insulation, reaching approximately 2,200 volts per mil at a paper density of 0.6 grams per cm^3 and 3,300 volts per mil at 0.8 grams per cm^3, in each case for 3-mil kraft paper. In oil-paper systems, unfortunately, the dielectric constant and power factor (dissipation factor) also both increase linearly with insulation density. This factor, combined with the square-law increase of loss with voltage makes extra-high-voltage cable design extremely difficult and prohibitive with normal kraft papers and oil. Low-loss-factor materials are needed for such designs.

1.15   Relation to Molecular Structure

A further examination of the functions, forms, and characteristics of insulating materials requires prior understanding of their molecular structure and its relationship to dielectric phenomena and properties. These subjects will be covered in the next chapter.

REFERENCES

(1) Moses, G.L., *Electrical Insulation,* McGraw-Hill, 1951.
(2) Stansbury, J.G., "Plastic Laminate Heat Shields," *Modern Plastics,* vol. 34, 1957, p. 188.
(3) Katz, I. and Goldberg, J., "Dielectric Properties of Reinforced Plastics at Elevated Temperatures," *Electrical Manufacturing,* Dec. 1958, p. 72.

CHAPTER 2 | INSULATION STRUCTURE AND PROPERTIES

2.1   Molecular Structure and Polarization

Atoms are generally considered to be composed of very small, but dense, positively charged nucleii surrounded by sparse clouds of negatively charged electrons.[1, 2] Only high-energy fields are able to remove the electrons; low-magnitude fields can only distort the electron clouds. Molecules contain two or more of these atoms in such close proximity that the bonded nucleii share electron clouds. Molecules range in size from two to many thousands of nucleii in strands or mesh-like structures such as exist in polymers much used in insulating materials.

In gases, the single molecules are free to move in any three-dimensional pattern governed by collision with other molecules and container walls. The molecules react at the slightest disturbance of an electric field but show absorption of electric energy only at resonance nodes. For gases, the nodes occur at extremely short wavelengths in the optical range and appear as spectral lines.

Liquids and solids are examples of increasing condensation with inherently greatly decreased intermolecular spacing that has marked effects on polarization, conduction, and dielectric breakdown. Polarization (the formation of electric-field-distorted molecules, dipoles, and the like) is treated in four recognizable effects: electronic, atomic, dipolar, and interfacial. The first three are bound-charge polarization effects; the fourth effect is due to charge carriers.

Electronic polarization occurs in the gas phase as the result of the distortion of the electron cloud on each atom by an electric field. The distorted electron cloud forms a dipole with the remaining atomic ion induced by the electric field. This dipole absorbs energy from the electric field as it reorients with each half-cycle of the alternating-current field.

Atomic polarization occurs in molecules that contain atoms of unequal binding forces. Under the electric field forces the atoms polarize to form an atomic dipole. Here, too, the dipole absorbs energy from the alternating field.

Dipolar polarization takes place in molecules that form permanent dipoles caused by an asymmetrical charge distribution between unlike partners in the molecules. These molecules are oriented in the electric field as if they were submicroscopic magnets and absorb energy at every alternating half-cycle.

Interfacial polarization, as previously noted, is due to charge carriers and not to bound-charge effects. These carriers (both electronic and ionic) are usually found in solid or liquid dielectrics. They are urged to travel in one direction within the dielectric by a particular half-cycle of the field and are trapped at an interface formed by a lattice impurity, defect, or the like, after having traveled for a relatively considerable distance. The distance is related to the size of the carriers. The entrapment of the charge carriers prevents their being discharged at the electrode, thus giving rise to a space charge within the dielectric. The presence of this space charge causes distortion of the electric field (macroscopic field) and a consequent increase in the dielectric constant and dielectric loss.

Total polarization is thus the sum of four additive components as described above, namely: electronic, atomic, dipolar, and interfacial polarization. Electronic and atomic polarization occur mainly in the optical regions—both visual and infrared. Vibration of atoms and segmental chain movements, such as rotation, are also found in the infra-red region and in the microwave region. Dipolar absorption takes place at power frequencies for some materials but is often observed at higher frequencies in the radio-frequency range and above. Interfacial polarization is essentially a low-frequency phenomenon that appears at 60 Hz and below. The occurrence of this phenomenon in the low-frequency region is due to the time required for ions to travel to and from interfaces.

Additional information on molecular structure and its relation to dielectric phenomena will be developed in subsequent sections of this chapter.

2.2   Insulators Versus Conductors

The classical description of a *metal* proposed by Lorentz[3] is a crystalline arrangement of positively charged metallic ions with free electrons moving about in the interstices. This representation of a rigid structure permeated with a conducting electron gas has been used effectively to explain the electrical, thermal, and mechanical properties of metals as well as their radiation resistance.[4]

On the other hand, the classical representation of an *insulator* is that of a medium of molecular units in which the electrons are tightly bound and are not free to travel within the material.

The band theory for solids suggests that there are three main energy levels or bands in which electrons can exist: (1) the conduction band of high-energy level, (2) the forbidden band of varying widths, and (3) the valence band of low-lying energy levels. Figure 2.1 presents a summary of the band positions.

Electrons are tightly bound in the valence band and are free to move in the conduction band. The size of the forbidden band and the possible overlap of bands, together with the partial filling of the conduction band, have much to do in determining the characteristics of materials.

Small forbidden bands (small in comparison with the temperature-energy

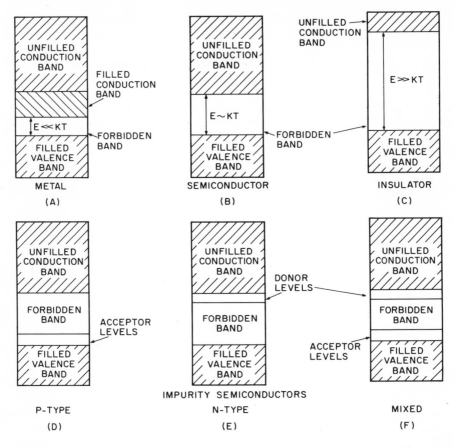

Fig. 2.1   Band theory for conduction in metals, semiconductors, and
insulators (After A. H. Wilson, *Proceedings of the Royal Society,*
A 133,458, 1931)

factor, KT), overlapping bands, and partially filled conduction bands are all indicative of metals. [See Fig.2.1(A).] On the other hand, large forbidden bands (very much greater than the energy factor, KT), unfilled conduction bands, and no overlapping of bands are indicative of insulators. [See Fig. 2.1 (C).] It may be concluded that the forbidden zone in insulators is too wide for easy bridging by thermal excitation. Impure insulators contain ions that carry conduction even if the electrons are tightly bound. Ions may also free electrons. A rising temperature will narrow the gap and so promote escape of electrons to the conduction zone. Radiation may also promote escape of bound electrons to the conduction zone.

2.3   Semiconductors

Semiconductors are a class of materials that occupy the region between metals and insulators. By again applying the band theory, we can represent semiconductors by a conduction band that is unfilled with electrons but is separated from the valence band by a forbidden gap approximately equal to the KT energy, as shown in Fig. 2.1 (B). The conduction band is accessible by thermal excitation, and an intrinsic semiconductor is obtained.

The n-type semiconductor is produced if only the electrons in the conduction band carry the current, as in Fig. 2.1(E). A p-type semiconductor results if the positive holes in the valence band carry the current, as in Fig. 2.1 (D). Mixed semiconductors carry current in both bands, as in Fig. 2.1 (F).

Impurities or lattice disturbances within some insulators tend to create intermediate localized energy levels and may lead to an impurity semiconductor. Impurities of an electronegative nature are added to potentially semiconducting materials to create electron acceptors in the valence band region of energy levels, as in Fig. 2.1 (D). Hole conduction thus results from the removal of the electrons as they emerge from the valence band and from the creation of vacancies in this band. This is known as p-type semiconduction [Fig. 2.1 (D)].

On the other hand, electropositive impurities form electron donor levels near the conduction band, supplying this band with electrons and establishing the n-type semiconductor [Fig. 2.1 (E)].

Both electronegative and electropositive impurities may be added, and the band picture is then that of an acceptor level near the valence band for producing p-type conduction and a donor level near the conduction band for producing n-type conduction [Fig. 2.1 (F)]. This type of conduction is typical of materials known as mixed semiconductors.

2.4   Conduction in Dielectrics

Direct-current conduction in dielectrics depends on (1) the number and kind of impurity ions present; (2) the mobility of these ions in the lattice governed by the nature of the dielectric, the temperature, and the environment; (3) the voltage gradient; and (4) the geometry of the dielectric and electrodes. Overriding influences that can add to d-c conduction occur when (1) the temperature reaches a level at which electrons can move freely from the valence band into the conduction band and thermal breakdown is imminent, or (2) the voltage reaches a level at which electrons are freed from the valence band and/or are interjected into the insulation from the electrodes so that disruptive breakdown becomes imminent. In the latter case, copious numbers of electrons stream through the dielectric, causing an avalanche build-up of electrons, carbonization of the dielectric, and molecular disintegration. Also active in controlling direct-current conduction is the space charge developed in the vicinity of the electrodes

It can be shown that in a vacuum the current density is limited by the Child-Langmuir Law[5]:

$$J \; = \; CV_0{}^{3/2} \tag{2.1}$$

in which:

J = current density
C = constant
V_0 = applied voltage

This relationship is unaffected by electrode geometry but is drastically reduced by the drag on the electron produced by liquid and solid media.

The dielectric absorption effects and space-charge build-up impose a conduction characteristic that is a function of time. Figure 2.2 shows how this conduction[6] develops in liquid dielectrics. Oscillographic studies have proved the build-up of a space charge in the liquid requiring only 3 to 4 milliseconds, which is followed by a plateau of conduction while existing ions flow to the electrode. An electrolytic cleaning sets in after 103 seconds, and in 10 minutes most of the conduction-producing ions have been swept out and the current reaches a very low level.

Convention has set the short-time d-c insulation resistance reading at 1 minute. This may be anywhere from two to more than ten times the duration of the "long-time" current. Solid dielectrics may not be polarized as rapidly as liquids, and the conduction current may be changing rapidly even after 1 minute.

Figure 2.3 reveals the two time effects in d-c and a-c conductivity measurements.[7] To illustrate the effects, pure Decalin (decahydro-naphthalene, $C_{10}H_{18}$) was bombarded by corona for several hours until the a-c conductivity attained the high and steady value of 9.0×10^{-12} mhos/cm, as shown by the horizontal line. After 60 minutes, the a-c source was replaced by direct current. Over a 30-minute period the application of d-c effected a cleanup of the Decalin and a reduction of d-c conductivity from 0.08×10^{-12} to approximately 0.04 mhos/cm, as shown. At the 90-minute point, a-c was reapplied, and in a sweeping curve a-c conductivity returned to its 9×10^{-12} mhos/cm level in approximately 150 minutes (the 240-minute test point). The d-c "clean-up" was faster, by 5 to 1, than the a-c recovery.

Moisture produces a more rapid polarization in solid dielectrics as well as a pronounced lowering of resistivity (ρ), both d-c and a-c. Rising temperature lowers resistivity in an exponential function, as shown in Fig. 2.4 for oils.[8] The characteristic effect is a straight line from $-75°$ to $-5°$ or $0°$ C and then a second straight

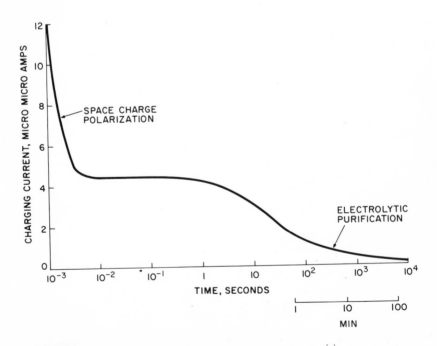

Fig. 2.2   D-C conduction in liquid dielectrics[6]

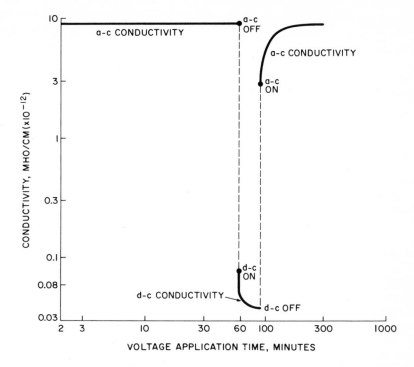

Fig. 2.3   Time related to conductivity for corona-bombarded Decalin a-c and d-c effects (After J. Sticher and J. D. Piper, *Ind. Eng. Chem.*, 33,1567, 1941)

line from this temperature range to $125\,^\circ$ C. The viscosity characteristic (η) is similar, breaking at $+50\,^\circ$ C

The form of the resistivity function is shown in the following equation:

$$log\ \rho\ =\ log\ \rho_0\ -\ (E\,/\,RT) \tag{2.2}$$

in which:

$log\ \rho_0$ = intercept at $T = \infty$
R = gas constant $/\ log\ e$ = 4.575
T = absolute temperature, $^\circ K$
E = activation energy, cal $/$ gm-mole

For the data in Fig. 2.4, the low-temperature portion (in the range of $-80\,^\circ$ to $-5\,^\circ$C) of ρ_2 is determined by the equation,

$$log\ \rho\ =\ 5.43\ +\ (2285\,/\,T) \tag{2.3}$$

and E equals 10,450 cal/gram mole. The high-temperature line (in the range of $-5\,^\circ$ to $+125\,^\circ$C) for ρ_2 follows the equation,

$$log\ \rho\ =\ 6.50\ -\ (2000\,/\,T) \tag{2.4}$$

and E equals 9,150 cal/gram mole.

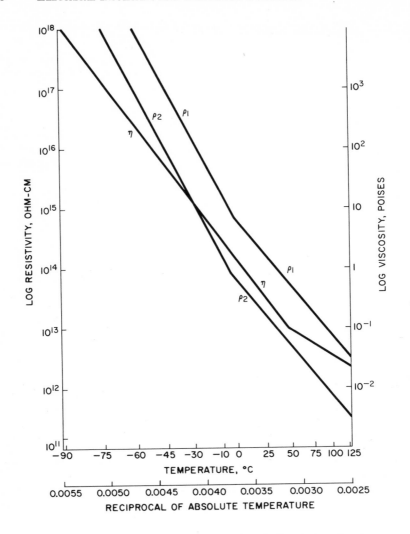

Fig. 2.4 Effect of temperature on resistivity of oil and viscosity changes with temperature[8]

The equations for ρ_1 have the same slopes and therefore the same activation energies, but the intercepts are each larger by one interval since the unaged oil was 10 times higher in resistivity than the aged oil.

From these values of activation energy it is seen that it is relatively easy to free electrons and ions in a liquid such as oil by increasing temperature. It is also shown that viscosity changes much as resistivity does.

A-C resistivity is equal to d-c resistivity only at temperatures above the point at which ions are free to migrate. When conduction is largely ionic, both conductivities are equal, as seen in Fig. 2.5. Below this temperature (which corresponds to the glass point in solid polymer resins[1], the a-c resistivity (due to polarization) is lower than the d-c.

Conduction in a-c depends on the frequency, the dielectric constant, and the dissipation factor (to be discussed later). The relationship is given by the equation,

$$\lambda = 0.556 \times 10^{-12} \, fk' \, \tan \delta \qquad (2.5)$$

in which:

λ = conductivity, in mhos / cm

[1] The glass point of a polymer is characterized by a sudden change in the plotted line for rigidity versus temperature. It represents a transition temperature at which the resin changes from a rigid solid to a rubber-like elastomer.

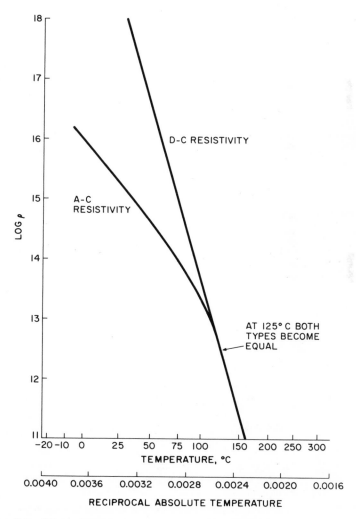

Fig. 2.5   Effect of temperature on a-c and d-c resistivity of asphalt-bonded mica insulation

f = frequency, in Hz
k' = dielectric constant
$tan\ \delta$ = dissipation factor

A-C resistivity is the reciprocal of Eq. 2.5.

There has been extensive inquiry into the proposition that there is a relationship between dielectric constant and d-c resistivity. Gemant has confirmed this tendency for correlation in a few isolated cases over a wide range of resistivity. (See pages 3 and 4 of Ref. 8.) However, only on an expansive scale is there any noticeable relationship of this kind. Figure 2.6 shows the size of the scale on which the relationship seems to exist. When insulators, semiconductors, and conductors (both metal and nonmetal) are examined and log resistivity extends from +20 (or more) to −8 (or lower), there is a continuing combination of linear relationships expressed by two main straight lines for good conductors as

$$log\ \rho\ +\ 3\ log\ k'\ =\ 5.7 \qquad (2.6)$$

and for poor conductors, semiconductors, and insulators as

$$log\ \rho\ +\ 12.5\ log\ k'\ =\ 22 \qquad (2.7)$$

The ferroelectrics are notable for their nonconformance to these relationships. Moreover, the good and poor insulators do not particularly conform in detail, as witnessed by the "shot gun" effect in that area. Ceramics tend to reside in the area to the right of the line of Eq. 2.7, and organic insulations tend to occupy the area to the left. Tap water conforms, but distilled water does not.

The superconductors listed have resistivities generally regarded as zero. However, silver and copper at 15° K both have measurable resistivities and are properly shown.

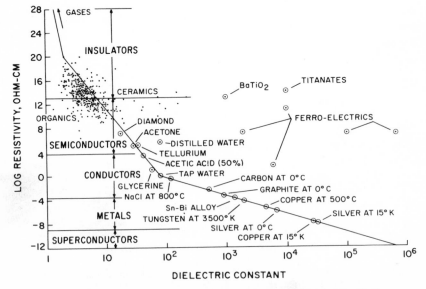

Fig. 2.6 Relationship between resistivity and dielectric constant

Gases show dielectric constants close to unity and resistivities of unknown but great magnitude. Ionized gas drops into the semiconducting region.

2.5 Dielectric Constant

In treating the a-c dielectric properties of insulating materials it is necessary to think in terms of loss currents and displacement currents. A displacement current is due to a capacitive charge formed around the insulation under measurement. This current leads the voltage by 90° and is plotted at right angles to the voltage vector. The amplitude of this current vector is given by the product of the susceptance, B, in mhos, and the voltage, V:

$$I_c = BV \tag{2.8}$$

$$B = \omega C = \omega K' C_0 \tag{2.9}$$

in which:

ω = $2\pi f$ radians / sec
f = frequency, in Hz
K' = relative dielectric constant
C_0 = geometric capacitance, in farads

K', the relative dielectric constant, is much confused in the literature with the real permittivity, ϵ', also called dielectric constant. In this book, the relative values will be represented by K' (relative dielectric constant) and K^* (complex relative dielectric constant), and the real values by ϵ' (real dielectric constant) and ϵ^* (complex real dielectric constant). The relative values will be referred to as "dielectric constant" and the real values as "permittivity."

K' is actually the ratio of the capacitance of the dielectric sample to the capacitance of the same geometric arrangement of electrodes but with the dielectric replaced by a vacuum. ϵ' is the relative value, K', multiplied by the permittivity of free space, ϵ_0 (8.8543 $\times$ 10^{-14} farads per cm). Permittivity of free space is related to the permeability of free space by the following equation:

$$\varepsilon_0 = 1 / \mu_0 C^2 \tag{2.10}$$

in which:

μ_0 = permeability of free space
C = speed of light

If C is 2.9979 $\times$ 10^8 meters/sec and $\mu_0 = 4\pi \times 10^{-7}$ newtons per ampere^2, then ϵ_0 becomes 8.8543 $\times 10^{-14}$ farads per cm.

Both the relative dielectric constant, K', and real permittivity, ϵ', become complex values when both the loss and displacement currents are considered as a vector system. Thus the complex real permittivity is given by the equation,

$$\varepsilon^* = \varepsilon' + j\,\varepsilon'' \tag{2.11}$$

in which:

ε^* = complex real permittivity
ε' = real permittivity
ε'' = real loss factor

and the complex relative dielectric constant by the equation,

$$K^* = K' + jK''$$ (2.12)

in which:

K^* = complex relative dielectric constant
K' = relative dielectric constant
K'' = relative loss factor

Most engineers work with the relative values of K' and K'' together with practical design factors. However, theoretical physicists continue to form expressions involving ϵ' and ϵ'' since they are based on units that allow for their ready interpretation in dielectric theory.

To dispel any further misinterpretation of the symbols and their evaluation, Tables 2.1, 2.2, and 2.3 were created. Table 2.1 lists 17 separate dielectric entities and shows how they are interrelated. In addition to the six major units already discussed, there are capacitive, conductive, and resistive units, and also geometric units. The "cell factor" unit is the equal of geometric conductance; it is used to convert resistance readings to resistivity. For example, a cell containing liquid dielectric can be used in such work.

The geometric capacitance is used to determine dielectric constant from a capacitance reading. Table 2.2 shows how these parameters are used. Dielectric constant (relative value) is also known as specific inductive capacity (S.I.C.).

Dielectric constant is affected by many conditions, as enumerated below:

Increasing temperature raises the dielectric constant of a material because of the freeing of ionic carriers and the resulting polarization.

Increasing moisture content raises the dielectric constant, because water itself has a high dielectric constant (80) compared to that of most dielectric materials in a normal (dry) condition.

Chemical changes brought about by curing processes, vulcanization, polymerization, catalysis, and the like, can cause marked changes in dielectric constant. The neutralization of acidic ingredients in a monomeric material by polymerization can lead to a large reduction in dielectric constant. The forming of a cross-linked polymer can also lower the dielectric constant of some materials.

Contamination by surface or internal impurities usually brings about an increase in apparent dielectric constant if the contaminant is conductive in nature.

Impregnation, in which air is replaced by a dielectric liquid — as for example, the impregnation of cellulose with oil — will increase the dielectric constant because the dielectric constant of air is practically unity whereas that of oil is 2.0 to 2.4. A similar effect is obtained when a resin is used as an impregnant, as in the potting of a coil.

Aging of solid dielectrics has several phases. Initially the dielectric constant may decline as curing proceeds. Then a reversal comes as the aging process begins to dehydrogenate and oxidize the dielectric. This process is speeded up as carbonization appears, and the dielectric constant rises rapidly.

Radiation (nuclear), as discussed elsewhere in this series,[2] has a marked

---

[2] The nature and effects of nuclear radiation are discussed in *Materials Science and Technology for Design Engineers,* Alex. E. Javitz, ed., Section II, "Critical Environments and Phenomena," another volume in the Hayden Book Company Materials Series.

Table 2.1 Dielectric Parameters Conversion

Obtain ↓ \ Given →	ϵ^*, $farad/cm$ (a)	K^*, (no unit) (b)	ϵ', ($farad/cm$) (c)	K', (no unit) (d)	ϵ'', $farad/cm$ (e)	K'', (no unit) (f)
Complex real permittivity, ϵ^*	a	bm	$\sqrt{c^2 + e^2}$	$\sqrt{d^2m^2 + e^2}$	$\sqrt{c^2 + e^2}$	$\sqrt{c^2 + f^2m^2}$
Complex relative permittivity, K^*	a/m	b	$\sqrt{(c^2/m^2) + f^2}$	$\sqrt{d^2 + f^2}$	$\sqrt{d^2 + (\rho^2/m^2)}$	$\sqrt{d^2 + f^2}$
Real permittivity, ϵ'	$\sqrt{a^2 - e^2}$	$m\sqrt{b^2 - f^2}$	c	dm	e/h	fm/h
Relative permittivity, K'	$1/m\sqrt{a^2 - e^2}$	$\sqrt{b^2 - f^2}$	c/m	d	e/hm	f/h
Real loss factor, ϵ''	$\sqrt{a^2 - c^2}$	$m\sqrt{b^2 - d^2}$	ch	dhm	e	fm
Relative loss factor, K''	$1/m\sqrt{a^2 - c^2}$	$\sqrt{b^2 - d^2}$	ch/m	dh	e/m	f
Capacitance, C	$n/m\sqrt{a^2 - e^2}$	$n\sqrt{b^2 - f^2}$	cq	dn	eq/h	fn/h
Dissipation factor, $\tan\delta$	$1/c\sqrt{a^2 - c^2}$	$1/d\sqrt{b^2 - d^2}$	e/c	f/d	e/c	f/d
Parallel resistance, R_p	$1/\omega q\sqrt{a^2 - c^2}$	$1/\omega n\sqrt{b^2 - d^2}$	$1/\omega chq$	$1/\omega dhn$	$m/\omega ne$	$1/\omega nf$
Parallel resistivity, ρ_p	$1/\omega\sqrt{a^2 - c^2}$	$1/\omega m\sqrt{b^2 - d^2}$	$1/\omega ch$	$1/\omega dhm$	$1/\omega e$	$1/\omega mf$
Parallel conductivity, λ_p	$\omega\sqrt{a^2 - c^2}$	$\omega m\sqrt{b^2 - d^2}$	ωch	ωdhm	ωe	ωmf
Parallel conductance, G_p	$\omega q\sqrt{a^2 - c^2}$	$\omega n\sqrt{b^2 - d^2}$	ωchq	ωdhn	ωqe	ωnf
Permittivity of free space, ϵ_0	$1/d\sqrt{a^2 - e^2}$	$c/\sqrt{b^2 - f^2}$	c/d	c/d	e/f	e/f
Geometric capacitance, C_0	$(g/cd)\sqrt{a^2 - e^2}$	$g/\sqrt{b^2 - f^2}$	gm/c	g/d	$m/\omega ie$	$1/\omega if$
Geometric conductance, G_0	$L/\omega\sqrt{a^2 - c^2}$	$(\omega h/k)\sqrt{b^2 - d^2}$	dn/c	dn/c	$1/\omega ie$	$1/\omega ifm$
Geometric resistance, R_0	$\omega i\sqrt{a^2 - c^2}$	$\omega im\sqrt{b^2 - d^2}$	c/dn	c/dn	ωie	ωifm
Cell factor, C.F.	$(g/mcd)\sqrt{a^2 - e^2}$	$g/m\sqrt{b^2 - f^2}$	dn/c	g/dm	$1/\omega ie$	$1/\omega ifm$

K' = Specific Inductive Capacity (S.I.C.)
= Dielectric Constant (D.C.)

$\omega = 2\pi f$ radians/sec (f = Hz)
$\epsilon_0 = 8.8543 \times 10^{-14}$ farad/cm
$1/\epsilon_0 = 1.131 \times 10^{13}$ cm/farad

Table 2.1 Dielectric Parameters Conversion (Cont'd)

Obtain ↓ ＼ Given →	C, farad	$\tan\delta$, (per unit)	R_p, ohms	ρ_p, ohm-cm	λ_p, mho/cm	G_p, mho
	g	h	i	j	k	L
Complex real permittivity, ϵ^*	$\sqrt{g^2 q^2 + e^2}$	$c\sqrt{1+h^2}$	$\sqrt{c^2 + \omega^2 i^2 q^2}$	$\sqrt{c^2 + j^2\omega^2}$	$\sqrt{c^2 + (k^2/\omega^2)}$	$\sqrt{c^2 + (L^2/\omega^2 q^2)}$
Complex relative permittivity, K^*	$\sqrt{(g^2/n^2) + f^2}$	$d\sqrt{1+h^2}$	$\sqrt{d^2 + \omega^2 i^2 n^2}$	$\sqrt{d^2 + j^2 m^2\omega^2}$	$\sqrt{d^2 + (k^2/\omega^2 m^2)}$	$\sqrt{d^2 + (L^2/\omega^2 n^2)}$
Real permittivity, ϵ'	g/q	e/h	$1/\omega q i h$	$1/\omega j h$	$k/\omega h$	$L/\omega q h$
Relative permittivity, K'	g/n	e/hm	$1/\omega n i h$	$1/\omega m j h$	$k/\omega m h$	$L/\omega q h m$
Real loss factor, ϵ''	gh/q	ch	$1/\omega q i$	$1/\omega j$	k/ω	$L/\omega q$
Relative loss factor, K''	gh/n	dh	$1/\omega n i$	$1/\omega m j$	$k/\omega m$	$L/\omega q m$
Capacitance, C	g	fn/h	$1/\omega i h$	$q/\omega j h$	$qk/\omega h$	$L/\omega h$
Dissipation factor, $\tan\delta$	fn/g	h	$1/\omega i g$	$1/\omega j c$	$qk/\omega g$	$L/\omega g$
Parallel resistance, R_p	$1/\omega g h$	$1/\omega q c h$	i	pj	$1/qk$	$1/L$
Parallel resistivity, ρ	$q/\omega g h$	$1/\omega c h$	iq	j	$1/k$	q/L
Parallel conductivity, λ_p	$\omega g h/q$	$\omega c h$	$1/iq$	$1/j$	k	L/q
Parallel conductance, G_p	$\omega g h$	$\omega q c h$	$1/i$	$1/pj$	qk	L
Permittivity of free space, ϵ_0	cn/g	e/dh	$1/\omega q f i$	$1/\omega f j$	$k/\omega f$	$L/\omega q f$
Geometric capacitance, C_0	g/d	$1/\omega i d h$	$1/\omega f i$	$\omega g j m h$	$\omega g m h/k$	$L/\omega f$
Geometric conductance, G_0	$\omega g h/K$	$1/\omega i c h$	$1/\omega i e$	$\omega f n j$	$\omega i g/k$	$L/\omega f m$
Geometric resistance, R_0	$K/\omega g h$	$\omega i c h$	$\omega i e$	$1/\omega f n j$	$k/\omega i g$	$\omega f m/L$
Cell factor, C.F.	$\omega g h/K$	$1/\omega i c h$	$1/\omega i e$	$\omega f n j$	$\omega i g/k$	$L/\omega f m$

Table 2.1 Dielectric Parameters Conversion (Cont'd)

Obtain ↓ \ Given →	ε_0, farad/cm	C_0, farad	G_0, cm	R_0, cm^{-1}	C.F., cm
	m	n	o	p	q
Complex real permittivity, ε^*	$\sqrt{m^2 d^2 + e^2}$	$\sqrt{(m^2 q^2/n^2) + e^2}$	$\sqrt{c^2 + (L^2/\omega^2 o^2)}$	$\sqrt{c^2 + (p^2/\omega^2 i^2)}$	$\sqrt{(g^2/q^2) + e^2}$
Complex relative permittivity, K^*	$\sqrt{(c^2/m^2) + f^2}$	$\sqrt{(q^2/n^2) + f^2}$	$\sqrt{d^2 + (L^2/\omega^2 o^2 m^2)}$	$\sqrt{d^2 + (p^2/\omega^2 i^2 m^2)}$	$\sqrt{(g^2/m^2 q^2) + e^2}$
Real permittivity, ε'	dm	mg/n	dn/o	dnp	dn/q
Relative permittivity, K'	c/m	g/n	oc/n	c/pn	cq/n
Real loss factor, ε''	fm	m/ωin	1/ωio	p/ωi	1/ωig
Relative loss factor, K''	e/m	1/ωin	1/ωimo	p/ωim	1/ωimq
Capacitance, C	cn/m	nc/m	ko/ωh	k/ωph	kq/ωh
Dissipation factor, tan δ	e/dm	en/gm	1/ωioc	p/ωic	1/ωiqc
Parallel resistance, R_p	m/ωen	1/ωnf	1/ωoe	p/ωe	1/ωqe
Parallel resistivity, ρ_p	m/ωenq	q/ωnf	1/ωfn	1/ωpfn	q/ωfn
Parallel conductivity, λ_p	ωenq/m	ωnf/q	ωfn/o	ωpfn	ωfn/q
Parallel conductance, G_p	ωen/m	ωnf	ωoe	ωe/p	ωqe
Permittivity of free space, ε_0	m	cn/g	n/o	np	n/q
Geometric capacitance, C_0	gm/c	n	mo	m/p	mq
Geometric conductance, G_0	n/m	n/m	o	1/p	q
Geometric resistance, R_0	m/n	m/n	1/o	p	1/q
Cell factor, C.F.	n/m	n/m	o	1/p	q

Table 2.2 Use of Dielectric Parameters

Characteristic	Formula	Units
Loss current, I_g	$\omega K'' C_0 E$	amperes
Charging current, I_c	$\omega K' C_0 E$	amperes
Total current, I	$\omega C_0 E \sqrt{(K')^2 + (K'')^2}$	amperes
Specific current, i_s	$\omega K' \sqrt{\tan^2 \delta + 1}$	amperes/volt-farad
Power loss, P	$\omega K' C_0 E^2 \tan\delta$	watts
Specific loss, P_s	$\omega K' \tan\delta$	mhos/farad
Power factor, $\cos\theta$	$K'' / \sqrt{(K')^2 + (K'')^2}$	none
$Cos\ \theta$ when $K'' \ll K'$	$K''/K' = \tan\delta$	none

Table 2.3 Geometric Factors

Geometric factor	Formula		Units
	Parallel plates	Concentric cylinders	
Geometric capacitance, C_0	$\epsilon_0 A/S$	$2\pi ML\epsilon_0 / ln(D/d)$	farads
Geometric conductance, G_0	A/S	$2\pi ML / ln\ (D/d)$	cm
Geometric resistance, R_0	S/A	$ln(D/d)/2\pi ML$	cm^{-1}
Cell factor, C.F.	A/S	$2\pi ML / ln(D/d)$	cm

A = area

S = spacing

$\epsilon_0 = 8.854 \times 10^{-14}$ farad/cm

$M = 1/ln10 = 0.43429$

$2\pi M = 2.729$

$2\pi M\epsilon_0 = 2.416 \times 10^{-13}$ farad/cm

$1/(2\pi M) = 0.3664$

ln = Naperian   logarithm

effect on dielectric constant since conductive components are generated by the penetration of the rays. Radiation usually increases the dielectric constant of polymers as damage builds up. If certain polymers such as polyethylene are subjected to mild radiation, there will be no significant permanent changes in dielectric constant.

Pressure has a consolidation effect that can increase the dielectric constant of composite insulations. But pressure also retards internal ionization, which, if present, is a factor in the apparent increase in dielectric constant with voltage. Thus pressure may reduce dielectric constant at high voltages.

Dispersion is a term used to denote the variation of dielectric constant that occurs at the frequency corresponding to a resonant peak in loss parameters. This phenomenon will be discussed further in Par. 2.8.

Ionic content of an insulation affects the dielectric constant when conditions permit the travel of such charged particles as ions. Dielectrics having a greater ionic content will show a greater rise in dielectric constant with rising temperature.

Corona effects on dielectric constant arise either from internal ionization at void sites or from external ionization of air in the vicinity of the electrode-insulation system. The higher the voltage, the more intense the corona and the higher the dielectric constant.

Molecular structure plays an important role in dielectric levels as dispersion regions are reached.[3]

---

[3]   Certain molecules, for example, contain nonrigid segments of chains that can rotate or vibrate under electrical stress. If the frequency is adjusted to the natural frequency of these segments, maximum energy absorption occurs, and at these maxima the dielectric constant moves rapidly with frequency to a new level, usually inversely (lower) with rising frequency. Moreover, molecules containing polar radicals that form dipoles under electrical stress also exhibit shifts in dielectric constant over wide frequency changes.

2.6   Macroscopic Permittivity and Polarizability

The macroscopically measured relative dielectric constant K' is related to the polarization vector, P, and the applied field, E, by the following equation:[7]

$$P = (K' - 1)\, \varepsilon_0 E \tag{2.13}$$

in which:

ε_0 = permittivity of free space
K' = relative dielectric constant

In molecular physics, the polarization vector P is expressed by:

$$P = N\alpha E' \tag{2.14}$$

in which:

N = number of elementary particles / unit volume
α = total polarizability
E' = local electric field

As seen previously, a is composed of four parts: electronic, a_e; atomic, a_a; dipolar, a_d; and interfacial, a_i.

Local field E' can be shown to be the result of two charge components: (1) The free charges at the electrodes, and (2) the free ends of the dipole chains within the cavity walls where the atomic and molecular activity occurs. A third component is present as a result of molecules within the cavity, but it can be neglected since there is a mutual cancellation of charges. The local field is thus represented by the equation,

$$E' = E_1 + E_2 \tag{2.15}$$

in which:

E_1 = field due to free charges
E_2 = field due to ends of dipole chains

The free charge at the electrodes is expressed by the applied field itself ($E_1 = E$). The second field formed by the dipole chains at the cavity wall is found by integration of the cavity wall and results in the equation,

$$E_2 = (E / 3)\,(K' - 1) \tag{2.16}$$

Thus the complete local field, also known as the Mosotti field,[9,10] is given by the equation,

$$E' = (E / 3)\,(K' + 2) \tag{2.17}$$

By equating Eqs. 2.13 and 2.14 and inserting Eq. 2.17 for the Mosotti field, we obtain the expression:

$$\Pi = \frac{N\alpha}{3\varepsilon_0} = \frac{K' - 1}{K' + 2} \tag{2.18}$$

in which Π = molar polarization (polarizability per unit volume).

For ideal gases, K' very nearly equals unity, and Eq. 2.18 reduces to

$$\Pi = N\alpha / \varepsilon_0 = K' - 1 \qquad (2.19)$$

in which N = the Loschmit number = 2.687×10^{25} molecules
per cubic meter at $0°$ C and 760 mm Hg.

For high-pressure gases, vapors, liquids, and solid dielectrics it is more convenient to discuss polarizability per mole. The N of Eq. 2.19 is transformed to Avogadro's number (molecules per mole). Thus,

$$N_0 = (NM / \delta) - 6.023 \times 10^{23} \qquad (2.20)$$

in which:

N_0 = Avogadro's number
M = molecular weight
δ = density, in kg per m^3

and the molar polarization, in meters^3, is

$$\Pi = \frac{N_0 \alpha}{3 \varepsilon_0} = \left(\frac{K' - 1}{K' + 2} \right) \left(\frac{M}{\delta} \right) \qquad (2.21)$$

Equation 2.21, the Clausius-Mosotti equation,[11] is useful in many cases, especially if dielectric constant K' is replaced by complex relative dielectric constant K^*, which is more representative of the general case. Also, in the optical frequency range, the relative dielectric constant can be replaced by the square of the refractive index, n^2, which itself can be replaced by the square of the complex refractive index, n^{*2}. These changes result in the final Clausius-Mosotti-Lorentz-Lorenz equation in complex values.[12, 13]

$$\Pi = \frac{N_0 \alpha}{3 \varepsilon_0} = \left(\frac{K^* - 1}{K^* + 2} \right) \left(\frac{M}{\delta} \right) = \left(\frac{n^{*2} - 1}{n^{*2} + 2} \right) \left(\frac{M}{\delta} \right) \qquad (2.22)$$

As before, a is composed of four fractions and the near field is completely neglected. It is expected that deviations from Eq. 2.22 in experimental results can be attributed to the influence of the near field in certain dielectrics.

Evaluation of the polarizability component of a can be made as follows:

$$\alpha_e = \varepsilon_0 4 \Pi r_0^3 \qquad (2.23)$$

in which:

α_e = electronic polarizability
r_0 = radius of undisturbed electron cloud sphere

The molor polarization, in meter^3, is therefore

$$\Pi = N_0 \alpha_e / 3 \varepsilon_0 = N_0 (4\Pi / 3) r_0^3 \qquad (2.24)$$

The polarizability per mole is apparently equal to the volume of the electronic sphere as defined by the Bohr model. Quantum mechanics, however, leads to a more extended electron cloud than the simple Bohr model of the monatomic gas. Thus measurements of dipole moment and dielectric constant do not closely satisfy Eq. 2.24. Spectroscopic methods of determining polarization come closer to the actual conditions within the atom under an electric field.

In the quantum treatment, the electrons assume fixed states or levels prescribed by quantum mechanics as the standing wave patterns of probability waves. The mutual interaction of the electrons is largely electrostatic in nature, but electron clouds may develop orbital magnetic moments and the electrons themselves may have magnetic spin moments.

2.7   Chemical Bond and Dipole Moment

By way of clarification, it becomes necessary to study the forces present when two atoms gradually approach one another. At a distance of several atomic diameters between electron clouds they come under the influence of Van der Waals attraction.[14] This apparent paradox (since the electron clouds of the two atoms are actually repelling each other) occurs because the electrons of the two atoms become correlated and a dipole-dipole coupling force develops because of induced and fluctuating moments. Van der Waals force between neutral atoms is consequently a short-distance force and decreases as the seventh power of the distance.

Then as the atoms come even closer, so that their electrons could overlap, a low coulomb attraction sets in as the nuclei begin to bind the electrons of the opposite atom. This is an intermediate force between Van der Waals force and the chemical bond.

Finally, depending on quantum mechanics, one of several mechanisms takes place as the nuclei lock into a firm chemical bond. (1) The atoms with unfilled orbitals may link together by contributing electrons from each partner and producing electron pair bonds with antiparallel spins (the normal covalent bonds of chemistry). (2) One atom may go over to an excited state. (3) The atoms may exchange electrons, thus forming ions of opposite polarity that attract each other by coulomb forces described by ionic bonds.

Ordinary collision of atoms is of insufficient energy to cause ionization, but such agents as heat, radiation, and catalysts can make it possible.

The Van der Waals, covalent, and ionic bonds described above are considered as completed reactions, and all degrees of intermediate states are found in practice, including elastic collisions and near misses where bonds are not completed.

Debye was the first to realize the significance of the permanent dipole moment that develops out of the different attraction two locking atoms have for electrons.[15] This attraction, for electrons, is called *electronegativity*. The unit of dipole moment is known as the *debye:*

$$1 \text{ debye } = 1 \times 10^{-18} \text{ e.s.u. } = 3.33 \times 10^{-30} \text{ coulomb meters}$$

Dipole moments have been measured for gaseous molecules by a quasistatic measurement of polarization as a function of temperature.[16] A statistical theory of orientation developed by Langevin for permanent magnetic moments of paramagnetic substances[17] was adapted to permanent electric moments by Debye.[10]

Molecules carrying permanent dipole moments are reoriented by an external electric field so that they are partially aligned in the field direction. Since the alignment is imperfect (because of thermal agitation of the molecules), a statistical equilibrium obtains. Although a calculation of the dipole moment can be made by rigorous mathematics involving the Langevin function, it is less time-consuming to work with measurements of the molar polarization at several temperatures and determine the moment by graphic means.

Thus the molar polarizability, Π, has a linear relationship with reciprocal absolute temperature, and the slope of this line is used to evaluate the dipole moment, μ, in debyes:

$$\mu = 12.7 \sqrt{B} \qquad (2.25)$$

in which B equals the slope of the Π vs $1/T$ line, using Eq. 2.19 for an ideal low-pressure gas.

Although polar gases give rising lines on these plots, nonpolar gases show horizontal lines of zero slope and therefore zero dipole moment.

Slope B in Eq. 2.25 is determined as follows:

$$B = \frac{\Pi_1 - \Pi_2}{(1/T') - (1/T_2)} = \frac{K'_1 - K'_2}{(1/T') - (1/T_2)} \qquad (2.26)$$

It is possible, by use of optical measurements, to separate molar polarization into its three components: electronic, atomic, and dipolar. For dipole moments greater than 1 debye, atomic polarization amounts to 10 percent of the electronic polarization. Electronic resonance spectra appear in the visible and ultraviolet light regions. Atomic polarization arises from vibrations between nuclei and is manifested by spectral resonance in the near infrared. Rotational action in molecules, including dipole orientation, often occurs in the far infrared and the microwave regions. Molecular spectroscopy is discussed by Herzberg.[18] Some dipole action can be measured at radio frequencies.

The potential energy curve for a diatomic molecule shows a minimum value at absolute zero with the two nuclei at a certain distance from each other. As the temperature is raised, this distance increases, but the potential energy also increases up to a level at which disassociation of the atoms takes place. If covalent-bonded, the two atoms are neutral. If ionic-bonded, they take the form of charged ions.

Atoms bond together in large aggregates to form liquids, crystals, metals and polymers. The two most used for dielectrics are liquids and polymers, although some crystals—for example, mica—are outstanding examples of very good dielectrics.

Polymerization of molecules to form plastics is considered an intermediate stage of condensation between liquids and crystalline solids. All levels of polymerization from sticky resins to resilient elastomers and tough cross-linked solid polymers are possible in this phase of molecular formation.

There are generally two main types of polymers: thermoplastics and cross-linked "thermoset" polymers. The thermoplastics are usually of the linear long-chain variety where molecules are formed of thousands of repeating atomic clusters that are often representative of the simpler monomer, gas, or liquid from which they were formed. Thermoplastics are often meltable below their decomposition temperatures and are moldable in this softened state. Some thermoplastics do not melt below their decomposition temperatures.

Thermoset polymers contain long chains of molecules just as the thermoplastics do, but they are also linked together along the chains by other chains of varying lengths. This network of polymer chains prevents true melting, although some softening can be observed below decomposition temperatures. Thermoset polymers often pass through a middle stage from liquid to rigid solid during which they can be molded. A final cure accomplishes the cross-linking. This curing is assisted by added catalysts, which may or may not take part in the final network of atoms and molecules.

Vibration and rotation losses, discussed for diatomic molecules, have added significance in polymers. Certain polar segments or branch chains or both can vibrate and rotate apart from the main chain or network. This behavior gives rise to dipole moments that are reflected in dielectric losses and dispersion of dielectric constant at lower frequencies than those for diatomic gases. In fact, it is this kind of absorption which allows efficient dielectric heating of some polymers.[4]

2.8   Dispersion and Absorption

Much information can be learned about the behavior of a dielectric and indeed its molecular and atomic structure by investigating the effects of frequency on both the dielectric constant and the dielectric loss, as well as by studying the spectroscopic response at infrared and optical frequencies.

Resonance bands occur from radio-frequency to optical wave lengths. These bands appear as peaking curves known as "resonance absorption." Absorption bands are narrow for pure gases, wider for liquids, and very wide for many solids. Increased width is caused by such conditions as thermal agitation, imperfections in the lattice, alternate amorphous and crystalline sections along polymer chains, foreign inclusions, mechanical stresses, electrical stresses, and many other factors. These factors are responsible for a form of the Stark effect in which the spectral lines or states are widened into continua and absorption bands become extensive.

As the dielectric loss approaches the resonant frequency, a change also occurs in the dielectric constant. With rising frequency, the dielectric constant, previously at a fixed level, starts to decrease gradually at first, then at a rapid rate, to a maximum slope at the resonance frequency. This rate then decreases until a lower constant level is reached at a higher frequency. The entire action is referred to as "anomalous dispersion." Figure 2.7 shows both absorption resonance and anomalous dispersion for polymonochlorotrifluoroethylene tested at frequencies from 30 to 10^6 Hz. Because of the polar nature of the molecule, the peak at 10^3 Hz comes at a low frequency compared to that found in many other polymers.

Because of the broad bandwidth for polymers in resonance (for example, see Fig. 2.7), Debye introduced the classical representation of dipoles as polar molecules rotating and vibrating in a medium of dominant friction.[10] This treatment completely neglects the inductive member of the usual electrical model of the LRC circuit used for gases. The resulting RC circuit yields instructive relationships that are useful in the treatment of dielectric loss and dispersion in polymers and liquid dielectrics.

2.9   Relaxation Time in Dielectric Theory

The absorption spectra of liquids and solids, including polymers, is much broader than those of gases. Much detail is lost in the process of studying the spectra, and interpretation is much less defined.

Thus Debye's parallel C_1 and R_2C_2 treatment (see Fig. 2.8) for dipole action

---

[4]   Dielectric heating is a unique method of getting heat into a plastic, plywood section, laminate, molded compound (or even food-products) without the use of heat conduction or convection. The heat is internal and more uniform than that obtained through conventional methods and is much more rapid in reaching the desired results.

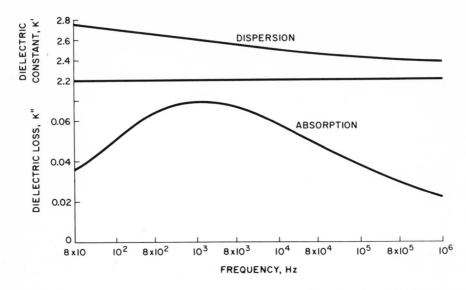

Fig. 2.7 Anomalous dispersion and resonance absorption for polytri-
fluoromonochloroethylene (After T. Nakajima and S. Saito,
"Properties of Polytrifluoromonochloroethylene," *J. Polymer
Sci.*, V 31, 423, 1958)

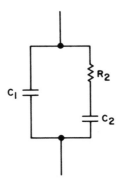

Fig. 2.8 Debye equivalent circuit for liquid and solid dielectric (After
P. Debye, *Polar Molecules,* Chem. Catalog Co., N.Y., 1929)

for electronic and atomic distortion results in an equivalent complex dielectric constant, K^*, as follows:[7,10]

$$K^* = K'_\infty + \frac{K'_s - K'_\infty}{1 + j\omega\tau} \tag{2.27}$$

in which:

K^* = complex relative dielectric constant
K'_∞ = optical dielectric constant = C_1 / C_0
K'_s = static dielectric constant = $(C_1 / C_0) + (C_2 / C_0)$

$$\tau = \text{relaxation time} = R_2 C_2$$
$$\omega = \text{angular velocity} = 2\Pi f$$
$$j = \text{complex vector operator}$$
$$C_0 = \text{geometric capacitance of } C_1 \text{ and } C_2$$

Relaxation time (τ) is best explained by observing the transient voltage relationship across C_1, previously charged to V_0, when suddenly shorted by the $R_2 C_2$ circuit:

$$V_1 = V_0 e^{-t/\tau} \qquad (2.28)$$

in which:

$$V_1 = \text{voltage across } C_1 \text{ at time } t$$
$$V_0 = \text{initial voltage across } C_1$$
$$t = \text{time, in seconds}$$
$$\tau = \text{relaxation time, in seconds} = R_2 C_2$$

The voltage thus drops exponentially with relaxation time, τ.

Equation 2.27 showed a replacement of the resonance spectrum by a relaxation time spectrum. The static and optical dielectric constants, K'_s and K'_∞, and relaxation time τ must be reinterpreted in terms of molecular quantities. K'_∞ represents the induced moment, μ_i, contributed by atomic and electronic deformation as follows:

$$\mu_i = (\alpha_e + \alpha_a) E' \qquad (2.29)$$

in which:

$$\alpha_e = \text{electronic polarizability}$$
$$\alpha_a = \text{atomic polarizability}$$
$$E' = \text{local field}$$

K'_s is represented by the sum of μ_i and an average dipole moment, $\bar{\mu}_d$, the latter determined as follows:

$$\bar{\mu}_d = (\mu^2 / 3kT) E' \qquad (2.30)$$

in which:

$$\mu = \text{dipole moment}$$
$$k = \text{Boltzmann constant}$$
$$T = \text{absolute temperature, in } °K$$

Use of the Mosotti field for E' and the polarizability per unit volume concept obtains the following relationship:

$$\frac{N\alpha}{3\varepsilon_0} = \frac{K^* - 1}{K^* + 2} = \frac{N}{3\varepsilon_0} \left[(\alpha_e + \alpha_a) + \left(\frac{\mu^2}{3kT}\right)\left(\frac{1}{1 + j\omega\tau}\right) \right] \qquad (2.31)$$

The static value of Eq. 2.31 is

$$\frac{K'_s - 1}{K'_s + 2} = \frac{N}{3\varepsilon_0} \left[(\alpha_e + \alpha_a) + (\mu^2 / 3kT) \right] \qquad (2.32)$$

The optical value of Eq. 2.31 is:

$$\frac{K'_\infty - 1}{K'_\infty + 2} = \frac{N}{3\,\varepsilon_0}\,(\alpha_e + \alpha_a) \tag{2.33}$$

By introducing Eqs. 2.32 and 2.33 into Eq. 2.31 and replacing τ by τ_e,

$$\tau_e = \tau\,(K'_s + 2)\,/\,(K'_\infty + 2)$$

we have:

$$K^* = K'_\infty + \left(\frac{K'_s - K'_\infty}{1 + j\omega\tau_e}\right) \tag{2.34}$$

Equation 2.34 is known as the Debye equation. It can be used to express the dipole orientation as a simple relaxation spectrum from which both dispersion and absorption functions can be evaluated.

The real and imaginary parts of 2.34 are as follows:

$$K' = K'_\infty + \left(\frac{K'_s - K'_\infty}{1 + \omega^2\tau_e^2}\right) = \text{dielectric constant} \tag{2.35}$$

$$K'' = \frac{(K'_s - K'_\infty)\,\omega\tau_e}{1 + \omega^2\tau_e^2} = \text{loss factor} \tag{2.36}$$

The dissipation factor is given by

$$\text{D.F.} = \tan\delta = \frac{K''}{K'} = \frac{(K'_s - K'_\infty)\,\omega\tau_e}{K'_s - K'_\infty\omega^2\tau^2} \tag{2.37}$$

Conductivity is found by multiplying Eq. 2.36 by ω:

$$\sigma = \frac{(K'_s - K'_\infty)\,\omega^2\tau_e}{1 + \omega^2\tau_e^2} \tag{2.38}$$

Equations 2.35, 2.36, and 2.38, in normalized form, are shown plotted in Fig. 2.9.[7] The frequency spread of the dispersion and absorption in Fig. 2.9 extends over approximately two decades of the logarithm.

2.10   Representation of Lumped-Circuit Constants

Although the macroscopic theory does not reveal what actually is going on in a dielectric under electrical stress, it is possible to derive many working relationships by considering the equivalent electrical-circuit lumped constants most descriptive of the dielectric under study.

Lumped circuits can be used over the frequency range from d-c (0 Hz) to about 200 MHz.[7]

At *low frequency* (d-c to 10 Hz), a voltmeter-ammeter method yields the shape of the current-time curve, which in turn gives information about the capacitance and resistance of the circuit.

At *medium frequency* (1 to 10^7 Hz), ordinary bridges can be balanced to yield dissipation factor and capacitance.[19] From these parameters loss factor and dielectric constant can be calculated.

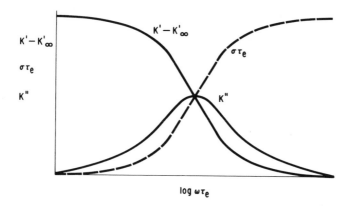

Fig. 2.9   Dielectric constant, loss factor, and conductivity of simple relaxation spectrum[7]

At *high frequency* (10^5 to 2×10^8 Hz), resonant circuits yield Q values[19,7] and capacitance; the dissipation factor can be calculated by the width of the resonance curve.

With reentrant cavity methods the upper frequency limit can be raised to 10^9 Hz.

The simplest forms of the lumped-circuit constant approach are the series and parallel RC networks shown in Figs. 2.10 (A) and 2.10 (B). For these circuits the dissipation factor, loss factor, and dielectric constant are given by the following equations:

Series R C	*Parallel R C*	
$tan\ \delta = \omega RC$	$tan\ \delta = 1/\omega RC$	(2.39)
$K'' = \omega R(C^2/C_0)$	$K'' = 1/\omega RC_0$	(2.40)
$K' = C/C_0$	$K' = C/C_0$	(2.41)

These circuits are useful in bridge-balancing techniques but do not necessarily represent a dielectric over a wide frequency span. For example, the series RC

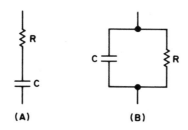

(A)   (B)

Fig. 2.10   Equivalent circuits for dielectrics used in bridge techniques: (A) series, and (B) parallel

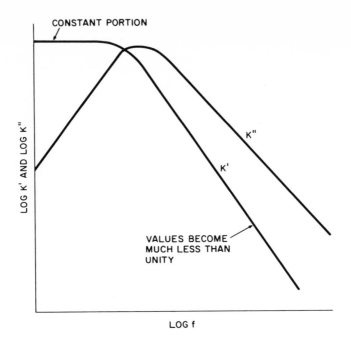

Fig. 2.11   Frequency response for series RC circuit

representation has the characteristic shown in Fig. 2.11 with a constant K' value only up to the vicinity of a peak in the K'' function. K' and the K'' function are given by the equations,

$$K' = C / C_0 (1 + A^2 f^2) \tag{2.42}$$

$$K'' = ACf / C_0 (1 + A^2 f^2) \tag{2.43}$$

in which:

$A = 2\Pi RC$

f = frequency, in Hertz

C = capacitance, in farads

R = resistance, in ohms

C_0 = geometric capacitance, in farads

The dielectric constant falls from the constant value as $A^2 f^2$ becomes comparable to or greater than unity.

The parallel response curves are shown in Fig. 2.12. The K' value is always a constant.

The decreasing K'' function is given by the equation,

$$K'' = A' / f \tag{2.44}$$

and

$$\tan \delta = B / f \tag{2.45}$$

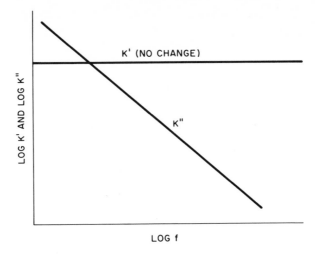

Fig. 2.12   Frequency characteristic of parallel RC circuit

in which:

$$A' = 1 \,/\, 2\Pi C_0 R$$
$$B = 1 \,/\, 2\Pi C R \qquad\qquad (2.46)$$
$$f = \text{frequency, in Hertz}$$

A true representation for dielectrics is not realized by simple series or parallel RC circuits. Several approaches for more meaningful representation are shown in Figs. 2.13 (A), (B), and (C).

Figure 2.13 (A) has been found to fit distilled water, whose tan δ dips to a very low level in the 10^7 to 10^8 Hz region.[1] A specific case with values of R and C inserted is shown plotted in Fig. 2.14. The parallel 53,000-ohm and 100-picofarad circuit shapes the low-frequency response, whereas the series 100-picofarad and 0.085-ohm circuit is responsible for the very high-frequency characteristics. All of the three circuit components affect the intermediate-frequency response. As seen in the dotted

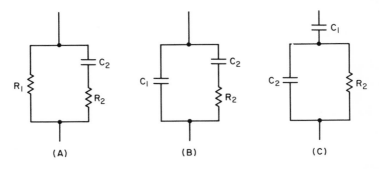

Fig. 2.13   Equivalent circuits more representative of dielectrics in terms
of frequency response

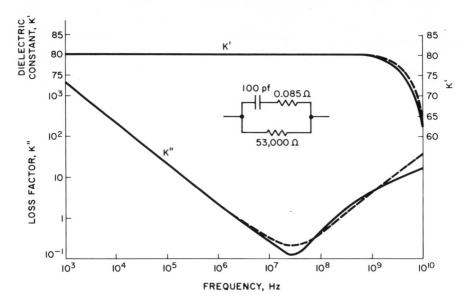

Fig. 2.14   Equivalent circuit response and actual measurements for water[1]

line, the circuit does not exactly fit but is sufficiently close for design use. The response for dielectric constant is also very close to actual measurements.

Figure 2.13 (B) has been found to fit Askarel (a chlorinated diphenyl).[1]

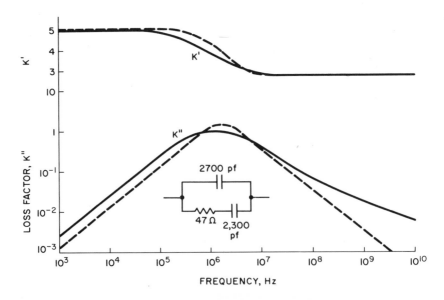

Fig. 2.15   Equivalent circuit response and actual measurements for a mixture of askarel chlorinated diphenyls[1]

Thus Fig. 2.15 gives the characteristic for this representation, using 2,700 pico-farads in parallel with the series 47 ohm and 2,300 picofarads. Again the fit is re-markable for both K'' and K'.

To show the varied possibilities, Fig. 2.13 (C) can also be made to fit the frequency response of Askarel with $C_1 = 5,000$ picofarads, $C_2 = 5,860$ picofarads, and $R_2 = 9.95$ ohms.[1]

2.11   Measurement of Dielectric Loss and Dielectric Constant

Because of the accuracy of definition of dielectric loss by RC circuits, means for measurement of the loss parameters of specimen dielectrics in the low- to medium-high frequency range have been developed using these equivalent circuits. These devices are called bridges but comparison networks, peak resonance circuits, and cavity resonators are included in this discussion.[5]

In a-c bridge networks the in-phase resistive component of loss is balanced by (1) resistive components in an adjacent leg (of a 4-arm bridge) or (2) a capacitive component in an opposing leg. The out-of-phase capacitive component is balanced by (1) capacitive components in an adjacent leg or (2) resistive components in an opposing leg.

The four-arm bridge is often used for these measurements, although added legs are used to balance out stray impedances. The Wagner ground is an example of this technique. Figure 2.16 illustrates the 4-arm bridge in which balance is obtained when the products of the opposing impedances are equal.

Given three of the four arms as R, C, and RC elements, the in-phase im-pedance, the out-of-phase impedance, and the phase angle can be calculated. By

---

[5]   The close similarity between the complex notation for impedance of an RC circuit and the complex nota-tion for dielectric loss based on real and imaginary vectors allows balancing of a bridge arm for both the dissipation factor (resistance) and dielectric constant (capacitance). When two bridge arms are involved, capacitance variation in each arm may balance each dielectric parameter, or the resistance in each arm may be used for balancing.

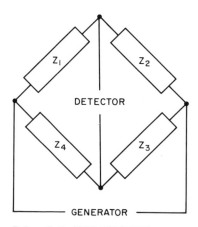

$$Z_1 Z_3 = Z_2 Z_4 \text{ (FOR BALANCE)}$$

Fig. 2.16   Impedance measuring bridge

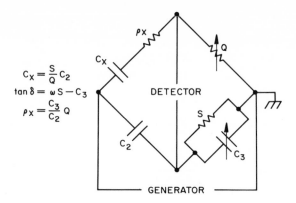

$$C_x = \frac{S}{Q} C_2$$

$$\tan \delta = \omega S - C_3$$

$$\rho_x = \frac{C_3}{C_2} Q$$

Fig. 2.17   High-voltage Schering bridge

selective use of components most bridges read out in terms of capacitance and dissipation factor.

Dissipation factor ($\tan \delta$) is the tangent of the complementary angle (δ) to the phase angle (ϕ).

Power factor, $\cos \phi$, is given by the equation,

$$\cos \phi = \tan \delta / \sqrt{1 + \tan^2 \delta}$$

For a $\tan \delta$ less than 0.1 (10 percent), the two are about equal.

Best-known in the power field for high-voltage measurements is the Schering bridge[19] (Fig. 2.17) dating back to 1920.[20] It was Semm[21] who first applied the bridge to high-voltage loss measurements.

Most of the high voltage is taken by the sample, C_x, and the standard, C_2, with the detector, Q arm, and the S-C_3 circuits at relatively low voltage levels.

The in-phase loss, ρ_x, is found by multiplying the value of Q by the ratio of C_3 to C_2 and the out-of-phase component, C_x, is found by multiplying C_2 by the ratio of S to Q.

The dissipation factor is $\omega S C_3$. S is chosen to make the quantity ωS a multiple of 10 so that C_3 in farads is a direct measure of $\tan \delta$, a dimensionless quantity. The loss factor K'' is given by $\omega S^2 C_2 C_3 / Q C_0$. Since C_x and C_2 are approximately equal, S and Q are also approximately equal.

The Schering bridge can be used with a grounded sample by remote operation of Q and C_3 or by using an ungrounded transformer secondary and grounding the junction of Q and ρ_x. Such an arrangement (grounded detector) requires a safety gap at the Q-S junction in case of failure of the specimen.

Three terminal measurements can be made with the Schering bridge by connecting the guard electrode of the sample on the shield circuit of the bridge. This and other methods on bridge design at high voltages are covered in References (19), (7), and (22).

For low voltages, the Schering bridge was modified by Field with a guard circuit, as shown in Fig. 2.18. The guard circuit is separately balanced to zero, like the Wagner ground. This bridge can be used to measure capacitors with very low loss.

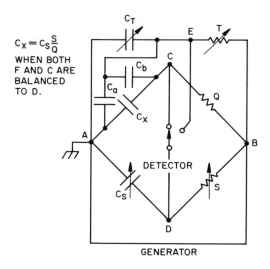

Fig. 2.18   Low-voltage Field bridge with guard circuit

For high-frequency measurements up to audio frequency, a conjugate Schering bridge is used, as shown in Fig. 2.19. For very high-frequency measurements, see References (7), (23), (24). At low voltage, power frequency, and grounded sample, the bridge shown in Fig. 2.20 has been found useful. It features simplified R_A and R_B arms.

2.12   Gaseous Breakdown Theory

Air breakdown theory, because it establishes the mechanisms for the avalanche theory of electrons, is a prerequisite to all breakdown theory—that of solids and liquids, as well as of gases.

In an air condenser at low pressure the plot of current against voltage, as

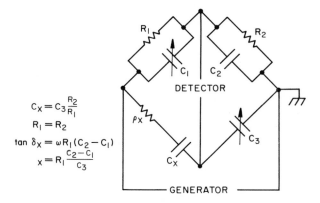

Fig. 2.19   Low-voltage conjugate Schering bridge, audio frequency

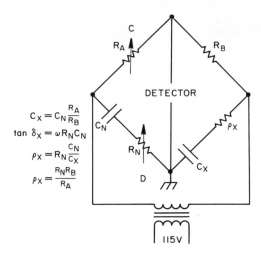

$$C_X = C_N \frac{R_A}{R_B}$$

$$\tan \delta_X = \omega R_N C_N$$

$$\rho_X = R_N \frac{C_N}{C_X}$$

$$\rho_X = \frac{R_N R_B}{R_A}$$

Fig. 2.20 Power-frequency bridge with low voltage and grounded sample

pioneered by Townsend[25] and repeated by many others, is shown in Fig. 2.21. The I_s value is a saturation current of extremely low magnitude which is caused by charge carriers in the gas formed at a constant rate. A rising current immediately prior to breakdown is caused by electron avalanches in which electrons collide with

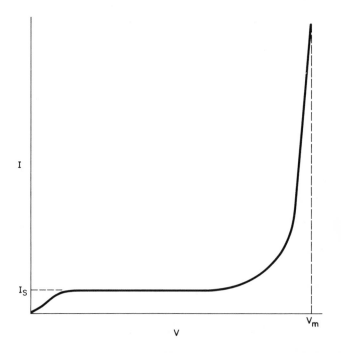

Fig. 2.21   Townsend current/voltage characteristic for low-pressure gas

molecules of gas, releasing other electrons which are accelerated by the field and collide in turn, releasing still further electrons. This process increases in intensity as the voltage is increased and forms ion avalanches in the opposing direction.

A condition is reached where the arriving positive ions produce electrons at the cathode with a probability that becomes a certainty. This is the condition for the breakdown of the gas, at which V_m is reached.

It is found that the only vital condition in air breakdown is the number of impacts across the gap of the air condenser. The number of collisions is proportional to the pressure times the distance. Thus in 1889 Paschen[26] stated his now famous law that breakdown voltage of a gas is a function only of the product, pd (pressure times spacing), and that one is replaceable by the other in a typical similarity law.

Figure 2.22 shows Paschen's Law for a-c voltages at a pressure of 760 mm and temperature of 25°C. The minimum voltage at a pd of 6.5 mm $\times$ mm is 240 volts (rms) for air. It is of some interest to note the empirical equations which fit the data in certain areas. The equation for zone A extending from a pd of 10 to 10^6 (and established by the dotted lines in Fig. 2.22) is

$$V = 10.9\, pd\,/\,(-0.55 + \log pd) \tag{2.47}$$

in which:

V = rms volts at breakdown
pd = pressure times spacing, in mm $\times$ mm

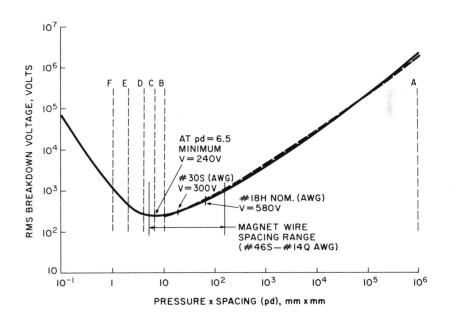

Fig. 2.22   Paschen's law for air breakdown converted to RMS, a-c volts, and assuming a uniform field (After A. Von Engel and M. Steenbeck, *Elektrische Gasentladungen*, vol. 2, Springer, Berlin, p. 45, 1934.

Equations for the other zones are as follows:

Zone	Range of pd	Equation for V
B	6.5-10	$V = 14.5\,pd/(-0.42 + \log pd)$
C	4-6.5	$V = 19.5\,pd/(-0.288 + \log pd)$
D	2-4	$V = 26.2\,pd/(-0.181 + \log pd)$
E	1-2	$V = 80.7\,pd/(0.067 + \log pd)$
F	10^{-1}-1	$V = 1,200\,pd^{-1.77}$

The Zone A relationship holds for most of the magnet wire spacings shown in Fig. 2.22. Therefore, if faults such as pinholes and cracks line up on adjacent turns of, say, AWG No.18H (heavy-build) wire, it would take 500 V to cause failure, whereas only 300 V would be needed for No. 30S (single) wire.

The IEEE No. 57 procedure for determining failure of samples aged for various periods of time at accelerated temperatures calls for 1,000 V on No. 18 wire in the form of twisted pairs.[6] Failure will occur when any two cracks are 5 mils or less from one another in adjacent wires. Some of the smallest wire insulated will approach or even go below the minimum point on Paschen's curve in terms of pd.

A more practical way of looking at air breakdown is in terms of a-c volts per mil (log scale) and spacing in mils (log scale), as shown in Fig. 2.23. This curve has

---

[6] IEEE No. 57, "Test Procedure for Evaluation of the Thermal Stability of Enameled Wire in Air," January 1959 (now ASTM D-2307).

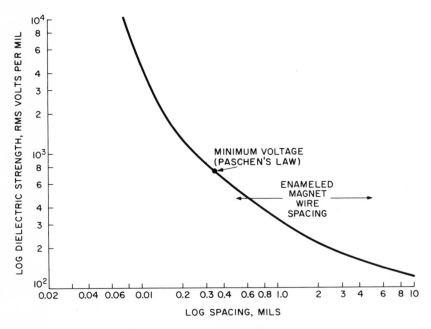

Fig. 2.23   Breakdown strength of small air films at atmospheric pressure (Combined data from F. W. Peek, Jr., *Dielectric Phenomena in High-Voltage Engineering*, McGraw-Hill, 1929, and W. O. Schumann, *Durchbruchsfeldstärke von Gasen*, pp. 51, 114, Springer, Berlin, 1923, for uniform fields)

wide sweeping limits extending from 0.075 mils to 10 mils with dielectric strength values from 10^4 to 10^2 volts per mil. This corresponds to the *pd* range of 1.45 to 193.

Gaseous breakdown takes various forms: arc discharge, spark discharge, glow discharge, and corona. In *arc discharge,* a high current at relatively low voltage exists. Frequently, the current reaches sufficient magnitude to evaporate metal. (Welding is an example of the use of arc-discharge conditions in practice.) In *spark discharge,* there are high-voltage gradients with relatively low-current passage and for extremely short periods of time. Lightning is one form of spark discharge.

A *glow discharge* is produced at low gas pressures (below 10 mm of Hg). It is characterized by lower voltage drops than those in spark discharge and at relatively low currents. A neon tube is an example of this type of discharge.

Corona is the result of partial breakdown of the air in the vicinity of sharp electrodes, series gas-solid insulations, and high voltage. Corona is visible in some states as a halo of low intensity light associated with the excited atoms present in a gas after electron avalanches have stripped electrons from the atoms of the gas. Corona is a hazard to solid insulation.

Certain gases known as electronegative gases have the property of absorbing the electrons that appear within them. If such gases are used alone or in conjunction with nonelectronegative gases, a marked change toward higher breakdown voltage is seen. Electronegative gases are used in certain high-voltage equipment where gas is the major dielectric medium.

The effectiveness of electronegative gases is shown by Table 2.4. Sulfur

Table 2.4 Relative 60-Hz Dielectric Strength of Electronegative Gases Compared with Air at N.T.P.

Source: References (7) and (27)

Gas	Formula	Vapor pressure, psi at 20° C	Relative dielectric strength at 20° C		
			pd = 9,650 mm × mm†	pd = 22,800 mm × mm††	pd = 38,600 mm × mm‡
Air	N_2-O_2		1.00	2.14	2.00
Freon 14	CF_4		1.01	2.31	2.56
Monochloro-monofluoromethane	CH_2ClF		1.03		
Freon 21	$CHCl_2F$	22	1.33		
Freon 22	$CHClF_2$		1.40		
Freon 13	$CClF_3$		1.43		
Monobromo-trifluoromethane	CF_3Br		1.60		4.32
Hexafluoroethane	C_2F_6		1.82	3.16	5.12
Octofluoropropane	C_3F_8		2.19	4.22	5.28
Sulfur hexafluoride*	SF_6	330	2.35	3.77	6.50‡‡
Freon 12*	CCl_2F_2	83	2.42		
Perchlorylfluoride	ClO_3F		2.48		
Perfluorocyclobutane	C-C_4F_8	40	2.80		6.93
Decafluorobutane	C_4F_{10}		3.08	5.39	7.50
Freon 11	CCl_3F	14	3.50		
Trichloromethane	$CHCl_3$		4.24		
Carbon tetrachloride	CCl_4	126	6.33		
Transformer oil**			5.37	6.50	9.47
High vacuum**			6.80	6.80	12.00
8% SF_6 + 92% N_2		26			5.2

* Most promising gases
†760 mm and 12.7 mm gap
‡‡100 psi and equivalent gap

** At equivalent gaps (for comparison only)
††1800 mm and 12.7 mm gap
‡760 mm and 50.8 mm gap

hexafluoride (SF_6), because of its high vapor pressure, chemical stability, and electronegativity, is considered the best of the gases shown. When taken to 100 psi (38,600 pd), it becomes a better insulating medium than transformer oil. Even when it is only a minor ingredient (8 percent) in nitrogen, SF_6 improves the dielectric strength markedly.

Investigations reported by Manion *et al*,[27] however, reveal that SF_6 is not completely stable in the presence of repeated arcing. In 30 minutes at 0.75-ampere arcing conditions, SF_6 had lost 25 percent of its original molar content, and in 30 seconds at 80-ampere arcing conditions, over 50 percent. In each case, however, the by-products of decomposition did not lower the dielectric strength, which remained at a high level. These by-products were listed as S_2F_2, SF_4, SOF_2, and SiF_4. The presence of the latter two molecules indicates attack on the glass container walls during the tests.

Other data on the corona instability of SF_6[28] also limits the use of this gas.

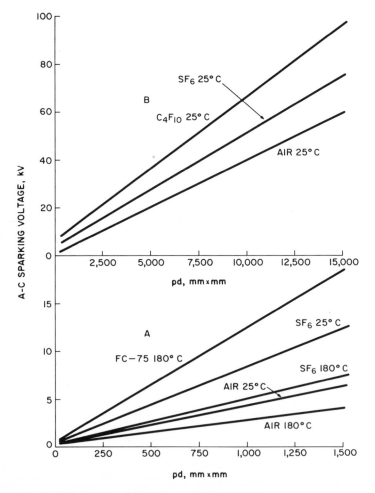

Fig. 2.24   Advantage of electronegative gases and vapors over air at elevated temperatures[29]

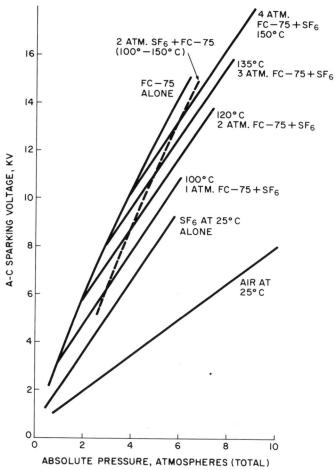

Fig. 2.25   Breakdown strength of mixtures of SF$_6$ and FC-75 in closed system with heat applied[29]

It is surpassed in this regard by BCl$_3$ (boron trichloride), which has both corona and arc stability and high dielectric strength. Unfortunately, the latter gas is hydrolyzed in the presence of water vapor.

A second gas, Freon 12 (DuPont trade name for one of its fluorocarbon refrigerants, CCl$_2$F$_2$) is also chemically stable and has a workable vapor pressure. The gas has been shown to withstand extensive corona and sparkover conditions with minor dissociation.

Figure 2.24 (A) shows the advantages of SF$_6$ and FC-75 (a perfluorocarbon vapor with the formula, C$_8$F$_{16}$O) at elevated temperature. In Figure 2.24 (B) still another fluorocarbon (C$_4$F$_{10}$) is compared to SF$_6$ over a pd range extending to 15,000 mm $\times$ mm. Figure 2.25 shows the result of mixing SF$_6$ and FC-75 in various proportions and using such mixtures at various temperatures. A closed system is used with a 10-mil gap ($d = 0.254$ mm). Temperatures from $100°$ to $150°$C produce a family of curves branching off the FC-75 curve. The dotted curve is for constant SF$_6$ and varying FC-75 pressures (temperatures), and the solid curves are for constant FC-75 pressures (fixed temperatures) and varying SF_6 pressures. SF_6 is also shown alone as is the curve for air at room temperature.[29]

Such action can be useful in apparatus under load where temperatures rise to the levels shown and the insulating gases and vapors increase in breakdown strength to protect the system.

2.13   Solid Dielectric Breakdown Theory

The theory for dielectric breakdown depends on the type of breakdown likely to occur under certain circumstances. The principal types of breakdown are (1) disruptive breakdown, (2) thermal breakdown, (3) conductive breakdown, (4) corona damage, (5) electromechanical breakdown, (6) electrochemical breakdown, and (7) bubble formation. A detailed discussion of each type follows:

1. In the *disruptive breakdown theory* (applied only to solid dielectrics), the effect of high-energy fields is physical or molecular disintegration for which visual evidence of the catastrophic mechanism is available. High-voltage insulation subjected to impulse testing gives evidence of this effect at breakdown. Even thin insulation tested at high field strength and low temperatures attest to this theory of failure.

No background of mathematical physics has been given for disruptive breakdown, and it may be that it is only a special case of conductive breakdown, but with higher energy release and consequently explosive results.

2. In the *thermal breakdown theory,* a nonuniform current distribution arises from the nonuniformity of electrical resistance existing in many heterogeneous solid dielectrics. Those sections exhibiting lowest resistivities will therefore be called upon to carry the larger part of the leakage current. If of sufficient magnitude, the heating effect of this localized current can lead to a lowering of the insulation resistance, which in turn causes the local leakage current to increase. This increased current raises the temperature, which further reduces resistivity. In this way a spiraling, cumulative heating condition finally results in thermal breakdown by electrical conduction.

Dielectric heating at high frequency is one form of this type of selective heating. The difference, however, is that the electrodes are removed from the specimen so that (1) they do not conduct heat away, and (2) complete breakdown is avoided. Thermal breakdown is a special case of the conductive breakdown mechanism.

3. In the *conductive breakdown theory,* dielectrics are assumed to contain ionic charge carriers that move in an electric field and produce a leakage current. The electric field itself produces additional ions by collision of electrons and molecules so that the current increases with increasing voltage stress. Eventually such a large flow of ions and electrons exists that the dielectric is transformed into a conductor at the weakest spot and complete breakdown follows.

Conductive breakdown lends itself to mathematical analysis, as represented by many studies in the literature both by physicists and electrical engineers. [See References (2), (7), (25), (26), and (30) through (35).] Conductive breakdown theory as developed by von Hippel (referred to as the *low-energy criterion theory*) is based on constant energy in electrons being thermally excited to the conduction band.[1,7] Frohlich postulated a distribution of energy for conduction electrons with some possessing very high energies.[34] His theory (known as the *high-energy criterion theory*) allows for ionization below breakdown. Both theories predict essentially no prebreakdown current.

Other investigators have modified the von Hippel and Frohlich theories by allowing for electron avalanches at field strengths 20 percent below breakdown values. This modification results in calculated values of breakdown field strengths that are approximately one-fourth of those predicted by von Hippel and Frohlich and that agree more closely with experiments on nonpolar crystals.

The results of an interesting study of the field strength of extremely thin crystalline mica plates by Austen and Whitehead are shown in Fig. 2.26.[34a] Frohlich's theory can be used to show approximate agreement with experimental data.

4. The *corona damage mechanism*[7] consists of (1) erosion caused by imping-

---

[7]   The theory of corona action is that whenever a sufficient a-c voltage stress is applied to a system of series solid or liquid dielectric and air or other ionizable gas, electrons generated by the field are accelerated alternatingly toward each electrode. In the half-cycles in which they speed toward the insulation-covered electrode, they collide with air molecules, causing ionization and dislodging several electrons. These new electrons take up the action and strip other electrons from other molecules in ionizing collisions. This action of electron production increases exponentially, forming what is known as an electron avalanche and the resulting air breakdown in the air gap adjacent to the insulation. This action continues as long as the voltage is applied and as long as the insualtion withstands the bombardment (in cycles) of millions of high-speed electrons impinging on the insulation surface. Erosion occurs in polymers, and eventually breakdown takes place. Damage also occurs when corona-producing square electrodes are placed in contact with an insulation film in air.

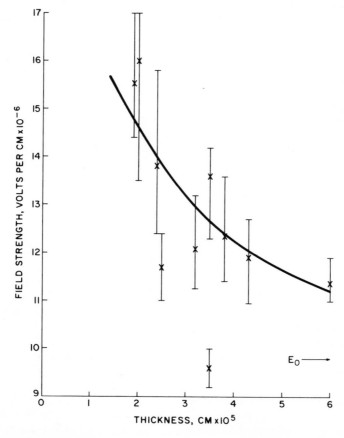

Fig. 2.26 Dielectric breakdown strength for thin plates of muscovite ruby mica[34a]

ing high-speed electrons generated in surrounding air or within void pockets in insulation at high voltages, (2) stress cracking (as in rubber under mechanical and electrical stress in air), (3) chemical attack caused by ozone formed during the corona process, or (4) surface tracking and leakage caused by the carbonization of insulation surfaces adjacent to electrodes in air at high voltages.

5. The *electromechanical breakdown* theory concerns itself with the compressive electrostatic force exerted on a dielectric by the electric field. As the thickness of material reduces to values below about 60 percent that of the original thickness (at very high voltages), the compressive force of the material can no longer balance the electrostatic force, and mechanical failure followed by electrical failure occurs.

This mechanism seems to be active in some soft polymer breakdowns at low temperature (polyisobutylene and polyethylene). It is not applicable to harder polymers such as polystyrene and polymethylmethacrylate.

6. *Electrochemical breakdown* occurs in devices such as capacitors and cables. It may happen in hermetically sealed refrigeration units and in encapsulated and potted or sealed systems. Usually ionic impurities, such as water, acids, amines, and the like, cause leakage currents, and over a period of time various reactive byproducts build up, some of which may be detrimental to the proper functioning of the dielectric. Examples are acids in oil, nascent hydrogen, water, hydrochloric acid, and organic acids in epoxy. These reactants can lead to thermal breakdown by causing a run-away build-up of heat and losses. This condition is more common in d-c devices, such as capacitors, than in a-c devices.

On a-c a point-to-plane electrode system is necessary to establish rectification for the electrolytic process. However, the general case of entrapped products of deterioration can cause a rapid system contamination even without the rectifying action.

7. *Bubble formation* occurs mainly in the liquid phase, as in entrapped solvents, moisture-contaminated solid insulation, and in partially deteriorated insulation components (for example, where depolymerization may lead to a monomer state). Internal voids are susceptible to expansion and to being filled by the liquid products of decomposition. The latter, under electric stress, are prone to bubble formation, thus leading to further expansion of voids.

2.14   Thin Film Breakdown

The data plotted in Fig. 2.26 and previously in Fig. 2.23 show the rapid increase in dielectric strength as film thickness or electrode separation is reduced. In air or gas this behavior is explained on the basis of the inability of electrons to form electron avalanches by collision since they have distances to travel which approach the mean free path in the gas. This path is the average unobstructed distance an electron can move at the pressure existing in the gas between molecules. It is a distance that is greatly reduced in liquid and solid dielectrics, but the increased dielectric strength with reduced thickness is still apparent in the condensed phases.

Not many dielectrics will attain the levels shown in Fig. 2.26 for a pure mica crystal. They will, however, reach astounding strengths even at the level of thickness of magnet-wire enamels. Values over 5,000 volts/mil (or 2×10^6 volts per cm) have been recorded for a 3-mil thickness of insulation between twisted pairs of enameled wire. Even higher values are obtainable when no air is present.

2.15   Liquid Breakdown Theory

Very little is known about the breakdown mechanism in uncontaminated insulating liquids. The most prominent theory is probably the bubble mechanism by which electric stress on the liquid molecules causes extreme agitation and bubble formation by heat or cavitation. The bubbles tend to align themselves with the lines of force of the field and allow avalanche stripping of electrons and ion formation in chains. These ions carry conduction currents which lead to heavy ionization and finally arcing through the liquid. (Bubble formation in relation to dielectric breakdown in solids was discussed in Par. 2.13.)

Once a liquid has experienced a breakdown, carbonaceous by-products may form and contaminate the liquid with mineral oils. Contaminated liquids may break down by a conduction mechanism.

Insulating liquids have breakdown strengths that are greatly dependent on contamination type and amount. Certain peroxide and acid formations during aging can grossly lower the dielectric strength of oils. Water in one form has a large effect whereas in another form the effect is not so evident. The contaminant that has the largest effect on raising conductivity of the liquid often will also be the one giving the greatest lowering of breakdown strength.

As an indication of the power of minute contaminants for increasing conductivity in oil, it has been shown that an increase of gilsonite from 0.01 to 2 percent by weight in liquid paraffin will raise the conductivity from 1.4 to 100 mho/cm ($\times 10^{-12}$). Such a rise of conductivity will foster a very large reduction in breakdown strength.

Liquid insulants are also very sensitive to effects of high voltage where internal ionization becomes apparent (shown by the rise of dissipation factor with voltage). The dielectric strength is seriously impaired by the development of a space charge built up within the liquids under electric stress according to Gemant.[8]

Gemant further states that the dielectric strength of oil is lowered "very considerably" when discrete water drops are present. His experiments showed that when these droplets, normally spherical, are exposed to an intense field they become ellipsoidal and may approach and join neighboring elongated droplets to form chains aligned with the applied field. Breakdown is concurrent with the completion of such chains.

Dry degassed oil may have a dielectric strength of 400 kV/cm, but with 0.1 percent of water the value can be as low as 40 kV/cm. With a water content of 0.5 percent, the dielectric strength is further reduced to 25 kV/cm. The normally acceptable lower level by the ASTM cup test is about 88 kV/cm.

The air content of oil is also a factor in its dielectric strength. Bubble formation is minimized by degassing the oil and the electrodes.

2.16   Voltage Endurance

Voltage endurance varies widely in different types of insulation. It is affected by the presence or absence of corona, and by the severity of corona. Data for voltage endurance have been obtained in the form of time-voltage plots for mica insulation,[36] polymers,[37, 38] and insulations under combined conditions.[39] Effects on rubber are also available in the literature.[40] A typical curve of test results is shown in Fig. 2.27 for polyester film. The voltage stress decreases

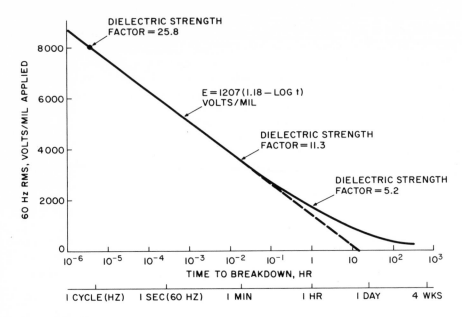

Fig. 2.27 Voltage endurance of 2-mil and 8-mil Mylar polyester film in air[37]

linearly at first with the increasing logarithm of the time and eventually curves to an asymptotic constant level when corona ceases to be a damaging factor.[37]

The equation for the linear part is included in the graph. The *dielectric strength factor* is over 25 for extremely short-time exposure to corona but drops to less than half that level by 1 min and in 1 hr it is 5.2.[8] The film is so severely attacked that it is totally useless in corona conditions.

Therefore, a film-insulated part must be protected by being embedded in resin, surrounded by oil, or used in an electronegative gaseous media. A possible means of protection would be the use of lower field strengths that do not produce corona. But this is a hazardous approach if made by increasing insulation thickness since corona damage can be a progressive effect. Thus, high-voltage insulation must be protected at its surface by corona-resistant materials. Semiconducting surface treatments to grade off stress at the electrode edges can be used for this purpose.

Internal corona also shortens insulation life. In this case, air voids present in the film-wrapped insulation develop corona which damages the film enclosing the voids and may eventually result in complete breakdown. This type of corona action is diminished by (1) filling the voids with resin or liquid impregnant. (2) using pressure to close the voids, or (3) controlling the gradient by inserting foil wraps at intervals in the insulation walls.

In some methods, initial pressure is used to reduce the size of voids. Application of heat then effects a seal that prevents the voids from becoming

8   The *dielectric strength factor* is the ratio of the dielectric strength of a material to that of air in the same spacing and electrode configuration. (A fuller explanation will be found in Chap. 3.)

large enough to support corona. There is some evidence (for example, corona-induced erosion on magnet wire) that points to a critical size of void for corona damage as 0.001 in. Smaller voids are not likely to support corona; larger ones will do so.[9]

A very useful relationship developed by Dakin *et al* concerns the corona threshold (corona initiation) as dependent on the ratio of insulation thickness to dielectric constant.[37] A nearly straight line exists between the log of corona threshold voltage and the log of the ratio of thickness to dielectric constant, thus:

$$log\ V = 2.858 + 0.46\ log\ t\ /\ K' \qquad (2.48)$$

Equation 2.48 applies approximately for insulations from polyethylene with a dielectric constant of 2.25 to alumina ceramic with a K' of 8.3 and for thickness from 1 mil to 250 mils. Thus, for example, to avoid corona damage at sharp edges to Mylar ($K' = 3.2$) at 1,000 volts, an insulation thickness of 6.45 mils is indicated by the equation.

The Starr-modified[38] Parkman[41] equation which included the discharge inception voltage in the form,

$$[(V - V_i)\ /\ V_i]^n\ t = C \qquad (2.49)$$

has been further modified by the authors using Dakin's data for nylon, Mylar, and Teflon as follows:[10]

$$[(E - E_i)\ /\ E_i]^n\ t = C \qquad (2.50)$$

in which:

E = applied field strength, in volts / mil
E_i = corona inception field strength, in volts / mil
t = time, in hours
C = constant

Table 2.5 gives the constants of Eq. 2.50 for the three materials studied (nylon, Mylar, and Teflon). Mylar, with a lower voltage endurance at high field strengths than nylon, has a greater endurance at lower field strengths (below 350 volts/mil). The low corona-inception voltage for nylon can be attributed to the greater dielectric constant of the sample tested.

---

[9]   Data from Anaconda Wire & Cable Co.
[10]   Mylar is Du Pont's tradename for polyester film and Teflon for polytetrafluoroethylene.

Table 2.5 Voltage Endurance (Eq. 2.50)

Material	E_i	n	C	Time (at E = 500 V/mil), hr
Nylon	126	1.79	362	60
Mylar	296	1.52	22.4	50
Teflon	286	1.79	6.89	11

REFERENCES

(1) von Hippel, A., *Dielectrics and Waves*, Wiley, 1954.
(2) von Hippel, A. *Molecular Science and Molecular Engineering*, Wiley, 1959.
(3) Lorentz, H. A., *The Theory of Electrons*, Stechert, N.Y., 1909.
(4) Linnenbaum, V. J., "The Effects of Radiation on Materials," *Insulation*, Jan., Feb., March 1960.
(5) Langmuir, I., *Physical Review*, vol. 2, 1913, p. 450.
(6) Whitehead, J. B., and Marvin, R., *Transactions of the AIEE*, vol. 48, 1929, p. 299.
(7) von Hippel, A., *Dielectric Materials and Applications*, Wiley, 1954.
(8) Gemant, A., *Liquid Dielectrics*, Wiley, 1933.
(9) Mosotti, O. F., *Mem. di Math. e di Fisica in Modena* (II), vol. 24, 1850, p. 49.
(10) Debye, P., *Polar Molecules*, Dover, 1945.
(11) Clausius, R., *Die Mechanische Wärmetheorie*, Vieweg, Braunschweig, vol. 2, 1879.
(12) Lorentz, H. A., *Ann. Physik*, vol. 9, 1880, p. 641.
(13) Lorenz, L., *Ann. Physik*, vol. 11, 1880, p. 70.
(14) Van der Waals, J. D., *Over de Continuiteit van den Gas en Vloeistoftoestand*, Leyden, 1873.
(15) Debye, P., *Physik Z.*, vol. 13, 1912, p. 97.
(16) Wesson, L. G., *Tables of Electric Dipole Moments*, Technology Press, Cambridge, Mass., 1948.
(17) Langevin, M. P., *J. Physique*, vol. 4, 1905, p. 678.
(18) Herzberg, G., *Molecular Spectra and Molecular Structure*, vol I: *Diatomic Molecules*, Van Nostrand, 1950; vol. II: *Infra-Red and Ramon Spectra*, Prentice Hall, 1945.
(19) Hague, B., *A.C. Bridge Methods*, Pitman & Sons, London, 1946.
(20) Schering, H., *Zeits f. Inst.*, vol. 40, 1920, p. 124.
(21) Semm, A., "Verlustmessingen bei Hochspannung," *Arch. f. Elekt.*, vol. 9, 1921, p. 30.
(22) Dunsheath, P., *High Voltage Cables*, Pitman & Sons, London, 1929.
(23) Bussey, H. E., *Cavity Resonator Dielectric Measurements on Rod Samples*, National Bureau of Standards, Boulder, Col.
(24) Dakin, T. W., and Works, C. N., "Microwave Dielectric Measurements," *Journal of Applied Physics*, vol. 18, 1947, p. 789.
(25) Townsend, J. S., *Electricity in Gases*, Clarendon Press, Oxford, 1915.
(26) Paschen, F., *Ann. Physik*, vol. 37, 1889, p. 69.
(27) Manion, J. P., Philosophos, J. A., and Robinson, M. B., "Arc Stability of Electronegative Gases," *IEEE Electrical Insulation*, vol. EI-2, No. 1, Apr. 1967.
(28) Eichberger, J. E., "Electrical Failure in Solids," *Electro-Technology*, vol. 77, May 1966.
(29) Sharbaugh, A. H., and Watson, P. K., "Breakdown Strengths of a Perfluorocarbon Vapor and Mixtures of the Vapor with SF_6," *IEEE Transactions*, Paper No. 63-922, 1963.
(30) Hurd, D. T., "An Introduction to the Mechanism of Dielectric Breakdown and Insulation Failure," *G. E. Res. Lab Rpt.* No. 1579.
(31) Frohlich, H., *Reports on Progress in Dielectrics*, The Physical Society, London, vol. 6, 1939, p. 411.
(32) Frohlich, H., *Nature V*, 1943, pp. 151, 339.
(33) *Encyclopedia of Physics*, vol. 17, "Dielectrics," S. Flugge, ed., Springer, Berlin, 1956.
(34) Frohlich, H., "On the Theory of Dielectric Breakdown in Solids," *Proceedings of the Royal Society*, London, Ser A 176, 1940, p. 33.
(34a) Austen, A. E. W., and Whitehead, S., "The Electric Strength of Some Solid Dielectrics," *Proceedings of the Royal Society* (London), A 176, 1940, p. 33.

(35) Clark, F. M., *Insulating Materials for Design and Engineering Practice*, Wiley, 1962.
(36) Moses, G. L., "A.C. & D.C. Voltage Endurance Studies on Mica Insulation for Electrical Machinery," *AIEE No. 51-127*, 1950.
(37) Dakin, T. W., Philofsky, H. M., and Divens, R., "Effect of Electrical Discharges on the Breakdown of Solid Insulation," *AIEE No. 54-70*, 1954.
(38) Starr, W., and Endicott, H., "Progressive Stress—A New Accelerated Approach to Voltage Endurance," AIEE No. 61-243, 1961.
(39) McMahon, E. J., "The Chemistry of Corona Degradation of Organic Insulating Materials in High Voltage Fields and Under Mechanical Stress," *IEEE Electrical Insulation*, vol. EI-3, No. 1, 1968.
(40) Starr, W. T., and Pomerantz, M., "Voltage Endurance Specifications for Butyl Rubber Insulated Pulse Cable RG-191/U," IEEE No. 31, 1966.
(41) Parkman, N., *British ERA Rpt. L/T*, 1957, p. 351.

CHAPTER 3 | APPLICATION OF DIELECTRIC PHENOMENA

The preceding chapter emphasized the theory of dielectric phenomena. This chapter will deal with the application significance of these phenomena, beginning with dielectric strength, a basic property demanded of insulation. The discussion will then turn to the dielectric constant, dielectric loss, dispersion and absorption, resistivity, corona, flashover, and tracking. The behavior of insulation in terms of each of these characteristics has an important bearing on the design of apparatus for both electrical and electronic service.

3.1 Dielectric strength

The *initial dielectric strength* of a material or system of insulation before the ravages of thermal aging, electrical fatigue, corona erosion, radiation exposure, and moisture attack is dependent on inherent and imposed factors such as the following:

1. Intrinsic breakdown strength
2. Original ion content (resistivity)
3. Moisture content
4. Consolidation of components
5. Insulation thickness
6. Test area
7. Electrode or conductor shape and type
8. Compatibility between adjacent dielectrics
9. Surrounding media
10. Test conditions
11. Design of insulation system
12. Frequency and wave shape of test voltage

3.1.1   Intrinsic breakdown strength

The intrinsic breakdown strength of an insulation or system is that inherent ultimate dielectric strength attainable only under special test conditions but responsible for that level of practical breakdown strength which becomes a fixed percentage of the maximum. Components selected on the basis of high intrinsic breakdown strengths are mica, glass, ceramics, nonionic resins and some hard-pressed laminates.

3.1.2   Original ion content

The original ion content of a resin or insulation can have a marked effect on its dielectric strength. The usual method of determining ion content is by d-c resistivity. High-resistivity components will usually show high dielectric strength levels under optimum test conditions. Examples of high-resistivity dielectrics are polystyrene, polyethylene, polycarbonate, polytetrafluoroethylene, mica, and glass.

Dissipation factor is another parameter used to estimate ionic content. But this characteristic involves other loss-producing phenomena and must be compared at conditions, such as high temperature and low frequency, where the ionic loss becomes the major one. Ion content and ion mobility are key factors in the level of the original dielectric strength determinations in insulating materials. The cure of synthetic resins often involves the conversion of ionic to nonionic components, the tying down of loss-producing chain members, and the elimination of volatile members, all of which decreases the mobility of the remaining ions.

3.1.3   Moisture content

The moisture content of an insulation or system plays an important part in determining its breakdown strength. For a given material, the wet breakdown strength is usually 50 percent lower than the dry strength, or may be even less. Moisture may be introduced into a material during processing or as a by-product of a resin chemical reaction. It may also be inherent in the original structure of the material, as in cellulose. To assure optimum electrical strength in an insulation system, any water present must be removed by suitable methods, such as the application of heat and vacuum.

3.1.4   Consolidation of components

In high-voltage insulation design, consolidation of constituents is necessary to attain high and long-lasting breakdown strength. This procedure is necessary because of the deleterious effect of any gas pockets or voids that may be present within the insulation. If not removed, these voids will ionize when the applied voltage is raised. This internal corona (from the ionized gas) will contribute to a gradual weakening of the insulation system and to an ever-decreasing dielectric strength. But the presence of these conducting areas within the insulation wall will immediately result in an initial lowering of the dielectric strength, as well.

Heat and pressure are usually applied together to achieve the highest degree of consolidation. Thermosetting laminates offer a good example. The laminating process calls for high pressure at relatively high temperatures to assure that all voids are closed and all trapped air expelled. The thermosetting resins are used since their cross-linked molecular structure formed under high heat will not permit relaxation during use conditions after removal of the heat and pressure.

As another example, taped insulation for form-wound coils is compressed to a high pressure under heat to yield high dielectrical strength. Thermosetting resins are again used for coating or saturating the tapes to prevent relaxation during use. In still another example, some systems, such as oil-filled cables, are consolidated by the continuous application of gas pressure that remains at high levels during the life of the system.

3.1.5 *Insulation thickness*

Insulation thickness plays an important part in the dielectric strength of thin films. The breakdown voltage rises linearly in a log-log plot of voltage gradient vs thickness in the thickness range of 0.1 to 100 mils, as shown in Fig. 3.1. This plot is for high-resistivity films such as terephthalate polyester and is extended to include nominal values for thick insulation (up to 10 in.) not necessarily of the polyester type.

The voltage gradient for solid insulation recedes with rising thickness values on the same log-log plot. The two empirical equations for voltage gradient and breakdown voltage in the range of *log t* from − 4 to − 1 (linear portion) are as follows:

$$G_s = 661 \ t^{-0.296} \tag{3.1}$$
$$E_s = 6.61 \times 10^5 \ t^{0.704} \tag{3.2}$$

in which:

G_s = voltage gradient, in volts / mil
E_s = breakdown voltage, in volts a-c
t = thickness, in inches

Similar curves are shown in Fig. 3.1 for air (uniform field). The voltage breakdown curve for solid insulation follows the Paschen Law shape, [1] reaching a minimum in the range of − 4 to − 3 *(log t)*. Extremely small spacings in solid insulation (thin films) theoretically would show very high breakdown strengths. Unsupported thin films made by modern methods achieve this excellence of dielectric strength to a marked degree. Coatings on metal conductors such as magnet wire, however, must overcome imperfections in the substrate before very high dielectric gradients are achieved. Thus whereas films 1 mil and thicker show very high dielectric strengths, those less than 1 mil thick are apt to fall far short of the theoretical level. (Note: In addition to metal imperfections, the presence of foreign impurities, such as oxides, dirt, oils, and moisture, also causes poor wetability and consequent nonuniformity of the thinner coatings with lower dielectric strength.)

The breakdown voltage curve for solids, then, is shown joining that for air breakdown at spacings less than 10^{-4} in. (tenths of a mil). The exact nature of this region is not covered by data, and the curve is therefore shown broken.

Dielectric strength factor   The foregoing discussion on thin films leads into the subject of the *dielectric strength factor,* which is a convenient and instructive means for dealing with the dielectric strength of films. The dielectric strength factor, defined as the ratio of the breakdown voltage of the insulating material under study to the breakdown voltage of air in a uniform field, is a factor similar to the dielectric constant when the capacitance ratio between insulation and air is evaluated. Since the characteristics of air breakdown in a uniform field are well known, use of the

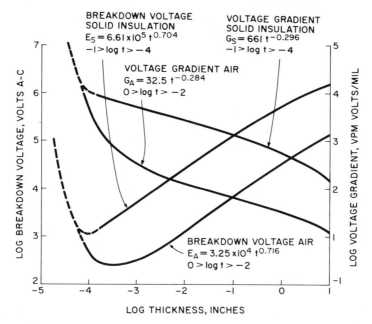

Fig. 3.1   Breakdown voltage and gradient for air and solid insulation related to insulation thickness (uniform field)

factor places the matter of the dielectric strength of solids on a firmer base than heretofore was possible. Figure 3.2 shows the dielectric strength factor (D.S.F.) for the insulation of Fig. 3.1 plotted against the log of thickness in inches.

The curve rises sharply from a low of 2 at 10^{-4} in. to a high of over 20 for spacings 3×10^{-3} to 10^{-1} in. Thereafter, the D.S.F. declines to a final value of 12 at a spacing of 10 in. These D.S.F. levels represent nominal values, with maxima as high as 20 to 30 percent over those shown in Fig. 3.2 and minima, of course, at unity or below. The "below-unity" phenomenon will be discussed presently.

Nominal enamelled-magnet-wire thicknesses (from 5×10^{-4} to 5×10^{-3} in.) yield D.S.F. values on the rising side of the curve (Fig. 3.2) with A.W.G. No. 18 heavy-build wire showing an expected value of 20. Actually, values up to 25 have been observed and, in poorly coated samples, as low as 2. But Fig. 3.2 represents a normal level of D.S.F. and shows the practical limits at each spacing or thickness.

Empirical relationships set up for the curve in Fig. 3.2 are used to determine D.S.F. by dividing the measured gradient for a given insulation by the gradient for air in Fig. 3.3 at the same spacing.[2,3] In effect, the D.S.F. value provides a factor of superiority, in terms of voltage gradient or breakdown voltage, for the insulation being evaluated when compared with air, the latter taken as unity. Values *below unity*, however, are possible and occur when the effects of creepage across contaminating surfaces are worse than those of an air-gap flashover. Values in the range of 1 to 5 indicate spacing insulation or porous enamel or film. D.S.F. levels of 5 to 8 indicate imperfect film and coating with cracks or porosity present. Normal barrier insulation (free from cracks) will yield D.S.F. values well above 10, in the 16 to 25

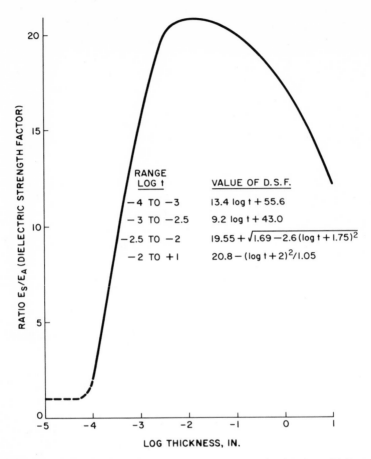

RANGE LOG t	VALUE OF D.S.F.
−4 TO −3	$13.4 \log t + 55.6$
−3 TO −2.5	$9.2 \log t + 43.0$
−2.5 TO −2	$19.55 + \sqrt{1.69 - 2.6 (\log t + 1.75)^2}$
−2 TO +1	$20.8 - (\log t + 2)^2 / 1.05$

Fig. 3.2   Dielectric strength factor (nominal values) related to thickness of insulation

region. Thickness will dictate the limit, as seen in Fig. 3.2. A 15 D.S.F. value for a 1-in. thick insulator will be good for this material and low for a magnet wire with a thickness (wire to wire) of 3×10^{-3} in. A value of 25 for D.S.F. represents 5,000 volts/mil gradient (2×10^6 volts/cm), which is approaching the intrinsic strength of the resin.

Although it is not difficult to reach high D.S.F. values for "heavy" and "triple" grades of wire in the larger sizes, it is quite difficult to attain D.S.F. levels over 5 to 8 for wire in the finer sizes (30–46 A.W.G.), especially in the "single" grade. One reason for this is the rapidly declining D.S.F. limit as thickness is decreased (see Fig. 3.2). Another reason, as mentioned before, is the degree of difficulty of coating fine wire with very thin coats.

Film thicknesses in the range of 10 to 100 mils enjoy maximum D.S.F. limits and make the best voltage barrier stock because they yield superior voltage breakdown levels when used in multiple layers.

Thicker insulation (up to 1 in.) shows a decline in the D.S.F. limit. Actually,

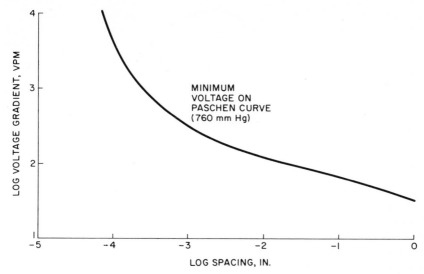

Fig. 3.3 Voltage gradient related to spacing for uniform field in air at atmospheric pressure (see Fig. 3.1)[2,3]

in practice, these limits are seldom obtained because of the statistical chance of imperfections in the thicker wall. This chance is intensified when even thicker insulation (up to 10 in.) is considered, as witnessed by the dropoff of D.S.F. to a value of 12. Electronegative gases and insulating oils are more useful at such spacings than solid insulation and with suitable barriers will yield higher D.S.F. levels.

The dielectric strength factor is a useful concept, allowing a direct comparison of all films and coating thicknesses in terms of dielectric strength. This had not been possible previously since dielectric strength was so dependent on thickness that direct comparison was not possible. The concept has broad applications. It can be useful in studying changes brought about by aging, moisture absorption, temperature levels, radiation exposure, voltage endurance effects, extent of impregnation and the like. D.S.F. is a material-oriented property as is its analog, the dielectric constant. It is dependent on prehistory, processing, and use of the insulating material involved. D.S.F. compares the barrier action of an insulating material against air as the standard.

3.1.6 Area of test

The test area of an insulation has a certain minor influence on the final dielectric strength result. On the basis of stastistics and the "weakest link" theory, it is understandable that the larger the area under test, the lower the dielectric strength and its dielectric strength factor (see section 3.1.5). Raising the general level of dielectric strength of the material by eliminating some weak spots will change the amount of this effect but will not remove the trend of data.[4]

3.1.7 Conductor shape and type

The electrode or conductor shape and type is more important to initial dielectric strength in vacuum and gaseous insulations than in liquid or solid types since dielectric breakdown voltage for a given gas is strongly dependent on the uniformity of the electric field applied. In liquids and solids the immediate effect is less pronounced but can become a serious factor over an extended period of time. Such parameters as electrode size, shape, surface condition, metal type, and conductivity all affect the dielectric strength of a given gas. Sharp edges produce corona at breakdown voltages, and this effect reduces the sparkover value. Surfaces with irregular dents, machine markings, die marks, slivers, dirt, oxides, and inclusions all reduce the breakdown voltage in gases. Under similar conditions, breakdown voltage is also reduced in extremely thin film dielectrics. Apparently, as previously pointed out, the behavior of such films is like that of gaseous films.

Electrode metals such as copper in oils can be catalytic, promoting oil deterioration and the eventual lowering of dielectric strength of the oil. Aluminum, on the other hand, is passive because of its oxide layer and will not act in this manner. Poor surface condition, however, may prevent good coatings of aluminum and thus lower the dielectric strength of film insulation. Conductivity may be a factor in the heating of insulation and cause a lowering of dielectric strength. Copper is often chosen over aluminum for its higher electrical conductivity and greater ease of drawing and shaping, especially for fine sizes of magnet wire, for example. Metals of low work function (such as stainless steel) show lower field emission, as a result of which gaseous insulation will exhibit a correspondingly higher dielectric strength.

Compatibility of insulation with the metal conductor is of major concern in encapsulants for insulation systems. Some 100-percent solids systems using peroxide catalysts (with styrene) can be inhibited because of their incompatibility with copper. Similar incompatibility is encountered in insulating compounds containing sulfur. Any incompatibility between the conductor and the insulation system will cause a lowering of dielectric strength. One method of making copper more compatible is to use a tin-plating or tin-dipping procedure before insulating.

Sharp edges and slivers should not be tolerated even for conductors within an encapsulated insulation system.

3.1.8   Intercomponent compatibility

Incompatibility with the constituents of an insulation system other than the conductors also poses a serious threat to high dielectric strength. Undercured resins, hydrolytic decomposition, unreacted high loss catalysts, presence of chlorine in some components, and internal voids caused by lack of internal cohesivity all contribute to loss of dielectric strength. Even loss of adhesion between insulation and conductor, as with a varnished enamel on magnet wire, can result in diminished dielectric strength. The long-time effects of incompatibility between components of an insulation system will be discussed in the next chapter.

3.1.9   Surrounding media

The surrounding media can have a great deal to do with the dielectric strength of a material or system. If a condition of corona exists within a gaseous medium, the dielectric strength will be lower than if the gas were replaced by a liquid insulant, a resinous potting compound, or an electronegative gas such as SF_6 (sulfur hexafluoride). Even good varnishing can improve the dielectric strength by raising the corona start voltage.

Conducting media must be avoided to assure adequate dielectric strength levels. However, if such media are unavoidable, as for example in immersion-type water heaters, or other devices that must operate in water, complete isolation of leads must be provided. Another approach is to provide metal isolation of the entire device. Still another successful approach is to pot the device in a water-repelling material, such as silicone rubber.

An organic insulating material placed in a medium of irradiated air or gas, or within a radiation field itself, can become nonfunctioning within a short period of time. Even irradiated materials removed from a radiation field when tested are also likely to exhibit greatly reduced dielectric strengths. This loss is caused by the induced ions set up internally within the material by radiation exposure. This ionic behavior, indicated by greatly reduced resistivity values, [5] can be relieved by leaving the material for an extended time at room temperature or for a short-time period in an oven at temperatures comparable with use temperatures. A space charge theory for the drastic changes in dielectric strength caused by radiation exposure is given in References (6) and (7).

Moisture in a material exposed to radiation will slow up the effects of penetration, but it may produce surface contamination that can promote early flashover or may tend to minimize high-stress areas, thus leading to higher corona-starting voltages.

The effect of altitude on the surrounding medium, air, is to reduce the breakdown voltage of the air as the altitude is increased. A point will be reached at which a glow discharge will be present at high altitude instead of flashover. To combat this phenomenon, all high-voltage air-borne equipment must be potted or otherwise isolated from the media.

At the other end of the scale, enclosed devices may be operated under pressures greater than atmospheric ones. This situation improves dielectric strength in several ways. It may close gas pockets within the insulation, and it may mitigate corona by raising the corona level above the operating voltage. In the latter case the dielectric strength will be improved because of the lower corona intensity, if any is present at all.

3.1.10   Test conditions

Test conditions include all of the discussion in Par. 3.1.9 above plus temperature considerations. Temperature of test is a very important condition. Moderately elevated temperatures, if over $100°$ C, will dry out a material or a materials system, thereby improving the overall dielectric strength. In addition, any ionic residue left from previous radiation exposure or corona may be eliminated.

Expansion effects caused by temperature, however, may develop mechanically stressed areas, delamination of poorly made systems, and rupture of embrittled insulation. All effects will lower the dielectric strength, and some effects may mean end of useful life (for example heat-shock cracking).

Temperatures raised to operating levels sometimes exceed the glass temperature (softening point) for resinous insulations. In this case, added ionic mobility will lower the resistivity and the dielectric strength. Usually, a 50-percent lowering of dielectric strength is observed and accepted as a workable level for a useful system. It must be pointed out, however, that even though the softening point is reached and sometimes surpassed by test conditions, the use of thermosetting resins precludes any "run out" of resin such as might be expected of thermoplastics. Although the thermosetting resins do become "rubberlike," they hold their shape.

3.1.11 Design of insulation system

The proper design of an insulation system in which several materials of varying dielectric constant are utilized in series can gain for the system a superior dielectric strength. The following example, limited to high-voltage power-cable design, is used to explain the principle of optimum use of materials. The next chapter will discuss the use of this principle in selecting thermal life and compatibility of various components with regard to their functions in the system.

In the following demonstration, a method is shown for combining three different types of paper insulation so that the overall dielectric strength will be increased by 75 percent over that of a single-paper insulation.

Starting with concentric cylinder electrodes (for a high-voltage cable), the dielectric breakdown voltage is given by the following equation:[2]

$$e = gr[(\log R \mid r) \mid M] \tag{3.3}$$

in which:

g = breakdown voltage gradient, in kV/cm
r = conductor radius, in cm
R = inner radius of sheath, in cm
M = log of the Naperian base

If any one of three available papers, each with a g value of 100 kV/cm, is used alone, the value of e in Eq. 3.3 becomes 75.3 kV. The dielectric constants of the

Table 3.1 Dielectrics in Series*

	Concentric cylinders				Parallel plates			
	3 Dielectrics		2 Dielectrics		3 Dielectrics		2 Dielectrics	
e	$\dfrac{g_1 r_1 K'_1 A}{M K'_1 K_2 K_3}$	(3.4)	$\dfrac{g_1 r_1 B}{M K'_2}$	(3.8)	$g_1 K'_1 C$	(3.11)	$\dfrac{g_1 D}{K'_2}$	(3.15)
g_1	$\dfrac{Me K'_1 K'_2 K_3}{r_1 K'_1 A}$	(3.5)	$\dfrac{Me K'_2}{r_1 B}$	(3.9)	$\dfrac{e}{K'_1 C}$	(3.12)	$\dfrac{e K'_2}{D}$	(3.16)
g_2	$\dfrac{Me K'_1 K'_2 K'_3}{(r_1 + t_1) K'_2 A}$	(3.6)	$\dfrac{Me K'_1}{(r_1 + t_1) B}$	(3.10)	$\dfrac{e}{K'_2 C}$	(3.13)	$\dfrac{e K'_1}{D}$	(3.17)
g_3	$\dfrac{Me K'_1 K'_2 K'_3}{(r_1 + t_1 + t_2) K'_3 A}$	(3.7)	—		$\dfrac{e}{K'_3 C}$	(3.14)	—	

* Unknowns in these expressions are as follows:

e = potential of high voltage electrode, in kV
g_1 = maximum gradient of t_1 cylinder, in kV/cm
g_2 = maximum gradient of t_2 cylinder, in kV/cm
g_3 = maximum gradient of t_3 cylinder, in kV/cm
K'_1 = dielectric constant of paper touching high-potential electrode
K'_2 = dielectric constant of paper in middle or touching ground-potential electrode
K'_3 = dielectric constant of paper touching ground electrode

t_1 = thickness of paper touching high-voltage electrode
t_2 = thickness of paper in middle or touching ground electrode
t_3 = thickness of paper touching ground electrode
r_1 = conductor radius, in cm
R = inner radius of sheath, in cm
M = $\log e = 0.4343$

$A = K'_2 K'_3 \log[(r_1 + t_1) \mid r_1] + K'_1 K'_3 \log[(r_1 + t_1 + t_2) \mid (r_1 + t_1)] + K'_1 K'_2 \log[R \mid (r_1 + t_1 + t_2)]$
$B = K'_2 \log[(r_1 + t_1) \mid r_1] + K'_1 \log[R \mid (r_1 + t_1)]$
$C = (t_1 \mid K_1) + (t_2 \mid K_2) + (t_3 \mid K_3)$
$D = K'_2 t_1 + K'_1 t_2$

three papers when oil-filled are $K'_1 = 5.4$, $K'_2 = 3.6$, and $K'_3 = 2.0$. The values of r and R are 0.5 cm and 2.25 cm, respectively. What is the optimum arrangement of the three papers and what is the best dielectric strength?

Reference to Table 3.1, for three concentric cylindrical insulations, will give the potential of the inner conductor by means of Eq. 3.4. The maximum gradients are also shown in Eqs. 3.5, 3.6, and 3.7 for the three concentric insulations. Each of these gradients can be assigned the 100 kV/cm value given. These maxima appear at the inner edge of each paper cylinder. Although any sequence of papers can be used, the best arrangement is the one in which the highest dielectric constant type is closest to the high potential conductor and the lowest dielectric constant type is adjacent to the outer sheath.

By trial-and-error methods, t_1, t_2, and t_3 are found to be 0.25, 0.6, and 0.9 cm, respectively. By applying these thickness values, the optimum dielectric breakdown voltage becomes 133 kV, a 75 percent increase over the 75.3 kV obtained with a single paper type.

Table 3.1 also contains equations (Nos. 3.8 through 3.17) for the two-concentric cylinder case and for the three- and two-insulation parallel-plate cases.

Two dielectrics in series do not always mean a higher dielectric strength than one alone. If we assume an air gap of 2 cm with an impressed voltage of 60 kV ($g = 30$ kV/cm), the air holds this voltage since sea level breakdown requires a 31-kV/cm gradient (62-kV breakdown voltage). The voltage is removed and a 0.2-cm thick piece of pressboard (with $K' = 4$) is placed in the air gap with a gap of 1.8 cm in series with the pressboard. The pressboard has a maximum breakdown gradient of 175 kV/cm. The 60-kV voltage is again impressed on the gap plus the pressboard. The following conditions exist:

1. The total capacitance of the system is increased.
2. The electric flux is decreased.
3. The air-flux density is increased.

The air-gap flux is given by:

$$\psi = C_a e_a = C_p e_p \tag{3.18}$$

in which:

ψ = air-gap flux
C_a = capacitance of air
C_p = capacitance of pressboard
e_a = voltage on air
e_p = voltage on pressboard

Then,

$$\left(\frac{K'_a A}{4 \Pi t_a}\right) e_a = \left(\frac{K'_p A}{4 \Pi t_p}\right) e_p \tag{3.19}$$

in which:

t_a = air gap thickness
t_p = pressboard thickness
K'_a = dielectric constant of air (1.0)
K'_p = dielectric constant of pressboard (4.0)
A = area of electrodes

Since,

$$e = e_a + e_p$$

$$\frac{K'_a e_a}{t_a} = \frac{K'_p (e - e_a)}{t_p} \tag{3.21}$$

Therefore,

$$e_a = \frac{t_a K'_p e}{K'_a t_p + K'_p t_a} \tag{3.22}$$

and

$$g_a = \frac{K'_p e}{K'_a t_p + K'_p t_a} \tag{3.23}$$

Substitution of the known values into Eq. 3.23 shows that the gradient is 32.5 kV/cm on the air component and that this component breaks down. All of the stress is thus placed on the pressboard with a gradient of 300 kV/cm, but the pressboard also breaks down. Thus, despite the insertion of a piece of better insulation, the air gap that previously held a certain voltage now cannot hold the voltage.

The above discussion makes evident that the utilization of dielectrics in series can both work for and against good design, depending on the circumstances. It also shows that where air is involved in the insulation system, particularly in a series arrangement, overstressing of the air can be a hazard to be avoided in best design practice. Internally trapped air should also be removed, as noted in Par. 3.1.4.

3.1.12   Frequency and wave shape

The frequency and wave shape of the test voltage can change the apparent dielectric strength of a system of insulation. Impulse testing for example, where a steep-wave-front single-impulse voltage is impressed on the system, will yield a higher dielectric strength for some materials, such as built-up mica-flake insulation for form-wound coils, than is seen in tests at 60 Hz.

Usually, higher-frequency testing has only minor influence on dielectric strength levels until a resonance occurs (called an *absorption band*). In this case, the loss-producing members (chain links, dipoles, ions, atoms, or electrons) resonate with the applied frequency and produce a higher loss than in the nonresonant frequency response. This loss frequently lowers the dielectric strength. Above the critical band the dielectric loss and strength return to "normal" values. This type of action can occur in the audio, radio, and microwave regions, and even in the infrared regions for vibration loss.

Many better high-frequency insulations used at the present time do not show sufficient absorption at resonant frequencies to seriously affect the dielectric strength in the manner indicated above. However, the threat is present for the use of materials known to show high loss at certain resonant frequencies.

A decrease in frequency below 60 Hz can also cause an increase in dielectric loss due to ionic activity and interfacial polarization. In some cases this loss may be sufficient to lower the dielectric strength.

When high-moisture-content insulations are subjected to frequencies on the order of 10 to 100 kHz, they will become internally heated (dielectric heating) and at this time will show low dielectric strength levels. Later, as the moisture is driven off by the heating, the temperature subsides, and a partial return to higher dielectric strength is observed.

Wave shape is important mainly because most voltage reading is done by rms meters calibrated to sine-wave voltages. If the nonsinusoidal wave shape contains sharp peaks superimposed on a sine wave, the meter responds only to the latter, and when breakdown occurs on the sharp peak, the meter will read a low value for the breakdown (apparent breakdown voltage).

3.1.13   Temperature effects

Temperature reduces the dielectric strength in a uniform manner for some materials. Figures 3.4 and 3.5 show linear plots of log dielectric strength vs reciprocal of absolute temperature for various ceramics and glass.

The form of the relationship in Fig. 3.4 is as follows:

$$log\ D.S. = (m\ /\ T_a) + K \qquad\qquad (3.24)$$

in which:

$D.S.=$ dielectric strength, in volts / mil
T_a = absolute temperature, in ° K
m, K = constants

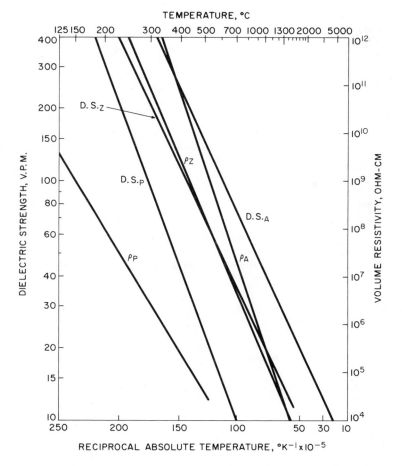

Fig. 3.4   Effect of temperature on dielectric strength and resistivity of ceramics[9]

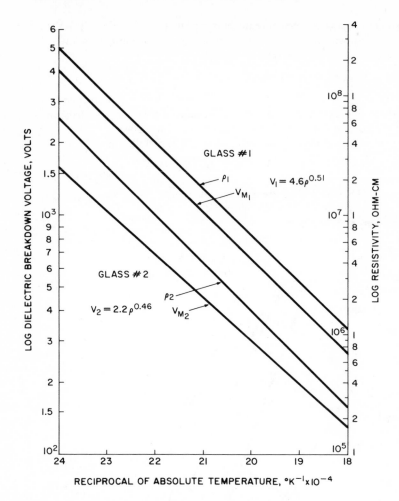

Fig. 3.5 Effect of temperature on dielectric breakdown voltage and resistivity of glass[10]

Typical values of m and K in Figure 3.4 are as follows:

	m	K
Alumina	1,170	0.7425
Zircon	1,062	0.471
Porcelain	823	0.060

The form of the breakdown voltage in Fig. 3.5 is as follows:

$$log\ V_m = (m\,/\,T_a) + K \qquad (3.25)$$

in which:

V_m = breakdown voltage, in volts
T_a = absolute temperature, in °K
m, K = constants

The values of m and K in Figure 3.5 are:

	m	K
Glass No. 1	2,000	−1.20
Glass No. 2	1,820	−1.17

From Eqs. 3.24 and 3.28 the following combined form can be evolved:

$$D.S. = K_T \rho^n \tag{3.26}$$

in which:

$D.S.$ = dielectric strength, in volts per mil
$K_T = [K_1 - K_2 (m_1 / m_2)]$ antilog
$n = m_1 / m_2$
ρ = resistivity, in ohm-cm
m_1, K_1 = constants in Eq 3.24
m_2, K_2 = constants in Eq. 3.28

Values for n and K_T are as follows:

	n	K_T
Alumina	0.156	6.152
Zircon	0.180	3.953
Porcelain	0.179	4.227

Governed by a very similar expression to that of Eq. 3.26, the breakdown voltage is also dependent on resistivity, with the following constants (Fig. 3.5):

	n	K_T
Glass No. 1	0.51	4.603
Glass No. 2	0.456	2.240

The K_T values in each table above differ according to the thickness value involved (in mils), but the exponents (n) are on an equal plane. Thus for ceramics, the dielectric strength varies as the $\frac{1}{6}$ power of the resistivity, approximately; for glass, the dielectric strength varies as the $\frac{1}{2}$ power of the resistivity (approximately as the square root of resistivity). Additional discussion on resistivity will be given in Par. 3.5.

Chapter 4 deals further with dielectric strength by considering the environmental/extended-time effects such as occur with exposure to thermal aging, voltage endurance, hydrolysis, nuclear irradiation, and the like.

3.2   Dielectric Constant

Unlike dielectric strength, which is always a vital factor in insulation, the dielectric constant is sometimes important and at other times of little significance. For example, it is extremely important for capacitor dielectrics to have a relatively high dielectric constant and to retain this value over a wide range of service conditions and for extended time periods. Many undesirable dielectrics show increased dielectric constant at higher dielectric losses. For capacitors, materials such as mica are chosen because they show high dielectric constant with low dielectric loss and maintain these advantages over a broad temperature range. Very high dielectric-constant materials are also in demand, such as barium titanate $(K = 1,500)$ and titanium dioxide $(K = 100)$, since they permit a reduction in size of the unit capacitors needed in electronic circuitry. Conversely, high-voltage cable requirements call for high dielectric strength but very low dielectric constant and dielectric loss.

Application areas for which the dielectric constant is of little importance are found in small power d-c motors (automotive) and small power transformers for low-frequency electronic power supplies. For most areas, excluding capacitors, in fact, electronic frequencies call for low-dielectric-constant insulation.

In electronic coils, the interloop capacitance is a major factor in determining the quality ("Q") of a coil. High "Q" coils result from low-dielectric-constant insulation plus some winding methods that offer as little turn-to-turn contact as possible. Low dielectric constant is also mandatory in communication cables to minimize attenuation of the signal. Air is frequently used as much as possible in communication cables of coaxial design. Use of low-dielectric-constant spacers is kept to a minimum to maintain the configuration within the cable.

When air is involved in power-frequency devices of moderately high voltages, the supporting insulation is usually made of a low-dielectric-constant material to avoid a mismatch that would concentrate all the stress on the air. When high-dielectric-constant materials are necessary, it is sometimes appropriate to "stress-grade" the surface of the solid insulation so as to eliminate air breakdown and corona conditions. An example is provided by high-voltage apparatus.

In high-voltage generator coils the stress can be graded by means of a semiconducting surface coating. This approach fits *every* high-voltage generator and motor situation. However, it may not fit where a feed-through problem exists with high-voltage bushings (sometimes referred to as "condenser bushings"), such as are used in transformers. To accomplish stress gradings in these bushings, a stepped dielectric constant is used. In effect, it is necessary that the parallel conductance of the surface be so designed that it increases the apparent dielectric constant of the insulation near the grounded stator iron.

A stepped dielectric constant is formed in the condensor bushings by the use of several types of paper with the one that possesses the highest dielectric constant placed closest to the conductor. This method also has been used in power-cable design.

Low-frequency operation (below 60 Hz) often produces a higher dielectric constant as well as higher losses. The raising of temperature of a material above its glass temperature (softening point) will also add to the dielectric constant by way of increased ionic behavior. Intrusion of moisture (which has a dielectric constant of over 80) into an insulating material will raise its overall dielectric constant, but it will also add to its losses.

Usually in cases of power applications such as motors and transformers, the increases in dielectric constant caused by temperature effects, frequency change, and moisture contamination are not sufficient to affect the operating characteristics of the device, but other long-term considerations such as hydrolysis may eventually shorten the life of the unit.

The effect of dielectric constant on dielectrics in series is covered in the discussion in Par. 3.1.11 on design of insulation systems. It is important that materials of highest dielectric constant be used at the position of highest voltage gradient and that successive materials be arranged in descending dielectric constant values as the voltage gradient declines.

Ceramics are often used in design of high-voltage insulators and bushings to resist weathering, effects of creepage, and flashover. The ceramics have dielectric constants far exceeding that of air, the usual medium where insulators operate. It is possible to design both electrode and insulator to minimize stress on the air. The common overhead line insulator meets this objective. It has an inverted saucer shape with a smooth round upper surface and a convoluted lower surface. The design is completed with well-rounded upper and lower electrodes (also functioning as fasteners). The larger ceramic bushings for transformer lead feed-through have a series of convolutions from tank to terminal with interspersed sections of air and ceramic in the flashover path.

3.3   Dielectric Loss

The dielectric loss of a material is evaluated by measuring its dissipation factor and dielectric constant on a capacitance bridge and by multiplying these parameters by the square of the applied voltage, the frequency, the ratio of area to thickness (for the parallel plate case), and a constant:[8]

$$W = 2\pi e_0 f E^2 \, (A \, / \, t) \, K' \, tan \, \delta \qquad\qquad (3.27)$$

in which:

$$W = \text{dielectric loss, in watts}$$
$$2\rho e_0 = \text{constant} = 0.556 \times 10^{-12} \text{ farads / cm}$$
$$f = \text{frequency, in Hz}$$
$$E = \text{applied volts}$$
$$A = \text{area, in cm}^2$$
$$t = \text{thickness, in cm}$$
$$K' = \text{dielectric constant}$$
$$tan \, \delta = \text{dissipation factor}$$

Equation 3.24 shows that the dielectric loss varies directly as the frequency, the square of the applied voltage, the area, the dielectric constant, and the dissipation factor, and inversely as the thickness of the insulation. These relationships explain why devices for high-frequency, high-voltage, and high-capacitance applications must use low-dissipation-factor dielectrics.

Low-loss ceramics and glass such as alumina, porcelain, low-alkali glass, quartz, and the like are used as high-voltage insulators. The materials are useful in several areas, such as high temperatures, exposure to weather, exposure to nuclear radiation, and where mechanical strength may be needed.

In less severe conditions, low-loss organic resins and compounds are available. The highly heat-stable polyimides (in enamel and sheet form) are outstanding in low-loss characteristics over a wide range of temperature. The imide linkage is so stable that it imparts a measure of stability and low loss to such dual molecular linkages as amide-imides and ester-imides as well as to the ester-amide-imides.

Polyamides in fiber form (DuPont's Nomex) have been improved to the point of being useful as phase and ground insulation for Class H and higher temperature motors.

Polyethylene has been used extensively in coaxial cable for high-frequency transmission because of its very low dielectric loss.

Polystyrene is a further example of a plastic with very low dielectric-loss characteristics. This plastic is used for structural purposes at high frequency because of its rigidity.

Polytetrafluoroethylene is a high-temperature polymer with extremely low dielectric losses over a wide range of temperature, frequency, and relative humidity. It also has high temperature stability. Its weaknesses are its vulnerability to attack by corona (affecting voltage endurance) and by nuclear radiation and its low resistance to mechanical flow. These characteristics will be discussed in greater detail in Chap. 4.

Thermoplastic polymers such as polyamide have true melting points at which they show phase changes from solid to liquid as the temperature is raised. Many thermoset polymers (cross-linked chain structure) do not actually melt but go through a transition temperature where some rigidity is lost and the material becomes more pliable and rubberlike. This transition is sometimes called the "glass point." It is often accompanied by increased dielectric loss as the temperature rises through the region. The explanation for this effect is that the ions and loss components within the material are "frozen in" at temperatures below the transition point but are more or less freed at this point and at temperatures above and produce increased dielectric loss. Some materials have transition temperatures so high that they actually are thermally deteriorated before the transition point is reached. Two such polymers are polytetrafluoroethylene and polyimides.

In tests where voltage is applied at rising frequencies above 60 Hz, the dielectric loss is affected in three ways. First, the frequency dependence is a direct effect and is extremely important at electronic frequencies. The second effect is the change in dielectric constant. (This change will be covered in detail in Par. 3.4. Briefly, the usual changes occur at loss peaks where polar molecules resonate with the applied frequency. Here the dielectric constant takes on a reduced value for rising frequency.) The third effect is that of the dissipation factor, which rises abruptly at or near the resonating frequency and recedes to normal above this frequency.

In exploring the sub-60 Hz region, some materials show rising dissipation factors and dielectric constants that may off-set the lowering of the frequency.

3.4   Dispersion and Absorption

3.4.1   Dispersion

Dispersion is the term used to describe the behavior of dielectric constant at resonance frequency in the dielectric-loss study of insulation for electrical and electronic devices. The dielectric constant suffers a reduction in value with rising frequency as the resonance band is approached and passed. The explanation is that a certain polarization of molecular sections of the polymer can occur at frequencies below the transition but not above it. Since polarization contributes to the dielectric constant, it is easily seen why the higher-frequency dielectric constant is of a lower value than the low-frequency value.

Dispersion may be a problem in electronic circuits where steady capacitance is required. Thus the choice of insulation for a capacitor would be based partially on the absence of a dispersion region in the range of frequency utilized. Dispersion might be used as a control or switch in electronic circuits.

Dispersion is not usually encountered in power-frequency apparatus.

3.4.2   Absorption

Absorption is the name given to the change in dissipation factor with frequency at the resonance point described above. In the vicinity of the resonance frequency the polar components in the insulation begin to show a movement which absorbs energy from the applied field. The dissipation factor (D.F.) begins to rise and then rises sharply to a peak with further increase in frequency before returning sharply to relatively the same level as at lower frequencies removed from the resonance band. The increased loss at the peak of the absorption produces internal heating within the dielectric which may itself raise the D.F. still further. If the applied voltage is sufficiently high at this point, a "runaway" condition, referred to as "thermal breakdown," can result. This possibility is the reason for the dielectric strength dip referred to in Par. 3.1.12 on frequency and wave shape.

Several absorption peaks may be observed for a given insulation in sweeping the frequency range from power to microwave and up to infrared frequencies. Each active component (polar molecule, dipole, polymer chain fragment, atom, or electron) will in turn, at specific frequency bands, resonate and absorb energy. The most important of the resonating parts is the dipole, a chain fragment which is polarized by the applied field and resonates somewhere in the audio, radio, VHF, or UHF regions. For best insulating properties, materials should be selected that avoid dipole absorption at the intended use frequency range.[1]

There are two other types of absorption regions. (1) Interfacial absorption, which occurs where a laminar barrier is present in a built-up wall of insulation. At a certain frequency (usually low) a maximum absorption of energy will be extracted from the electric field as ions and electrons within the insulation are stopped by the barrier and then removed as the field reverses. (2) Low-frequency absorption is shown in ions by conduction, even with no interfacial action. However, this latter type of absorption does not peak but continues to rise as the frequency is lowered. Absorption is minimized by using relatively pure, nonpolar, organic plastics and resins, ceramics, glasses, mica, and some liquid insulants such as transformer oil.

3.5   D-C Volume Resistivity

The volume resistivity of an insulating material is dependent on the ionic or electronic content of the material and the ease of moving these charged particles through the insulation under the applied voltage. Both the magnitude of the voltage and the temperature of the material are important. To these factors should be added the history of the insulation, including exposure to such environments as moisture, corona, and nuclear radiation.

Polystyrene is an example of a material with a very high d-c resistivity. It has a very low ionic content when made without contamination, and at room temperature it is very rigid. This rigidity ties down the movement of any chance ionic impurities present. As the temperature is raised, a slow rate of change in resistivity is

---

[1] Dipole absorption, however, is used to advantage in dielectric heating processes, in which a material is heated by absorption of energy when placed in a high-frequency field. The process is used, for example, in plywood production where the adhesive resin is heated in order to solidify the individual plies rapidly to the desired strength.

observed until the transition temperature is reached. (For a discussion of transition temperature, see Par. 3.3.) At the transition temperature the rigidity gives way to a semiflexible state in which ions are released. The ions are able to move under the applied field, thereby imparting increased conductivity and reduced resistivity to the material.

Many glass and ceramic materials show a linear semi-log plot between log resistivity and reciprocal absolute temperature. They also show a linear relationship between log dielectric strength (volts per mil) and reciprocal absolute temperature. On simultaneous plots using different log scales (at a ratio of two to five cycles of resistivity to one of dielectric strength), the two characteristics can be made to run

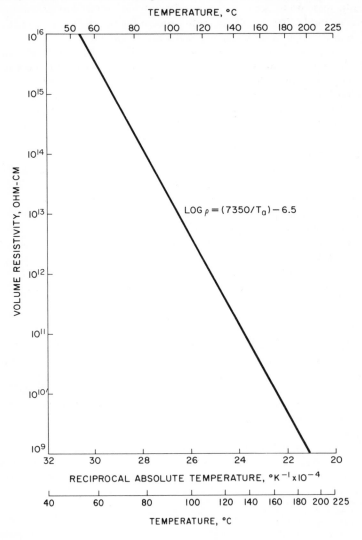

Fig. 3.6   Effect of temperature on resistivity of an epoxy resin[11]

in nearly parallel lines for certain ceramics and glasses. (See Figs. 3.4 and 3.5)[9, 10]. An interesting combined form of dielectric strength and resistivity is shown in Eq. 3.26 in Par. 3.1.13.

Resistivity can thus be a controlling influence on dielectric strength, at least at elevated temperatures. At low temperatures the dielectric strength appears to be independent of resistivity within the practical range of useful resistivity values for good insulation. Other conditions (such as moisture or radiation exposure) that lower the resistivity are also known to lower the dielectric strength of organic insulation. This aspect of the discussion will be covered in Chap. 4.

Returning to Figs. 3.4 and 3.5, which show both resistivity and dielectric strength for zircon, alumina, and porcelain and for two glass types, an empirical relationship for resistivity is obtained as follows:

$$log \, \rho \, = \, (m \, / \, T_a) \, + \, K \qquad (3.28)$$

in which:

ρ = resistivity, in ohm-cm
T_a = absolute temperature, in °K
m, K = constants

Figure 3.6 also shows the same characteristic for an epoxy resin.[11]

Typical values of m and K are as follows:

	m	K	Figure No.
Epoxy resin	7,350	−6.5	3.6
Glass No. 1	3,940	−1.05	3.5
Glass No. 2	4,000	−1.80	3.5
Porcelain	4,600	3.16	3.4
Alumina	7,500	−0.3	3.4
Zircon	5,900	0.7	3.4

Moisture, although it is an environmental factor rather than an operational one, shows similar effects to those of temperature.[12] The relationship between moisture and resistivity for nylon is indicated in Fig. 3.7. The semi-log plot for log volume resistivity vs percent moisture yields the straight line shown. It seems that the introduction of moisture releases ions within the nylon much as rising temperature does.

The relationship between resistivity and percent moisture is given by the equation:

$$log \, \rho \, = \, K \, + \, m \, (\text{P.M.}) \qquad (3.29)$$

in which:

ρ = volume resistivity, in ohm-cm
P.M. = per cent moisture
K, m = constants

As shown in Fig. 3.7, the values of K and m are 15.12 and − 0.748, respectively.

Volume resistivity is often affected by surface resistivity, the two being difficult to separate unless guarded electrodes are used. Thus, in design considerations it is as important to provide sufficient surface creepage distance as to use materials with high volume-resistivity levels. Insufficient creepage distance plus contamination can lead to either flashover or tracking, both of which will be covered in Par. 3.7.

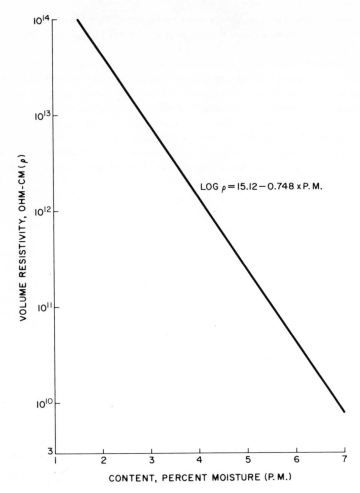

LOG $\rho = 15.12 - 0.748 \times$ P. M.

Fig. 3.7   Effect of moisture on resistivity of nylon[12]

3.6   Corona

Corona was described in Par. 2.16 of Chap. 2 as a partial breakdown of an electrical system involving the air or gas in the immediate vicinity of an electrode in series with a solid or liquid insulation extending to the opposite electrode. The avalanche theory provides the explanation for the wide-spread ionized molecular glow observed while corona is in action. Both external and internal corona can occur. Internal ionization occurs in voids within the insulation wall, and pressure is used to control it. External corona occurs at the insulation surface where the gaseous medium is over-stressed. Both types of corona will produce surface degradation in nearby insulation (if it is solid). This degradation, if not checked, will lead to eventual dielectric breakdown or surface tracking.

A study of the complete insulation wall after dielectric failure from corona attack shows that surface erosion does take place but that a considerable amount of wall remains with no indication of physical damage. In fact, the remaining insulation can be expected to hold the voltage applied at breakdown. A recent study of this condition showed that a very low resistivity exists during and following the failure by corona attack. A theory has been proposed that the insulation seemingly left uneroded is in fact penetrated by a space charge effect which extends well into the remaining wall and becomes an extension of the conducting corona cloud that is present around the insulation. Such a penetration would explain the very low dielectric strength observed after very short time periods with corona-producing voltage. This resistivity change caused by corona is quite similar to the effect of radiation exposure, and both conditions can be cleared up by heating the insulation in an oven for a short time.

The various means for control of external corona are as follows:

1. Removing air or gas (other than electronegative types) from critical areas by varnish, impregnation, potting, oil-filling, or embedment of electrodes

2. Changing the gas to an electronegative type (such as SF_6) so that electrons may be trapped before they form a corona cloud

3. Removing sharp edges on electrodes

4. Increasing the gas pressure

5. Using semiconducting components or surface treatment materials to grade the stress below the corona inception voltage

6. Using stress rings to alter the electric field in the vicinity of the gas-insulation-electrode area

7. Avoiding the use of high-dielectric-constant materials adjacent to air.

In removing the air or gas, the process must be thorough since any gas pockets or cracks and poor adhesion areas will ionize even more readily and the system will still be in danger. Often vacuum impregnation is needed to insure complete fill. Electrode embedment is usually expensive and requires the use of material similar to potting.

Electronegative gases, such as SF_6, and fluorocarbon gases require an enclosure as does the increased gas pressure method. These effective gases are used in transformer design.

The removal of sharp edges on the electrodes is a prerequisite to all corona-free systems. The effect of edge sharpness on both the ionizing distance (the distance necessary before an avalanche can start and sustain corona) and the ionizing voltage is shown by the wire radius in the following equations for parallel wires. The first equation is

$$d_i = 0.301 \sqrt{r} \qquad (3.30)$$

in which:

d_i = minimum ionizing distance, in cm
r = radius of wire, in cm

Thus, a 0.01-cm diameter wire requires a distance of 0.0213 cm (0.0084 in.) for ionizing air (corona start) whereas a 0.05-cm diameter wire requires 0.0477 cm (0.0185 in.), twice the distance of the smaller wire. Then,

$$g_i = 29.8\left[1 + (0.301 / \sqrt{r})\right] \tag{3.31}$$

in which:

g_i = minimum peak ionizing voltage gradient for corona start, in kV/ cm
r = radius of wire, in cm

In this case, the peak gradient needed for corona on the 0.01-cm wire would be 156.5 kV/cm (398 V/mil) whereas that needed for the 0.05-cm-wire would be 86.3 kV/cm (219 V/mil). The expression for e_i becomes

$$e_i = g_i \, r \, ln(S / r) \tag{3.32}$$

in which:

e_i = minimum a-c peak voltage for corona, in kV
r = wire diameter, in cm
S = wire spacing (wire-to-wire), in cm

For very small spacings (adjacent wires), small wire diameters (magnet wire), and thin insulation, the value of S is replaced by d_i, the minimum distance for corona. Thus the minimum peak a-c voltage for corona on the 0.01-cm wire is 1,130 volts and on the 0.05-cm wire, 1,390 volts.

For temperature and pressure effects, air density (δ) is used:

$$\delta = 3.92b / (273 + t) \tag{3.33}$$

in which:

δ = density of air
b = barometric pressure, in cm
t = temperature, in °C

The density is unity when $b = 76$ cm and $t = 25°$ C.

For surface irregularities on the conductors, a factor m_v is used as a multiplier, its value ranging from 0.25 to 1.0. Therefore, Eq. 3.32 becomes

$$e_i = 29.8 \, m_v \, \delta \left[1 + (0.301 / \sqrt{r})\right] r \, ln(S / r) \tag{3.34}$$

The value of m_v is 0.5 in outdoor conditions when water, sleet, or snow are present. Oil and insulation have very little effect if the conductor is smooth. The factor m_v is 0.82 for stranded wire. Current in the conductor produces heat and lowers the value of δ.

Corona takes energy away from a system, and, if that system is a transmission line, the fair-weather corona loss, in kW/km, will be:

$$P = (241 / \delta) \sqrt{r / S} (f + 25)(e - e_0)^2 \tag{3.35}$$

in which:

δ = air density
r = radius of conductor, in cm
S = spacing between conductors, in cm
f = frequency, in Hz
e = 1 / 2 line voltage for single-phase, in kV, or 1 / 3 line voltage for three-phase, in kV
e_0 = 1 / 2 breakdown voltage for single-phase, in kV, or 1 / 3 breakdown voltage for three-phase, in kV

Furthermore,

$$e_0 = m_0 g_0 \delta r \, ln(S/r) \text{ kV to neutral}$$

in which m_0 is the irregularity factor (1.0 for smooth wire, 0.98 to 0.93 for roughened wire, 0.87 to 0.83 for 7-strand cables) and g_0 equals 21.1 kV/cm (disruptive gradient for air at 25° C and 76-cm pressure).

For small-diameter wires (less than 0.2 cm) the e_0 value, in kV, is

$$e_0 = m_0 g_0 \delta \, r \, ln(S/r) \left[1 + \left(\frac{0.301}{\sqrt{r\delta}} \times \frac{1}{(1 + 230r^2)} \right) \right] \qquad (3.36)$$

That corona is reduced by higher pressures or replacement of air with SF_6 gas is shown in Fig. 3.8, where the corona-initiating voltage rises with pressure increase for both air and SF_6 in nonuniform fields[13].

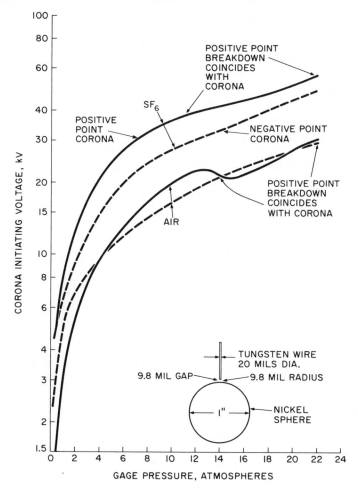

Fig. 3.8   Effect of pressure on corona voltage for air and SF_6 for non-uniform field[13]

Both positive and negative d-c points to a spherical electrode were used. The three- or four-to-one advantage in breakdown voltage of SF_6 over air is reduced to about two-to-one when positive point corona is considered. (Breakdown curves were omitted for simplicity in Fig. 3.8.) The negative point corona for SF_6 is only about 50 percent better than the same polarity for air and is below the positive point breakdown for air in some areas.

This study indicates that if the use of pressure and SF_6 gas is contemplated, field distortion must be minimized by careful rounding of all edges and corners to secure the full benefit in corona abatement.

3.7   Flashover and Tracking

Both flashover and tracking are surface phenomena. Flashover is a breakdown of the air or gas in parallel with solid or liquid insulation. Tracking is a disruption of a solid insulation surface between two electrodes. The latter results in a carbonized path which eventually shorts the electrodes.

Flashover is controlled by creepage distance, insulator surface design, and many of the methods suggested for corona abatement in Par. 3.6. In fact, the reduction of corona is a first step in improving flashover voltage levels. The familiar convolutions on the porcelain stand-off insulator and feed-through bushing greatly improve the flashover resistance of these components. By inserting multiple alternating rings of air and ceramic, the corona-start voltage is raised and the flashover voltage is also raised. Even on straight-sided ceramic bushings of the "solder-in" type, the flashover voltage can be improved by using semi-conducting glaze to grade off the stress at the electrodes and ground rings.

Flashover voltage is affected by humidity, type of voltage, spacing, dielectric constant of solid insulator, contamination, and uniformity of field, among other things. Repeated flashover on an organic resin-filled laminate can lead to tracking. ASTM test D-495, which measures the arc resistance of laminates under voltage stress (15,000 V) and arcing conditions, has the following sequence of conditions:[14]

Current, ma	Time cycle, sec	Step time, sec	Total time, sec
10	¼ on–1¾ off	60	60
10	¼ on–¾ off	60	120
10	¼ on–¼ off	60	180
10	Continuous	60	240
20	Continuous	60	300
30	Continuous	60	360
40	Continuous	60	420

Laminates are listed below in order of increasing time as revealed by the above test:[9]

Laminate	Time, sec	Laminate	Time, sec
Phenolic-paper	5–10	Alkyd-glass	175–185
Phenolic-fabric	5–10	Silicone-glass	180
Phenolic-asbestos	5–10	Teflon (TFE)-glass	180
Phenolic-glass	5–10	Asbestos	180
Phenolic-nylon	5–10	Mica	180
Epoxy-glass	100–130	Glass	180
Polyester-glass	140–160	Ceramics	180
Melamine-glass	175–185		

Tracking resistance[15] is measured by the electrolyte drop test.[16] Aqueous NH_2Cl is allowed to drop in 30-sec intervals on an area containing electrodes resting on the horizontal laminate at 4-mm gap spacing. A current of 0.5 amp flowing for 1 sec will trip the relay, but a short-circuit current of 3 amp is possible. The maximum number of drops before failure at each of three voltages (250,300,500 volts a-c) determines the resistance.

Laminates and other insulators are classified as follows by the drop test:[9]

Class A (30 drops at 500 V)[2]

Ceramics
Glass
Mica
Asbestos
Melamine formaldehyde
Urea formaldehyde
Epoxy
Polyester
Alkyd
Polymethyl methacrylate
Polystyrene
Polyethylene

Polyvinyl chloride
Polyamide
Cellulose esters

Class B (7 drops at 300 V, 15 at 250 V)

Phenolic (mineral loaded)
Phenolic blended with melamine
Phenolic-paper
Phenolic-asbestos

Class C (7 drops at 300V, 15 drops at 250 V)

Phenolic-wood flour
Phenolic-glass reinforced

Saito[17] has introduced a dry-tracking method of determining track resistance by using a third (trigger) electrode above the two conventional slanted-point electrodes in contact with the laminate. A truer picture of actual track resistance under dry conditions is claimed for this test since no contaminants are present to mask the results. The trigger method allows a dry test at relatively low test voltage (1,000 V—main electrodes). To establish the arc, a 5,000-volt condenser is discharged from trigger to the grounded main electrode. The resulting ionized air causes the main 1,000-volt condenser to discharge across the insulation surface. Failure occurs when the cyclic arcing current (at 0.45 Hz) rises from 12 to 20 ma. The number of discharges to failure (NDTF) is taken as the tracking resistance figure of merit. Some values of NDTF are shown below:

Material	NDTF at 1,000 V[3]
Silicone rubber	E*
Butyl rubber	10
Chloroprene rubber	1
Phenolic resin paper	3
Epoxy resin paper	12
Molded polyester premix	4
Polyethylene carbon black	1
Polyvinyl chloride	2
Polystyrene	2
Polymethylmethacrylate	E*
Polyethylene (cross-linked)	E*

This chapter has presented important design considerations pertaining to the behavior of insulation in electrical and electronic devices in each of several fundamental characteristics from dielectric strength to surface tracking. Chapter 4 will cover the behavior of insulation under environmental conditions and long-time aging processes.

---

[2] Not to be confused with IEEE Thermal Index Ratings.

[3] E* means erosion type—no carbon formation.

REFERENCES

(1) Paschen, F., *Ann. Physik,* vol. 37, 1889, p. 69.
(2) Peek, Jr., F. W., *Dielectric Phenomena in High-Voltage Engineering,* McGraw-Hill, 1929, p. 382.
(3) Schumann, W. O., *Durchbruchsfeldstarke von Gasen,* Springer, Berlin, 1923, pp. 51, 114.
(4) Hill, L. R., and Schmidt, P. L. "Insulation Breakdown as a Function of Area," *Transactions of the AIEE,* vol. 67, 1948.
(5) Boeck, W., "Effect of Space Charges from Corona on Plastic Films," *Elektrotechnische Zeits,* PtA, vol. 88, No. 26, 1967.
(6) Weeks, R. A., and Binder, D., "Effects of Neutron and Gamma Irradiation on the Dielectric Constant and Loss Tangent of Some Plastic Materials," *Transactions of the AIEE,* vol. 78, Pt. II, 1959.
(7) Dakin, T. W., Philofsky, H. M., and Divens, R., "Effect of Electrical Discharges on the Breakdown of Solid Insulation," *AIEE* No. 54-70, 1954.
(8) Greenfield, E. W., "Refresher for Dielectric Calculations," *AIEE Electrical Engineering,* Oct. 1943.
(9) Clark, F. M., *Insulating Materials for Design and Engineering Practice,* Wiley, 1962.
(10) Inge, L., and Walther, A., "Penetration of Glass in Homogeneous and Non-Homogeneous Electric Fields," *Arch. Elektrotech,* vol. 19, 1928, p. 257.
(11) Delmonte, J., "Electrical and Mechanical Properties of Epoxy," *ASTM Bulletin* (TP 152), vol. 32, 1957.
(12) "Nylon for Electrical Insulation," *Insulation,* vol. 1, 1955, p. 16.
(13) Pollack, H. C., and Cooper, F. S., "Effect of Pressure on the Positive Point-Plane Discharge in N_2, O_2, CO_2 SO_2, SF_6, CCl_2F_2, Ar, and He," *Physical Review,* vol. 56, 1939, p. 170.
(14) *ASTM D 495–61T,* 1961.
(15) Nichel, K., and Hillenkamp, M. H., "Determination of the Tracking Resistance of Organic Insulating Materials," *Bull. ASE* (Switz.), vol. 50, 1959, p. 601.
(16) "IEC Drop Test," *ASTM D 2132-62T,* 1962.
(17) Saito, Y., and Hino, T., "A Dry Tracking Resistance Test Method Using a Trigger Discharge," *IEEE Transactions on Electrical Insulation,* VEI-3, No. 18, 1968.

BIBLIOGRAPHY

A. H. Sharbaugh and P. K. Watson, "Breakdown Strengths of a Perfluorocarbon Vapor and Mixtures of the Vapor with SF_6," *IEEE Transactions,* No. 63-922, 1963.
E. J. McMahon, "The Chemistry of Corona Degradation of Organic Insulating Materials in High Voltage Fields and Under Mechanical Stress," *IEEE Electrical Insulation,* VEI-3, No. 1, 1968.
J. P. Manion, J. A. Philosophos, and M. B. Robinson, "Arc Stability of Electronegative Gases," *IEEE Transactions on Electrical Insulation,* VEI-2, No. 1, 1967.
C. N. Works and T. W. Dakin, "Dielectric Breakdown in SF_6 in Non-Uniform Fields," *AIEE Transactions,* Pt. I, vol. 72, 1953.
G. L. Moses, *Electrical Insulation,* McGraw-Hill, 1951.
L. J. Frisco, "Frequency Dependence of Electric Strength," *Electro-Technology,* Aug. 1961, p. 110.
J. B. Whitehead, D. L. Birx, and C. F. Miller, "The Electric Strength of Air in Non-Uniform Fields at Radio Frequency," *AIEE* No. 53-162, 1953.
J. J. Chapman, J. W. Dzimianski, C. F. Miller, and R. K. Witt, "Behavior of Insulating Materials at Radio Frequency," *Electrical Manufacturing,* July 1951, p. 107.
F. W. Peek, Jr., "The Law of Corona," *Transactions of the AIEE,* 1889, 1911.
A. H. Sharbaugh, "Dielectric Constant and Loss—Their Significance in Insulation Selection," *Electro-Technology,* Feb. 1962, p. 111.

CHAPTER 4 | ENVIRONMENTAL FACTORS IN DESIGN

The preceding chapter dealt with the practical significance of the dielectric characteristics in electrical and electronic design. These characteristics were treated as instantaneous phenomena that determine initial design parameters. However, dielectric materials are not indestructible. They age. They are changed by weather and moisture, by heat and cold, by radiation, by shock and vibration, among other conditions. In short, to function at all times a given design must have built-in safeguards to offset or limit such environmental hazards. This chapter, therefore, will look at those thermal effects that involve aging, both at high and low temperatures, and over extended time periods; it will examine the effects of radiation on good insulation and make comparisons with the effects of corona; it will discuss mechanical stresses, corrosive effects, ultraviolet and ozone exposure, weathering, fungus contamination, ultrahigh vacuua, and the various combined effects that materials encounter. Finally, those processing effects that may be definitely harmful will also be discussed.

4.1 Thermal Aging

Thermal aging produces different effects in different materials. For those solid insulations prone to oxidation, a "burning up" of the insulation occurs where volatile gases and vapors are driven off and the remaining member gets more and more like carbon with a tendency to crack and conduct electricity. Cellulose, rubber, polyvinyl chloride, polyvinyl acetal, acrylic resins, and polyamides are attacked by oxygen in this manner. Liquids like oils used in transformers and cables are oxidized under the conditions that form conducting acids and peroxides. These materials should not be used at temperatures greatly exceeding $105°$ C so as to limit the oxidation mechanism and extend the life of the device of which they are a part.

Many new resins of the last two decades, however, raise this operating temperature by 100 or more degrees. All are more resistant to oxidation than the above-mentioned materials. Among them are terephthalate polyesters (with and without cyanurate modification), polyesterimides, polyimides, polyamideimides, silicone resins, polytetrafluoroethylene, silicone rubber, silicone fluids, high-temperature polyamide fibers, fluorinated liquids, and Dowtherm.[1]

Along with oxidation there are several other thermal aging mechanisms: polymerization, depolymerization, hydrolysis, delamination, and evaporation. While these are active in some cases, the major aging factor is oxidation.

With the new oxidation-resistant materials available it is now possible to design electrical and electronic units in any one of six or eight different thermal classes. A description of thermal classification schemes is given in Par. 4.2.

4.1.1   High-temperature effects

The effects of high temperatures on basic dielectric characteristics (dielectric strength, dielectric loss, and dielectric constant) have already been examined in Chap. 3. Physical changes (such as melting, boiling, creep, or excessive flow) can usually be prevented by proper design precautions. Thus, for the most part, the more dimensionally stable thermosetting polymers (with their cross-linked chains) are preferred to the thermoplastics. However, if the latter have to be used because certain of their properties are desired, the melting points should be well above the anticipated use temperatures. Nylon molding compounds provide an example of the wide usage of thermoplastics in electrical design because of their properties (for example, high impact strength), but the use temperatures are safely below the melting point of these materials. In liquid dielectrics, boiling is avoided by the use of high-flashpoint oils as well as liquids with high boiling points.

Sudden expansion brought about by rapid exposure to heat can produce ruptures in an insulation system. Large electrical machines such as generators and motors are permitted to expand in their slot coils when re-energized after a long shut down period. These coils will move out of the slot a perceptible amount. The tying of such coils must be in such a manner as to maintain this freedom of movement. The same amount of contraction must be permitted on shut-down to avoid a "ratchet action" that would eventually destroy the insulation.

Insulation must provide for heat conduction so that the point of highest temperature will not rise above the maximum level for that material. Good heat conductivity is often accompanied by good electrical conductivity in nature. We need good thermal conductivity but also good electrical resistivity. Some materials are outstanding in this regard. Mica and glass, inorganic-filled-phenolic and -epoxy are examples of materials with good electrical resistivity but high thermal conductivity. Plastic foams, on the other hand, have very poor thermal conductivity but usually good electrical resistivity. Conventional polymeric films and enamels may have good electrical properties but only fair thermal conductivity. Built-in air ducts, water cooling devices, and heat sinks are some of the methods used for cooling. A thorough job of 100-percent solids impregnation of a unit (by vacuum techniques) can improve the heat removal greatly over that of a normal system containing trapped air.

Under given thermal conditions, the hottest spot temperature will determine

---

[1] Dow Chemical Company's tradename for various compounds that provide heat-resistant properties.

the long-time life of a given device or piece of equipment. Thermal classification systems (see Par. 4.2) also provide suitable rating systems for various electrical products—a transformer, say, can be rated as a "Class H" (180 °C transformer).

Thermal hazards are present during the processing of insulation. For example, a form of heat shock (solvent shock) can take place during the application of insulating varnishes. .Such shock can crack previously applied and cured varnish films in adjacent areas. Paragraph 4.11 summarizes various processing effects.

4.1.2   Cryogenic temperature effects

The range of temperatures identified as "cryogenic" extends from nearly absolute zero (−273 ° C) to about −173 ° C, or in the Kelvin absolute scale, from 0 ° K to 100 ° K. The most important development in the cryogenics area has been the application of the phenomenon of superconductivity (the apparent absence of resistivity in a given material at its critical transition temperature) in the design of various new devices, such as the superconducting magnet, cryogenic memory units, and superconductor cables and wires. Detailed discussion of such developments is not within the scope of this work, but we *are* concerned with the effect of cryogenic temperatures on the properties and reliability of electrical insulation. Such concern arises from the fact that many electrical devices are required to operate in or near tanks and lines containing cryogenic fluids and so are directly or indirectly exposed to the cryogenic environment. Frequently the normal characteristics of insulating materials are drastically changed by such exposure.

The cryogenic behavior of organic electrical insulation can be examined through studies of physical and mechanical effects as well as dielectric effects. Quite often, insulation fails through physical or mechanical deterioration *before* any severe loss in dielectric properties takes place. Many organic insulations exhibit an inherent brittleness at very low temperatures which can become an aggravated problem when combined with such mechanical effects as contraction caused by thermal expansion differences with adjacent metal components. Such conditions can occur in encapsulated coils, various types of electronic modules, and the like. The presence of shock or vibration may trigger failure. No generalization can be made, however, about the response of organic insulations to the conditions described. Various tests (usually mandrel flexibility tests in a liquid helium medium) show a diversity of behavior. The flexibility of a given insulation appears to be affected by a number of factors, including its basic structure and microscopic properties.

Various other studies of mechanical properties of organic polymers under cryogenic conditions have led only to generalities or inconclusive data. However, for thermoplastics it appears that tensile strength and impact resistance increase as temperatures decline and then become relatively stable at the lowest point.

Based on still other tests conducted in cryogenic media (such as tests on standard twisted-pair wire samples) the dielectric properties of insulations range from good to excellent. However, the value of such properties can be vitiated if unsatisfactory physical or mechanical effects are present. The criteria for selection of insulation for service under cryogenic conditions therefore requires consideration of *all* factors involved, including a study of the possible effect of the surrounding cryogenic medium in which the insulation is to serve. Some media (such as liquid nitrogen, which is a good dielectric) may increase the dielectric strength of a given material; another medium (such as liquid helium, which has a low dielectric constant and low breakdown voltage) may cause a reduction in the dielectric strength of the material.

Despite the various problems of electrical insulation in cryogenic environments, some materials have evidenced satisfactory performance records. For example, several formulations of magnet wire enamels survive cryogenic exposure and return to room-temperature conditions without apparent damage. Polyurethane and polyvinyl acetal have been used successfully at $4\,°K$ ($-269\,°C$).

The selection of a suitable insulation for superconductive wires imposes particular problems because the wire carries a large current in relation to its size and is always subject to the danger that it will suddenly become resistive and burn out. Even though the wire is protected by being imbedded in copper of larger cross section or by being clad or plated so that, in effect, a circuit shunt is provided, the insulation may be subjected to sudden temperature surges of $200\,°C$, to $300\,°C$. Design of superconducting wires must therefore provide for unrestricted expansion and contraction and the use of an insulation capable of resisting the tremendous forces generated.

As an overall guide to cryogenic applications, it is recommended that all required data for candidate materials be obtained under actual cryogenic use temperatures, and that materials compositions be selected (or modified if necessary) to meet the requirements at hand.

A growing body of information on the effects of cryogenic temperatures on materials is being developed and appears in the literature. References are given at the end of this chapter.[1-6]

4.2   Thermal Classification Systems

A thermal classification system has been established by the IEEE, as listed in Table 4.1. Related tests to determine the classification of materials and materials systems had been established by the American Institute of Electrical Engineers (AIEE) prior to its merger with the Institute of Radio Engineers (IRE) to establish the IEEE. The classification system of Table 4.1 gives applicable materials in the last column.[7]

Table 4.1 IEEE Thermal Classification System

Class	Temperature, °C	Applicable materials
Class O	90	Unimpregnated cellulose
Class A	105	Polyvinyl acetal, polyurethane, impregnated cellulose, polyamide
Class B	130	Epoxy, polyester glass (styrene type), phenolic glass
Class F	155	Terephthalate polyester, modified silicone
Class H	180	Silicone, silicone glass, silicone rubber, ester-imide, cyanurated terephthalate polyester
Class 200*	200	Amide-imide overcoated ester-imide, amide-imide overcoated polyester
Class 220**	220	Amide-imide, polyimide, high-temperature polyamide fibers†
Class C	250+	Polytetrafluoroethylene, glass, ceramics

* U.S. Navy-designated as Class K
** Industry-designated as Class M
† Du Pont's Nomex

The tests employ models that are exposed to several environments at several temperatures and are checked periodically for failure by voltage. The failures, linked to exposure time, are treated statistically to determine the constants of the best straight line in the equation:

$$log\ L = A + B\ /\ T \tag{4.1}$$

in which:

L = life, in hours
T = absolute temperature, in °K
A,B = constants

This equation is a form of the Arrhenius equation for chemical reaction rate studies that were first applied to insulation by T. W. Dakin in 1948.[8]

A relation exists between the *activation energy* of insulation and its resistance to the aging process. The reason for this relationship is that the activation energy is the amount of energy needed to oxidize a gram molecular weight of the insulation. It follows that when the activation energy is low, the insulation is easily destroyed by heat-induced aging in air, whereas a high value for the energy indicates a greater resistance to aging.

In Eq. 4.1 above, the constant B contains the activation agent parameter for the aging process, thus:

$$E = B(R\ /\ M) = 4.574B \tag{4.2}$$

in which:

E = activation energy, in cal / gm-mole
R = gas constant = 1.98 cal / °C / gm-mole
M = reciprocal of natural log of ten = 0.4343

For a polyvinyl acetal system (to take an example), the value of B is 5,110 and E = 23,350 cal/gm-mole. This value represents a rather low activation energy in comparison with those of most insulation materials, for which the range is from 20,000 to over 50,000 cal/gm-mole. Because of its low activation value, polyvinyl acetal is considered to be a low-temperature material and is placed in the 105°C class. A list of temperature classes and corresponding activation energies follows:

Temperature, °C	Activation energy range, cal/gm-mole
105	20,000–23,400
130	23,500–26,300
155	26,400–29,400
180	29,500–32,200
200	32,300–35,100
220	35,200–40,300
250	40,400–47,300
300	47,400–55,000

It should be noted that for ceramic insulations (which are in the higher temperature classes, 250° C and 300° C) the above activation energies do not apply, since the ceramics do not age in the same manner as do the polymeric insulations.

Data from aging tests at accelerated temperatures can be fed into a computer that is programmed to read out the values of constants A and B of Eq. 4.1 and the temperatures at which plotted lines for aging temperature vs life cross specified life levels, such as 5,000, 10,000, and 20,000 hr. These temperatures are known as "temperature indices" since they characterize an insulating material for a given set of heat-aging conditions. If the conditions are altered, a new group of indices would be expected.

Figure 4.1 shows a plot of life lines for materials, all fulfilling the form of Eq. 4.1. The indices can be read from the intercepts of the plotted lines with the

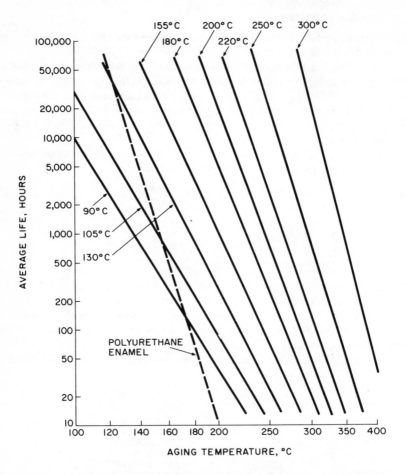

Fig. 4.1 Aging temperatures plotted against average life for electrical insulating materials and systems (Note: Intercepts at the 20,000-hr level conform to temperature indices for classes from 105° to 300°C, the intercept for class 90°C falling outside the graph; the line for polyurethane enamel illustrates occasional divergence from the usual relationship between activation energy and resistance to aging)

20,000-hr line, a level chosen to represent continuous service at full load at the hottest spot in a winding (which is the basis of the standard temperature classes previously listed). Load cycling and intermittent service would greatly extend the life beyond the 20,000-hr level.

The plotted lines in Fig. 4.1 also indicate that insulating materials can be used at higher temperatures if shorter lives can be tolerated. For example, if a material with a temperature index of 155 °C at 20,000 hr is used at 200 °C, it will have a 1,000-hr life. This fact is useful in the design of short-mission devices or whenever down time is a large part of the anticipated service life of a piece of equipment.

It should be noted that occasionally a material will depart from the relationship between activation energy and aging resistance described above. Such a departure from the norm is shown in the dashed line in Fig. 4.1, which plots relationships for polyurethane wire enamel. At the higher temperature level (200° C, for example, the enamel has only a 10-hr life, but at 130°C (only 70° lower), that life extends to 15,000 hr. This behavior makes it necessary to place polyurethane enamel in the Class A (105°C) category.

The behavior of polyurethane enamel may be explained as follows. In a typical set of data for this material, the value for constant B of Eq. 4.1 was shown to be 8,350 and that of the activation energy, 38,200 cal/gm-mole, which would normally locate the material in the class-200° C category. However, the value for constant A was $-$ 16.55, greater than that found for the materials with the best thermal properties. As a consequence, the behavior shown in Fig. 4.1 occurs, and the material falls into the Class A group. In actual service, polyurethane enamel is used in a solderable form, and there is a large degree of evaporation at high temperatures (for example, when hot solder is used at about 250° C). The process of evaporation takes precedence over the mechanism of oxidation, thus leaving little if any significance to the activation energy of the material.

In summary, thermal classification systems offer design engineers the advantage of selecting materials that have a proven capability of meeting life service requirements for predetermined operating temperatures. It should be understood, however, that if some constituents of an insulating materials *system* are not subject to the maximum predetermined temperature, alternative materials in lower temperature categories may be used. This procedure may be followed in cases where insulation is used in close proximity to cooling elements or must perform structural functions as well as insulating, or in other similar situations.

4.3 Radiation Aging

A second form of aging that has become important in recent years is that caused by electromagnetic and nuclear radiation. *Electromagnetic radiation* is related to light waves. Thus gamma rays and X-rays are types of such radiation. They are, in fact, at the short-wave end of the extensive electromagnetic spectrum that starts with electrical power frequencies and advances through radio frequency, television, microwave, infrared, visible light, and ultraviolet before it ends up in the X-ray and gamma-ray regions. *Nuclear radiation* refers to the radiation (corpuscular in nature) that results from alpha particles, beta particles (electrons), neutrons, charged ions, and even molecules—all accelerated in high-energy streams.

4.3.1   Electromagnetic radiation

Gamma rays are similar to hard X-rays in their resemblance to light. They can generate much heat in organic insulation and, because of their wavelike character, they can penetrate deeply into an insulation wall. Gamma rays also cause such damage as chain-fracture in the molecular structure of insulation; moreover, these rays leave a path of ionized fragments that greatly diminishes insulation resistance and dielectric strength. In addition, the air in the vicinity of the insulation may be ionized, as a result of which flashover voltage is reduced and surface tracking and breakdown are increased.

Gamma rays are often present simultaneously with nuclear radiation. Predictably, the combined effects intensify the severity of thermal damage and general deterioration in an insulation under attack.

The degree of electromagnetic radiation is measured in terms of megarads.[2] An indication of the magnitude of the megarad in terms of insulation deterioration is given by the behavior of polyimide film which can be destroyed by a gamma flux of 3×10^3 megarads per hour over a relatively short period of 3 hr. This dosage is approximately the equivalent in its effect to 20,000 hr of steady heat in air at $250\,°C$. Equivalent destructive effects will occur at a much milder flux of 3×10^{-1} megarads per hour, but over an extended period of 10,000 hr.

4.3.2   Nuclear (corpuscular) radiation

Radiation of the nuclear type is corpuscular in that discrete particles are accelerated to extremely high speeds (approaching the speed of light) and impinge on the surface of the insulation under exposure. Two deleterious effects can take place: (1) The particles can cause erosion in the insulation by dislodging molecules from the chains that form the surface, the result being *mechanical damage;* and (2) the particles can also penetrate the insulation wall to leave an ionized path that extends well below the surface, the result being *electrical weakening.* These two forms of damage are very similar to the effects of corona exposure during which electrons are the impinging particles. (Discussion of corona appears in Par. 4.4.)

As with electromagnetic radiation, the deteriorating effects of nuclear radiation are determined by the amount (dose) and its duration. For the same total dose, the same degree of deterioration will be experienced, regardless of duration of exposure.

4.3.3   Radiation effects on components

Radiation may be damaging to electronic parts, as well as to the insulation on which they may be mounted. In a study of radiation effects, a nuclear flux of 10^3 neutrons per cm^2 combined with a gamma-ray dose of 10 megarads was applied to several thousand components. Included in the study were transistors, diodes, capacitors, resistors, switches, transformers, photomultiplier tubes, fiber-optic discs, and cadmium sulfide cells. Temperature, vacuum, and radiation cycles were used to simulate long periods of service in deep space. Thus 10,000-hr periods were monitored in various combinations of $100\,°C$ temperature, 10^{-5} torr vacuum, and radiation exposure as described above.[9]

---

[2]   The *rad* is a unit of absorbed radiation dose. It is equal to 100 ergs per gm.

Up to 75-percent degradation in current gain was suffered by the medium- and high-power transistors as a result of radiation, and all other transistors suffered some permanent damage. The observed damage occurred within as relatively short a period as 100 hr, with not much more damage inflicted within 10,000 hr. Some recovery was seen only after the 100-hr exposure period. In low-power transistors, current-gain decreases and saturation-voltage increases were observed. However, fast-switching transistors were not seriously affected. Capacitors, diodes, resistors, relays, switches, transformers, and connectors were not affected by the radiation received.

The photomultiplier tubes, fiber-optic discs, and cadmium sulfide cells were exposed only to gamma rays at a rate of 10^4 rads per hour for 200 hr. The photomultiplier tubes experienced rapid increases in "dark" current, and the tests were stopped after 100 hr. The fiber-optic disks also showed rapid deterioration, the light transmittance dropping to below 1 percent in 200 hr. The cadmium-sulfide cells decreased in cell resistance to 50 percent in low light and 10 percent in bright light.

4.4   Corona effects

A theoretical explanation of corona phenomena has already been given in Chap. 2 (Pars. 2.13 and 2.16), and the design significance of this phenomena has been analyzed in Chap. 3 (Par. 3.6). In essence, corona causes both erosion and internal ionization akin to the disruption caused by the radiation effects described above.

In the erosion process the electrons that reach the insulation surface as a result of ionization dislodge molecules of the insulation and thus initiate a progressively weaker mechanical structure. Eventually, the process leads to cracking and dielectric failure. But often before the erosion process continues to the point of failure, there is a second and more subtle process that takes place simultaneously and forces final disruption by dielectric failure alone. The second process is that of internal ionization caused by penetrating electrons as they strike molecules deep within the internal structure. The ionization promotes greater conductivity in the insulation wall, with the effect that the thickness of the wall is progressively reduced until failure occurs. The erosion and ionization mechanisms, since they take place simultaneously, contribute jointly to the ultimate demise of the insulation.

Such demise may take place quickly, or it may require an extended period of time, depending on the applied voltage. In fact, corona resistance may be properly described as *voltage endurance*. An indication of the effect of voltage on the time to failure is given in Fig. 4.2. for magnet wire. Fortunately, magnet wires are not called upon to sustain voltages even remotely on the level of their breakdown strengths. Such voltages would cause failure in a matter of minutes. However, there are certain conditions that are capable of producing corona and long-time failure of magnet-wire insulating films even under system voltages that are only 5 percent of the breakdown values (beyond the limits of Fig. 4.2). Conditions leading to such failure are (1) poorly applied varnish or aged varnish, (2) faulty ground or phase insulation, (3) improperly placed or damaged sleeving, and (4) loosened cross-over wires.

A loose or open surface is one of the effects of corona erosion. A surface so impaired attracts and retains contaminants such as dirt, moisture, and the like, and so creates a condition that promotes surface tracking and conduction to ground. The

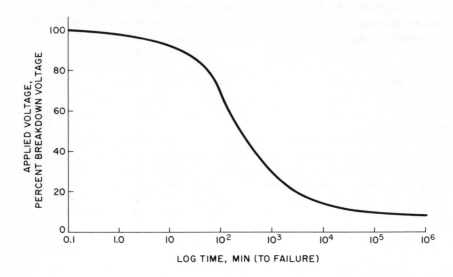

Fig. 4.2   Voltage endurance for organic film insulated magnet wire

surface effects described here may occur even with built-up mica, in which puncture by voltage is a remote possibility.

A different form of corona effect is experienced by rubber. Here, molecules of oxygen are attracted together in the ionization process to form ozone, a powerful oxidizing agent. Nitrous oxides are also formed. These gaseous products envelop the rubber insulation and attack it chemically. The simutaneous presence of mechanical stress would accelerate and intensify the chemical effect until the rubber cracks and becomes useless as an insulator. Silicone rubber and certain specially compounded rubbers, resistant to chemical attack, are used where corona must be tolerated. In special compounds, an ozone-attack inhibitor can be included in the formulation to reduce corona effects.

Generally, corona damage in organic insulations can be minimized by good varnishing techniques and can be eliminated by means of oil and resin impregnation as well as by high-pressure or electronegative gases such as SF_6.

4.5   Mechanical Effects

Deleterious effects on insulating materials are also experienced in the presence of mechanical stress. Fatigue, creep, cold flow, and stress cracking all occur during any period of time in which tensile force, compressive force, shock, vibration, abrasion, or other mechanical stress is encountered.

As electrical machinery is energized and called upon to do work or carry current, it heats up to some predetermined operating temperature. Thermal expansion forces set in and maintain a continual stress on the insulation system. The latter, therefore, must be designed to sustain these forces through the anticipated service life. Moreover, in repeated cycles of machine down-time and energized time, the insulation must not exhibit any fatigue failure.

Short-circuit conditions will impose large mechanical forces on certain types of electrical apparatus, for example, on transformer windings. It is necessary to design the equipment so that the insulation neither cracks nor flows nor suffers abrasion. Proper use of coil-blocking in the design will help achieve this goal.

Vibration of equipment can result in abrasion damage unless all components are securely tied or cemented in place by encapsulants or potting compounds. Ultrasonic vibration can introduce heating effects as an additional source of damage.

Metal fatigue studies are common in which millions of cycles of tensile and compressive forces are applied to determine failure levels. Such studies are not usually conducted on insulation except where insulator parts are also expected to sustain both continuous and intermittent, large mechanical loading. Such structural insulating parts are fiber-reinforced, as in laminates or the like, so as to provide an ample margin of safety in withstanding these stresses.

Insulation does fail by fatigue, however, with the aid of thermal degradation. The insulation is embrittled by the aging process to the extent that it can no longer endure the stress imposed by the expansion forces and so it cracks. If the extent of damage is critical, dielectric failure follows. Often, a certain amount of such cracking can be sustained without catastrophic results if the system is backed up by ground insulation, as in a motor that will continue to run even though some windings become partially cracked. However, operation under such conditions is precarious since a vulnerability to contamination is present that may lead ultimately to the end of useful service life.

The study of creep is likewise metal-oriented. For example, much work has been done in the ability of metal conductors to resist continual stress creep. Under similar conditions, insulation, too, can fail as a result of cold flow or hot flow. However, as previously mentioned, the possibility of such failure can be minimized if there is a proper choice of insulation. For example, composite (built-up) insulations that contain resin plus mica or glass are prone to failure if the resins are thermoplastics; on the other hand, the use of thermosetting resins results in insulations that are essentially immune to creep and flow problems.

Abrasion problems in various equipments are related to the original coil-winding methods and to the service conditions. Windings in which turns have not been properly secured may suffer abrasion during any period of operational vibration. Also a factor is the original handling of wire during manufacture of a given coil. Proper tension, smooth guides, careful positioning of coils after winding, and application of protective tapes or varnish dips, or both, all help prevent the initial abrasions. Long-time abrasion can be minimized by various standard practices, such as coil-tying, impregnation, and potting. Metallic armor can be used on some cables and wires to protect against abrasion. In other types of wires, tough plastics such as nylon and PVC are used as jacketing.

Stress-cracking is quite often associated with other environmental conditions such as radiation, corona, ultraviolet light, sunlight, infrared, solvent attack, and hydrolysis, any of which can occur concurrently with mechanical stresses. Thermal aging may also contribute to this form of cracking. (In effect, we have here an illustration of combined environmental effects, such as discussed further in Par. 4.10.)

A classic example of a test for stress cracking is the heat-shock test used in magnet wire practice. A piece of enamel-insulated wire is elongated by 15 percent and then wound on itself. This sample, with its greatly stressed areas, is then placed in an oven and required to withstand without cracking a sudden exposure to a temperature exceeding its normal operating level. For successful application, a wire enamel must pass this test.

4.6   Corrosive Effects

The materials (including insulation) and the components of electrical and electronic equipment can be severely affected by the presence of corrosive elements. Such elements may be natural atmospheric pollutants, man-made atmospheric contaminants, or they may be generated during the operation of any given equipment through electrical discharge, chemical reactions, solvent release, and the like.

Corrosion attack is often directed at the conductors in insulated wires. The metal may be slowly destroyed with time. The insulation, however, may be adversely affected at an earlier point. Corrosion in the conductor may impair the effectiveness of the adhesion between it and its insulating coating, thus leading to the deterioration of the latter, and even possible failure. Such conditions could be caused, for example, by copper oxidation in the presence of a very high temperature (250 °C), in an enriched-oxygen atmosphere (20 percent O_2), or in the presence of ozone (O_3).

Chlorine is a highly corrosive element when released either as a gas or in a combined form such as HCl or as a phosgene ($COCl_2$). As indicated above, wire conductors would be the first to be attacked. However, chlorine emission from any components adjoining insulation materials can also gradually affect the latter, and degradation will occur that is not unlike that caused by hydrolysis.

Fluorine as a gas and in some compounds is catastrophic in its effects on all equipment parts, even on glass. However, fluorine-based refrigerants are not extremely corrosive unless fluorine is released through electrical discharge. Such behavior is also true of sulfur hexafluoride (SF_6). Provisions should be made for removal of any free fluorine should a discharge take place. Fluorocarbon plastics are normally chemically inert but can generate corrosive elements under severe conditions.

Of the other chemicals, sulfur dioxide and sulfur trioxide in the presence of moisture can form acids which can seriously affect the life of electrical and electronic equipment. Ammonia fumes, solvent vapors, and oil sprays may have a softening effect on insulation and so bring on creep and flow.

Explosion hazards always exist in the presence of combustible fumes. The best method to combat such hazards is to specify explosion-proof motors, switchboxes, lights, controls, and other components. Protection against liquid contaminants can be provided by suitable design such as splash-proof motors.

4.7   Ozone and Ultraviolet Exposure

Natural rubber is a particularly good example of high sensitivity to attack by ozone. It is also vulnerable to ultraviolet rays. As might be expected, natural rubber ages rapidly under outdoor service conditions where it is exposed to sunlight. Compounds such as chlorinated rubber (neoprene), however, can provide protection from these elements.

Organic insulation of various types are similarly degraded, in varying degree, by ozone and ultraviolet. Usually, materials such as vinyls, cellulose derivatives, and polyamides are attacked only after long exposure.

4.8 Humidity, Moisture, Weathering, Fungus

Moisture in any form is unwanted in electrical equipment. At high humidity, exposed insulation may absorb moisture when equipment is not operating and then be unable to eliminate it when operation is resumed. Trapped moisture may explode in blister or void formation that may cause hydrolysis (polymer breakup through water intrusion). Moisture as a surface contaminant promotes surface leakage, tracking, flashover, and dielectric breakdown by bridging weak spots.

Weathering, aside from the moisture effects, includes contamination from pollution, possible salt deposits, heat and cold, wind and sand abrasion, ice and snow loading, and sunlight. Outdoor units, except for insulators and bushings, are often totally enclosed because of the extreme environments mentioned. The insulators are designed to present a maximum of surface area between opposite potentials with as much shielded area as possible. Weather-resistant wires are made with waterproof impregnants and heat-resistant reinforcing materials that are all able to withstand snow and ice loads and wind forces.

Fungus forms in areas of excessive moisture, such as the tropics and subtropics. If the insulation materials contains nutrients for the fungi, the latter will grow until the insulation is so weakened that it fails mechanically or electrically. As a preventative, fungicides are incorporated in its basic resins. Such additives are able to inhibit the growth of fungi even in very high humidity environments. Fungicides are formulated from mercurial compounds, phenols, and chlorinated materials, among other constituents. Chlorine-based products are often preferred because chlorine does not affect the chemical structure of the polymers as some of the other fungicide ingredients may do.

The susceptibility of various insulating materials to fungus attack varies. In fact, some materials may be classified as fungi-resistant ("funginert") because they possess inherent fungicidal properties, or because they offer no nutrients that support fungus growth. Insulating materials that may be so classified include mica, asbestos, polyamides, fluorocarbons, silicone resins, and polyvinyl chloride.

Fungus attack and consequent failure can be affected by the type of moisture conditions present. The severity of the effects may be influenced by whether the moisture is of the absorption surface type or the wicking type and also by the wetting and drying cycle. Another relevant factor is the presence of other corrosive conditions, such as direct chemical action, galvanic action, pitting, stress-cracking, silver migration, and surface migration. High-temperature, as would be expected, is at all times a contributing factor to degradation. The design of fungus resistant equipment has received renewed attention in the last few years.[11]

4.9 Ultrahigh Vacuum

The effects of ultrahigh vacuum, on the order of 10^{-5} torr, are of particular concern in the operation of electronic systems designed for space vehicles. Exposures in deep space to such pressures may extend to tens of thousands of hours. All types of electrical/electronic materials and components specified for such service must be capable of surviving ultrahigh vacuum without impairment of functions.

Of itself, ultrahigh vacuum has practically no effect on the functioning of such components as transistors, diodes, photosensitive tubes, capacitors, and relays. This conclusion was arrived at as part of a study summarized in Par. 4.3.[9] However, when heat and radiation are also present, and, in particular, when exposure to these conditions is lengthy, considerable degradation is observed (especially in photo-tubes and high-power transistors).

In its effects on electrical insulation, high vacuum (when heat is present) will always act on internal voids and cause delamination and blistering.

If the vacuum is extremely good, flashover and breakdown problems will be minimal. But if it becomes less "hard" and the glow discharge region is approached or entered, both low flashover voltage (long-glow discharge) and low breakdown voltage can be expected.

Ultrahigh vacuum is itself a very good insulating medium, and devices have been made, such as X-ray transformers, that use it as such. An extremely reliable leakproof case is paramount, and continuous pumping is usually considered neces-sary to maintain the low pressure needed.

4.10   Combined Environmental Effects

Having covered thermal, radiative, electrical, mechanical, chemical, light, weather, and vacuum environments as separate sources of degradation, we must now consider them in combined situations. Almost always two or more environmen-tal effects will reinforce each other's separate effects and speed up the final break-down of insulation of an organic type. Thus, if radiation and thermal degradation run concurrently, the material will not last as long as the thermal environment alone would dictate.

The two effects on lifespan are similar in analysis to two resistances in paral-lel. The effective lifespan is equal to the product of the two lives determined sepa-rately divided by the sum of the lives. When one environment grossly overshadows the other, the life approximates that allowed by the more powerful environment. Thus, if temperature exposure would normally end the life of an insulation at 20,000 hr, but a radiation dose rate is imposed that would end it at 1,000 hours, the com-bined life will be 950 hr. But if the two environments allow equal lives separately, such as 20,000 hr apiece, the combined life will be 10,000 hr. To plan for this effect one must derate the system until both environments are yielding 40,000 hours so that the combination will make an acceptable 20,000-hr life.

Figure 4.3 points out the relationship between radiation damage and thermal degradation plotting dose rate (in megarads per hour of gamma radiation) and the reciprocal of absolute temperature (in Centigrade degrees).[10] Along the line shown in the figure, equal time periods of exposure in either direction will produce failure for a particular insulation. The expression for the line is:

$$\log{(mr / hr)} = 6.8 - 3.2 \, (1{,}000 \, / \, K) \qquad (4.3)$$

By means of this equation Fig. 4.4 was constructed, the log of life plotted against

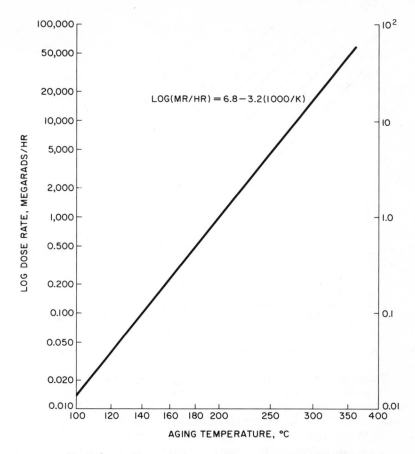

Fig. 4.3   Radiation damage related to heat degradation

log of dose rate for various classes of insulation. It is a useful chart for selecting insulation to withstand known radiation levels for required times. Example: If Class B insulation is adequate for the thermal aging alone, but a gamma dose rate of 0.01 megarads/hour is superimposed, the insulation should be improved to Class F to increase both aging indexes to 40,000 hr so that the combined life will be 20,000 hr.

Other exposures also cut into combined life. Corona effects are similar to radiation damage. However, life is drastically reduced by corona exposure, as shown in Fig. 4.2. Apparently the dose rate of electrons, in equivalent megarads/hr, is about equal to 300 at 600 volts per mil stress on a magnet wire twisted pair. And at 300 volts per mil the dose rate equivalent is 0.3 megarads/hr with a life of 10^4 hr.

Mechanical stress in combination with heat and voltage shows up as stress-cracking effects and loss of dielectric strength sooner than anticipated.

Thus all exposures should first be evaluated separately and then fed into a combined relationship similar to that for a number of paralleled resistors so that the combined life may be calculated.

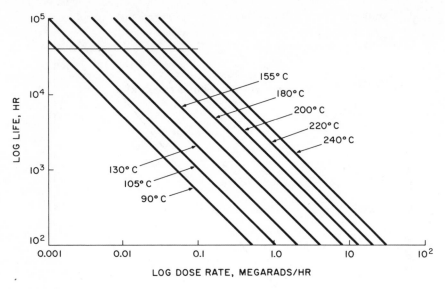

Fig. 4.4   Effect of radiation on life of various classes of electrical insulation (90°, 105°, 130°, 155°, 180°, 200°, 220°, and 240°C)

4.11   Processing Effects Considered as Environment

Before a given material or materials system is ever exposed to natural or man-
made environments, it must survive the hazards of initial processing. In making
electrical tapes, winding tension must not injure the material. In making magnet
wire, the winding must not scrape the conductor through the enamel, tape, or fiber
insulation. Wound coils may have to be pushed into motor slots, formed by massive
bending, crushed to conform to existing space, or twisted to form desired shapes.
All operations are exacting. They require materials with a high level of mechanical
strength to assure freedom from cracks. The materials must also be strong enough
to remain securely in place as applied.

After a given unit (say a coil) has been formed, it may be dried under high
heat and in vacuum. These processing conditions may cause shrinkage cracks, heat
shock, internal gas evolution, and embrittlement. However, choice of proper materials will minimize such effects, and the unit will be ready for varnish dip, encapsulation, potting, or impregnation. Solvent shock, incompatibility, softening, and stress
cracking can occur in this phase. These hazards can be minimized by proper choice
of materials (with emphasis on excellent chemical resistance), use of optimum potting techniques, avoidance of harmful activators, and like measures.

The finished units must be capable of withstanding applicable electrical tests
(which may be destructive or nondestructive). Such tests as those for impulse voltage, high potential, insulation resistance, dissipation factor, and induced voltage are
used to investigate the entire insulation system and to evaluate the ultimate functioning of the unit. The results can also be used to assess the marketing possibilities of
the unit and customer acceptance.

In summary, the processing effects call for maximal thermal, mechanical, chemical, and electrical properties for components in conjunction with careful forming and treating techniques and the least destructive testing that will weed out the weak units.

Various aspects of critical environments and phenomena that affect materials are examined in detail in another volume of this Series.[12]

REFERENCES

(1) Mathes, K. N., "Electrical Insulation at Cryogenic Temperatures," *Electro-Technology*, Sept. 1963, p. 72.

(2) Milek, J. T., "Behavior of Dielectric Materials and Electrical Conductors at Cryogenic Temperatures (A Bibliography)," *Interim Report No. IR-5*, Electronic Properties Information Center, Hughes Aircraft Co., Aug. 1965.

(3) Mathes, K. N., "Dielectric Properties of Cryogenic Liquids," *IEEE Transactions on Electrical Insulation*, Vol. EI-2, No. 1, Apr. 1967, p. 24.

(4) Saums, H. L., "Section 3–18: Magnet Wire, Strip, Hollow Conductors, and Superconductors," *Insulation Directory/Encyclopedia*, June/July 1970, p. 264.

(5) Laverick, C., "High-Field Superconducting Technology," *IEEE Spectrum*, Apr. 1968, p. 63.

(6) Garwin, R. L., and Matisoo, J., "Superconducting Lines for the Transmission of Large Amounts of Electrical Power over Great Distances," *Proceedings of the IEEE*, Apr. 1967, p. 538.

(7) IEEE No. 99, "A Guide for the Preparation of Test Procedures for the Thermal Evaluation of Insulation Systems for Electric Equipment," 1957.

(8) Dakin, T. W., "Electrical Insulation Deterioration Treated as a Chemical Rate Phenomenon," *AIEE Transactions*, vol. 67, 1948, p. 113.

(9) Klippenstein, E., Hamman, D. J., and Hanks, C. L., *Radiation and Electronic Parts*, Battelle-Columbus Report, 1970.

(10) Pendleton, W. W., "Advanced Magnet Wire Systems," *Electro-Technology*, Oct. 1963.

(11) Lascaro, C. P., "Design of Tropical-Resistant Electronic Equipment," *IEEE Transactions on Parts, Materials, and Packaging*, June 1970, p. 60.

(12) *Materials Science and Technology for Design Engineers*, A. E. Javitz, ed., Part II: "Critical Environments and Phenomena," Hayden Book Co., New York, 1972.

CHAPTER 5 | APPLICATION PROBLEMS IN DESIGN

5.1 Basic Problem Causes[1]

Electrical insulation—by its definition and function—may be considered as the common denominator of all categories of electrical/electronics design. As such, it should be expected that its application problems are of major significance. The very advances in electrical insulation—the proliferation of new materials, the improved standards for processing and fabrication, and the increasingly exacting specifications for end-product performance and reliability—have, in fact, served to intensify the demands on insulation capabilities and have thus served to uncover many problems not heretofore encountered.

What are the sources of these problems? These may be summarized as follows:

1. Defects in processing and fabrication procedures, with resultant materials properties degradation

2. Gaps between performance requirements and the *realistic* capabilities of commercially available materials

3. Environmental conditions not adequately considered that affect materials beyond their inherent endurance limits

4. Excessive variability in batch-to-batch properties of materials as supplied by producers under established specifications

5. Inability, or disinclination, of producers to reduce the cost of new, and otherwise satisfactory materials, to practical levels

6. Failure of design engineers to apply materials *systems* concepts in design, resulting in incompatible combinations of insulating and associated materials

7. Malfunction or failure of the insulation caused primarily by defects in the design or manufacturing of the end-product in which the insulation is applied

---

[1] This chapter has been contributed by Alex. E. Javitz

8. Lack of the "team" approach, essentially a lack of interdisciplinary aware-ness between the design engineers and the others involved in the design process—the materials engineer, the chemists, physicist, metallurgist, and others

9. Inadequate testing of insulation, particularly a lack of functional life test-ing of the insulation as incorporated in the end product.

The importance of the various possible causes of insulation problems shifts from category to category of end-product and is also dependent on the relative re-quirements for performance levels and service reliability. Broadly, the many cate-gories of electrical/electronics product groups can be consolidated as follows:

1. Rotating machinery
2. Power transmission and distribution equipment
3. Communication systems
4. Electronics (other than aerospace and defense)
5. Aerospace and defense devices and systems
6. Appliances (household, commercial, and industrial)

Another approach is to categorize electrical/electronics products on the basis of design parameters:

1. Advanced parameters (such as are encountered in aerospace and defense complex electronics, data processing machines, electromedical devices, communica-tions systems, and the like) in which high performance and exacting environmental conditions are over-riding considerations.

2. High-performance commercial parameters (such as are of concern in special-purpose industrial machinery, heavy-duty motors, power-generating appa-ratus, instruments and controls, special processing equipment, and the like) in which severe operational conditions, such as extreme temperatures, shock and vibration, and power surges, are the basic factors and long-life reliability is required.

3. Conventional commercial and industrial parameters (such as govern the design and manufacture of household appliances, consumer electronics, general-purpose electrical tools, and the like) in which performance reliability has to meet the requirements of competitive costs.

To examine the entire spectrum of insulation application problems within the space of this one chapter is obviously not feasible. The discussion here will there-fore concentrate on several selected examples of major interest and importance. The selection has been based on opinion and data provided by insulation authorities in industry, engineering schools, and government.[1]

An excellent source of information is also found in the Proceedings of the IEEE-NEMA Electrical Insulation Conferences. A Bibliography of selected papers from the last three Conferences, arranged by subject, appears at the end of this chapter.

5.2   Aerospace

Electrical insulation and dielectric materials are critical elements in many constituents of complex aerospace electrical/electronic systems. The miles and miles of insulated wire and cable in a modern aircraft alone pose a taxing problem on the selection of insulation capable of withstanding the environmental and operational conditions involved. Several problem areas will be discussed here.

5.2.1 High-temperature-resistant wire insulation

Increasingly severe service conditions, such as aerodynamic heating and higher engine temperatures, have made necessary the development of new and improved high-temperature wire insulation able to function up to 700°F. The particular application was for multistrand connector wire. Comparison of the properties of the new experimental insulation (polyimide-polyimidazopyrrolone) with the "state-of-the-art" MIL-W-81381 polyimide-FEP Teflon insulation is given in Table 5.1.[2,3]

The insulation is suitable for mass production on commercial wire-wrapping machines and can be prepared in the same forms (single, double, or triple-wrapped) as MIL-W-81381 polyimide/FEP Teflon wire. It is anticipated that this improved insulation will find use in future Air Force applications requiring high-temperature, lightweight, insulated, multistrand connector wire.

The development of this new material is believed to represent a significant advancement in the state-of-the art and the material will be a prime candidate for use in advanced Air Force systems areas where temperatures caused by aerodynamic heating effects or radiant heating from the engine cause temperatures to exceed 400°F. The development work was accomplished at the Hughes Aircraft Co. under contract to the Elastomers and Coatings Branch, Nonmetallic Materials Division, Air Force Materials Laboratory.

5.2.2 Integrated circuitry

Another aerospace problem arose in dielectric substrates for integrated circuitry. The advantage of dielectric substrates as opposed to silicon-on-silicon devices lies in their capability for higher circuit density because of reduced leakage.

Table 5.1   Experimental High-Temperature-Resistant Wire Insulation Compared to a "State-of-the-Art" Material*

Source: References (2) and (3)

	State-of-the-art Spec. MIL-W-81381 Polyimide-FEP teflon insulated wire	Newly developed experimental polyimide-polyimidazopyrrolone insulated wire
Max. temp. for continuous indefinite service	392°F	550°F
1,000-hr service	Slightly over 400°F	600°F
Short term (several hours service)	480°F	700°F
Cold bend, −65°F	Passes	Passes
Life-cycle test, 28 days, 185°F, 95 percent RH	Passes	Passes
Cut-through resistance	Passes (at 400°F)	Passes (at 700°F)
Insulation breakdown	Passes	Passes

* Topical Report, Air Force Materials Laboratory, September 22, 1970

The dielectric substrates must be single crystal to permit epitaxial deposition of the single crystal semiconductor circuits. Sapphire and magnesium aluminate spinel are the two most advanced single-crystal insulating substrate materials. The elimination of the substrate-epitaxy isolation junction always present in conventional silicon-on-silicon integrated circuits also eliminates the isolation junction parasitic capacitance that causes shunting of conventional circuits. Integrated circuits deposited on dielectric substrates are capable of higher operating frequencies and speeds.

Results of work on single-crystal spinel substrates supported by the Air Force Materials Laboratory at RCA are given in the referenced reports.[4]

5.2.3   Electronic modules

The encapsulation or potting of electronic modules has been the source of problems. Such modules have either welded or soldered joints. In one study, the best potting resin was found to be an epoxy compound cured with chlorinated anhydride and benzyldimethylamine as a catalyst. The insulation resistance requirement was 10^{12} ohm-cm or higher after MIL-E-5272 humidity cycling. Diallylphthalate cases were selected as the best receptacles for containing the module and resin. The epoxy was filled with silica or other mineral filler to achieve a compromise of low shrinkage, low expansion coefficient, and low viscosity. The main problem encountered was that of internal voids in the cured resin. It proved to be difficult to process a module without forming some voids.

Two serious quality control questions arose: How many voids were permissible? What size voids were permissible? (The use of radiographic techniques was ruled out because of excessive cost on a 100-percent screening basis.)

These questions were never completely answered, but, of course, if voids were found to be contiguous and could provide a moisture path between conductors, this was immediate cause for rejection or for rework. Another problem area was that of plastic shrinkage stresses that caused solder or weld joints to open. The controversy that arose was that if the joint was a good reliable one to begin with, it would not break open and that the epoxy shrinkage would merely point out the marginal joints. Predictably, the controversial positions were influenced by the specialities of the engineers involved—the electronic and metallurgical engineers would blame the epoxy resin, whereas the plastic engineers would blame the metallurgical joint.

5.2.4   Gyros and inertial instruments

Instruments such as gas-spin-bearing gyros employ electrical insulation in a number of critical locations. Insulation is used for motor stator impregnation or encapsulation, stack lamination bonding, bearing shaft insulation, flexible circuitry, and some other minor uses. Because of the ultra-precision and high dimensional stability requirements of these instruments, particles as small as 50–100 microinches generating from the plastic may be serious. Potting compounds and adhesives must be highly filled to closely match expansion coefficients of metals and ceramics of which the gyro is fabricated, yet these filler particles must not dislodge and migrate within the instrument. Another problem is that of outgassing of the insulation during gyro operation. Carbon dioxide, water, ammonia, and hydrocarbons have all been shown to emanate from cured epoxies. These gases can either affect spin bearing performance due to changes in gas viscosity or can interact with other impurities to form solid particles.

5.2.5   *Transformer encapsulation*

Stress cracking of the encapsulating compounds used with transformers is a major problem, especially during thermal shock or thermal cycling. However, this problem can be obviated through certain precautions in design, such as avoidance of stress risers, 50-deg angles, and the like. Encapsulating compounds can be screened by means of appropriate tests such as the 3M Company washer test or the hex-nut test, but ultimately functional tests on encapsulated transformers should be performed. Specially developed flexibilized epoxy compounds exhibit superior noncracking properties and so provide still another solution to the problem.

5.3   Computer Wire and Cable

The use of flexible flat conductor aluminum cable in computers offers obvious advantages—space saving, light weight, and adaptability to special configurations and stress. The specific problem in the selection and application of the insulation for such cable is the retention of the inherent insulation resistance and dielectric strength under service conditions. These conditions involve (1) permanent folding of cables through angles of 180 deg, (2) possible penetration of insulation by metal parts in routing channels, and (3) aging of the insulation. Examples of the flat conductors investigated are shown in Fig. 5.1.

An experience report based on extensive tests by IBM indicates that polyvinylchloride (PVC) insulation has the most acceptable characteristics. The advantages are resistance to stress cracking at folds, resistance to shrinkage with aging, and a favorable cost factor.

Limitations of PVC are associated with temperature rise when individual conductors are stacked or built up in layers. However, this limitation is not considered serious at this time. Also, some "shrink-back" from cut ends occurs.

The flat cable that has been worked with was 0.012×1.00 in. The PVC

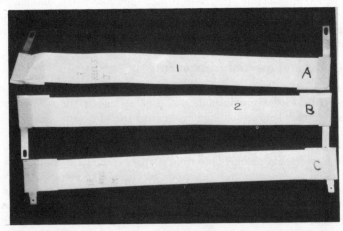

Fig. 5.1   Examples of flat aluminum conductors developed for computer use

Fig. 5.2   Stress cracking in flat conductor cables used in computers (*Courtesy,* IBM, Rochester, Minn.)

insulation used in the tests was 0.010 in. thick and was applied by an extrusion process. Also tested was a polyethylene insulation, 0.008 in. thick, applied by an edge-lamination process.

The stress-cracking hazard at the 180-deg folds occurs when the cable routing direction changes in the machines. As noted above, it is required that the insulation resist such cracking and maintain its inherent electrical characteristics. Stress-cracking tests using a solution of Igepal-630 (a nonionic surfactant) at a 50° C temperature resulted in cracks at fold locations when polyethylene was used, as shown in Fig. 5.2. The same test performed with PVC showed no cracking of the insulated cable at the fold locations.

Another disadvantage of the flat cable tested with polyethylene insulation was that the cable tended to wrinkle because of shrinkage. This condition was evidenced when the insulation was subjected to 80° C temperatures with uncontrolled humidity and also during the following temperature and humidity cycles: 4 hr at 75° C, 90–95 percent RH; 4 hr at 25° C, < 50 percent RH; 16 hr at 75° C, < 50 percent RH. Shrinkage of PVC occurred, but not to the extent of that of polyethylene.

Although the test program established the advantages of PVC insulation, it would still be desirable to obtain an insulation with a temperature rating higher than that of PVC (105° C), provided it would meet the required processing and service conditions as well and would be available at equivalent cost.

5.4   Motors and Motor-Generator Sets

In this area, environmental protection is seen as the primary design problem, particularly if motors are exposed to severe moisture conditions, aggravated by other contaminating elements.

The failure of shipboard motors due to insulation contamination caused by moisture has long been a Naval problem. A survey of shipboard motor failures over several years has shown that approximately half of all failures were caused by bearing failures and the other half by insulation failure, with the latter primarily caused by moisture. The combination of condensate, steam leaks, splash, and salt spray presents a truly severe environment. A number of continuing studies have been conducted for the purpose of evaluating this problem and developing methods to minimize or eliminate it.[5, 6] The use of encapsulated motors was the first significant improvement. The use of solventless varnish treatments also provided relief from the moisture problem. However, some reservations still exist in regard to both the encapsulation techniques and the use of solventless varnishes. Currently, the existing problems may be summarized as follows:

1. Round magnet wire with heavy, triple, or quadruple polyester or polyamide insulation must be used based on space limitations.

2. Limitations are in varnish treatment of windings to provide bonding and filling of interstices.

3. Unless the varnish supplier can come up with improved products the magnet wire supplier must add the bonding agent to his wire.

4. Encapsulated windings will provide protection for severe environments but encapsulation presents other problems such as heat transfer, difficulty in applying mold casts in some cases, and need for special handling. Also a problem is the limited number of suppliers. Solventless varnish treatments offer some advantage, but materials and processing techniques need to be improved. Bonded magnet wire seems to offer the most promise and is being evaluated at present by the Navy in a motor test program.

5.5   Distribution Transformers

The most commonly used insulation systems for these transformers consist principally of cellulosic material (kraft or rag paper and pressboard) impregnated with, and immersed in, mineral oil. Synthetic enamels are used as conductor insulation in smaller transformers, but not ordinarily in larger units. The oil-impregnated cellulosics have some important advantages in transformer appplications, but also some disadvantages, among which is a need for improved thermal endurance.

The principal advantages of oil-impregnated cellulosic materials are the following:

1. The high electric strength available, particularly under short-duration (impulse) stresses.

2. Ability to withstand fairly high (250° C) temperature for short times, with no greater deterioration than results from exposure to lower temperatures for longer times. (See referenced guide for loading mineral oil-immersed overhead-type distribution transformers.[7])

3. Reasonable cost.

The chief *disadvantages* of cellulose are the following:

1. Tendency to absorb water from its surroundings.

2. Reduction of insulation strength and thermal life as moisture content increases.

3. Evolution of water as a decomposition product in thermal aging.

4. In consequence of (1), (2) and (3), the necessity for very careful drying of the transformers during manufacture, sealing of transformer cases to prevent entry of atmospheric moisture, and care in design accumulation, in highly-stressed regions, of water released by thermal aging.

Much development effort has been expended in modifying cellulose to improve its thermal endurance. Use of such "thermally upgraded" insulation has made practical the design of distribution transformers with 65°C rated load rise, with substantial savings to both manufacturers and users.[2]

Numerous synthetic films and papers made from polymeric fibers have been evaluated as possible replacements for cellulose in oil-immersed transformers, but thus far have generally been considered unsatisfactory because of inferior thermal performance at high temperature, relatively low impulse strength, or cost. One polyamide paper, however, has shown outstanding thermal and electrical properties and is in limited use at this writing. Its present high cost seems likely to restrict its use to special situations where the cost can be justified.

A laboratory evaluation of polyamide papers has been followed by life tests on small power transformers. On the basis of extensive tests, it has been suggested that the special properties of the polyamides may permit design improvements that may offset their higher costs. With projected increased use there is also the prospect of cost reduction and, so, an availability for a wider field of applications.[8]

Test data show significantly better properties for the polyamide papers than for the cellulosics, as shown in Fig. 5.3, which compares both in respect to the percentage of original bursting strength against aging time in oil. In Table 5.2 the

[2] For example, Westinghouse Insuldur stabilized cellulosic insulation in its present improved version.

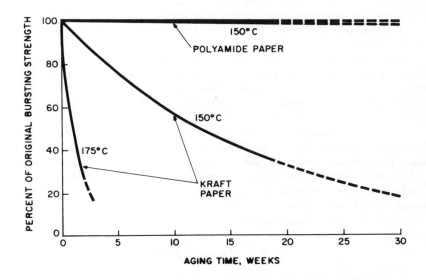

Fig. 5.3   Bursting strength related to aging time in oil for 10-mil polyamide paper and 10-mil Kraft paper[8]

Table 5.2   Comparative electric strength of polyamide papers, cellulosic papers, and synthetic fiber mats

Source: Reference (8)

	Impulse test	60 Hz, rapid rise test		
	Puncture strength, volts/mil			
Materials	Oil	Air	Oil	Inerteen*
Polyamide papers:				
3-mil calendered, 3 layers	3,060	580	1,750	2,030
5-mil calendered, 2 layers	3,100	680	1,630	2,060
6-mil uncalendered, 2 layers	2,650	240	1,780	1,650
Cellulosic papers:				
3-mil rope-kraft, 3 layers	2,370	170	1,560	1,680
5-mil kraft, 2 layers	2,840	180	1,500	2,000
Synthetic fiber mats, range of 8 grades tested (8 to 12 mils total specimen thickness)	1,040 — 1,470	100 — 140	790 — 1,300	660 — 990

*Proprietary liquid dielectric

electrical strength of polyamide papers is compared against cellulosic (kraft) papers and synthetic fiber mats.[8]

Liquid dielectrics used in transformer insulation systems provide their own set of problems. Primarily, these problems lie in the selection of materials with optimum properties of chemical stability, corrosive effects, gas solubility, viscosity, insulation strength, breakdown voltage, and the effect of arcing on the chemistry of the liquid.

Oils that are generally referred to as transformer oils (those made up of parafinic and napthenic constituents) and also the oils known as askarels are the normal solution to the above problems. However, in specific instances, purer liquids (for example, hexane and parafin) may provide better experimental and practical solutions.

Askarels are widely used in practice, but from an experimental and research viewpoint leave much to be desired. They are more difficult to handle and are less stable than their naturally occurring counterparts. Additionally, their shelf life is limited and their cost considerably higher. From a dielectric standpoint, their temperature constancy leaves room for improvement.

In many applications, it would be desirable to use a solid insulant if one could be obtained without voids. Alternatively, an ideal research objective would be a liquid insulant that would combine the best of the two worlds of transformer oils and askarels: longer life than the transformer oils, less corrosiveness and less harmful effects on the environment than the askarels, and a combination of the insulation strength and flame resistance of both.

A discussion of liquid dielectrics will be found in Chap. 12.

5.6   Magnet Wire

Although magnet wire *per se* cannot be categorized as an end-product in the same sense as an electronic device, electrical apparatus, or an appliance, it *is* a critical component in a broad spectrum of end-products. The problems that affect its optimum application therefore engage constant investigation and development activity.

At the core of magnet wire problems is its *insulation,* since it is the insulation that protects the conductor that carries the electric current that establishes the magnetic field. The conductor metal or alloy may also pose some problems. Of course, if special requirements of elevated temperature resistance, corrosion resistance, high physical strength, or other factors are of concern. Such problems, however, are not within the scope of this work and are treated in a companion volume in this Series dealing with conductive functions and materials.[9]

The problems associated with the selection of magnet-wire insulation (that is, enamels and various other coverings) range from the mechanical (windability), through the electrical (maximum dielectric strength), chemical (compatibility), and thermal (resistance to heat aging), to the physical properties such as smoothness, lubricity, thinness, and flexibility. No other constituent of a device encounters higher temperatures than such enamels and coverings. Critical thermal points commonly described as "hot spot" temperatures must be endured by these wire coatings for periods extending through maximum load cycles throughout the life of a given device or apparatus.

Magnet-wire insulation is combined with varnishes, impregnants, encapsulants and supporting insulation to complete the system of which it is a member. The wire insulation must match these materials and other solid, liquid, or gaseous members and media in a compatible "inner world."

Exposures to heat, extreme cold, radiation, corona, humidity, refrigerant liquids, gases, corrosive atmospheres, contaminants, centripetal forces, vibration, mechanical shock, expansion forces, and even the vacuum of outer space are some of the severe conditions that are met by magnet-wire insulation in a variety of power and electronic applications.

For example, the most important aspects of magnet wire insulation in *motor design* are (1) toughness and windability; (2) thermal stability (both short time and long time); (3) chemical stability in relation to associated materials in the wire and also in relation to various media, such as cryogenic fluids or refrigerants, that the insulation may come in contact with; and, not the least, (4) mechanical stability to provide extended resistance to vibration, shock, fatigue, thermal expansion, abrasion, and the like. Tables 5.3 and 5.4 list suitable magnet-wire insulations for open motors.

Table 5.3 Suitable Magnet-Wire Enamels for Open Motors

Thermal class, °C	Description of enamels
220	Polyimide
220	Polyamide-imide
200	Polyamide-imide-overcoated terephthalate polyester
200	Polyamide-imide-overcoated polyester imide
200	Polyamide-imide-overcoated polyester amide-imide
155	Polyamide-overcoated terephthalate polyester
130	Polyamide-overcoated polyurethane
105	Polyvinyl formal
105	Polyamide-overcoated acrylic
105	Polyamide
105	Polyamide-overcoated polyvinyl formal

Table 5.4   Suitable Coverings for Open-Motor Magnet Wires

Thermal class, °C	Description of covering
220	Polyimide film (tape)
220	Aromatic polyamide fiber
220	Glass fiber with polyimide bond
220	Glass fiber with silicone bond over polyimide film
200	Glass fiber with silicone bond
200	Glass fiber with silicone bond over Class 200 film
180	Glass and polyester fibers with silicone bond
155	Glass fibers with phenolic bond
155	Glass and polyester fibers with phenolic bond

In *dry-type transformers,* a high degree of thermal stability for wire coatings is demanded as well as resistance to mechanical and electrical shock from short circuits and power surges. For *liquid-cooled* types, the insulation must be resistant to softening caused by solvent action and must be compatible with the other constituents of the materials system. Tables 5.5 and 5.6 list insulations suitable for these transformers, and Table 5.7 for electronic transformers and small dry-type units.

Hermetic motors must contain magnet-wire insulations that are not softened,

Table 5.5   Wire Enamels and Coverings in Dry-type Transformer Applications

Thermal class, °C	Description of enamels and coverings
220	Polyimide enamel
220	Polyamide-imide enamel
200	Polyamide-imide-overcoated terephthalate polyester, polyester imide, or esteramide-imide
220	Polyimide film (tape)
220	Aromatic polyamide paper
220	Glass fiber with polyimide bond
200	Glass fiber with silicone bond
180	Glass and polyester fibers with silicone bond
155	Glass fiber with phenolic bond
155	Glass and polyester fibers with phenolic bond

Table 5.6   Wire Covering for Liquid-Cooled Transformer Applications

Thermal class, °C	Description of covering
130	Epoxy enamel
105	Polyvinyl formal enamel
220*	Aromatic polyamide paper
105	Cellulose paper

* This maximum temperature may be greatly reduced by the maximum temperature of the coolant (see Chap. 12 on liquid dielectrics).

Table 5.7   Suitable Magnet-Wire Enamels for Electronic Transformers and Small Power Dry-type Transformers

Thermal class, °C	Description of enamels
155	Solderable polyester
105	Solderable acrylic (with and without polyamide)
105 & 130	Solderable urethane (with and without polyamide)
90	Solderable polyamide
105	Solderable cellulose acetate
105	Solderable urethane with polyvinyl butyral overcoat (bondable and solderable); also with polyamide coat

Table 5.8   Magnet Wire Enamels Suitable for Hermetic-Motor Application

Thermal class, °C	Description of enamel (including fiber combination)
220	Polyimide enamel
220	Polyamide-imide enamel
200	Polyamide-imide-overcoated terephthalate polyester
200	Polyamide-imide-overcoated polyester imide
200	Polyamide-imide-overcoated polyester amide-imide
105	Polyurethane-modified polyvinyl formal
105	Acrylic copolymer
155	Glass fibers with phenolic bond and with or without hermetic enamel undercoat

embrittled, nor attacked by the refrigerant used. In addition, the insulation must possess all the properties demanded in application for open motors. Table 5.8 lists some suitable enamels for the hermetic motors. (Paragraph 5.7 discusses hermetic motor problems in greater detail.)

The question of a "universal" magnet wire is frequently posed. The present best all-around enamel combination is the *amide-imide* overcoating on a suitable base (see Tables 5.3, 5.5, and 5.8 for specific combinations). The costs of this type of enamel is less than that of a polyimide enamel and it is resistant to hydrolysis as well.

In comparison, the *polyimide* materials (both enamels and those in tape form) have two limitations to universal use, these being relatively high cost and a tendency to hydrolyze in sealed systems with moisture present. The high-temperature-stable *polyamide* papers are limited only in respect to spacing insulation. The dielectric strength is restricted to that of the media in which the polyamide is used. A similar limitation is faced by glass-fiber materials.

Development of new magnet-wire insulations should point to lower-cost enamels with enhanced resistance to such media as refrigerants, high-temperature liquids, and askarels.

Since many windings are encapsulated for better protection against environmental conditions, the problems of optimum selection of encapsulating compounds is certainly a companion factor in magnet-wire selection. Some basic encapsulating compounds are described in Chap. 8. Also, the Bibliography at the end of this chapter lists several excellent reports of various performance studies.

A detailed discussion of magnet wire will be found in Chap. 9.

5.7   Hermetic Motors

Application problems in hermetic motors affect a considerable range of end-products in which such motors are required to operate efficiently with long-life reliability. The most common examples are the household and commercial refrigerator-freezer, various types of air-conditioning systems, and refrigerating units for food processing, transportation, and the like. The discussion here will take up first some specific problems in magnet wire for these motors and will then review some aspects of motor and compressor design, construction, and operation that must be considered in relation to the failure of either the insulation system or the *entire* system, or both. These factors form a subject of extensive investigation and involve some distinctly controversial viewpoints.

5.7.1   Hermetic magnet wire

A summary of major problem areas in magnet wire for hermetic motors follows:

1. Among the design problems in current practice and experience is the need for a blister-free hermetic magnet wire. During the past 15 years, several stages of improvement have been made in hermetic wire insulation. Most manufacturers used Formvar as their first film insulation and from it came the Formvar-urethanes and acrylics. Most recently, the amide-imide overcoated modified polyester wires and the polyester coatings, modified with amide and imide linkages, have proven to be the most windable and solvent resistant insulations. One technical problem with these newest materials is that their high resistance to solvents causes a diffusion barrier to rapidly escaping refrigerants. Under certain types of air-conditioner field conditions, where the wire has had a long soak time in the refrigerant, followed by a rapid heat rise in the motor windings, interlayer blisters may form in the magnet wire coating. These blisters have not significantly affected the operating dielectric protection between turns, but they represent an undesirable surface phenomenon.

2. The best material to prevent refrigerant blistering within controllable winding temperature excursions is the polyamide-imide. This polymer possesses outstanding solvent resistance and toughness and possesses unusually attractive high temperature capability, thermoplastic flow resistance, and excellent overload or burnout protection.

3. One major limitation is cost, and intensive research efforts are being made to bring the cost down. Processability is another technical roadblock that must be overcome.

4. Insulation development should be aimed at a lower cost polyamide-imide with good coatability characteristics.

Looking to the future, the magnet-wire industry feels that the next generation of hermetic motor grades should keep pace with the expected advances in motor design and manufacture, including extensive life-testing practices, quality-control standards, and the like. Hermetic grades should be developed that possess full technical capability to fulfill all service requirements, and, in particular, be significantly upgraded in those properties that assure longer-life reliability than presently achievable. The development aims for the hermetic grades, it will be noted, are essentially similar to those for magnet wire generally, as indicated in Par. 5.8.

5.7.2   Hermetic motor construction

Several elements in motor design and construction can be appraised for their effect on performance failures:

Ground or slot insulation   "Hipot" tests on new motors cause rejections because of phase-to-ground failures. Such failures may be blamed on the insulation used (usually terephthalate polyester) or on the processing, depending on the point of view expressed. Concededly, proper motor construction would not normally split or crack ground insulation. On the other hand, a sufficiently resilient ground insulation would not crack or split under present processing conditions.

Also to be considered is the possibility that loss of "charge" in compressors that have only external protective devices can also lead to ground failures. Such failures have been found in laboratory tests. Test temperatures were high, 375° to 400°F, but these may be experienced in actual service and loss of charge can go undetected for extended periods.

Still another point of view is that the problem of ground-insulation failure is really a current leakage problem that may constitute a safety hazard. On the turn-to-ground test, voltages are high, and failure to ground may occur, in fact, only because of defects in the enamel. In field service, therefore, it may be expected that the wire will deteriorate first at a spot adjacent to the slot insulation. In such a case, there is a danger that current leakage to ground occurs *prior* to a motor failure. The interval between the deterioration of the wire and motor failure to ground is potentially a safety hazard even though precautions are taken for proper grounding. This problem alone may require a more dependable ground insulation.

Defects in rotors   Open rotor bars or improperly cast rotors, rather than the insulation, may be a cause of motor failures. Aluminum conductors may be defective, for example, because of porosities in die castings. Even though tests are used to detect and reject rotors with open conductors, some faulty ones do slip through and may be used in motors. The generated excess heat is transmitted to the adjacent areas and parts such as bearings and stator slot liners.

Defects in stators   Latent defects in stators may show up in reparied rejects that have passed final inspections. It is possible, in the opinion of some, that *internal* damage may persist and cause failure in service.

Variability in uniformity or concentricity of enamel   The batch-to-batch uniformity of enamels produced in Europe is apparently superior to that produced in the United States, although the costs are higher. It is suggested that improvements in U.S. enameling techniques should be studied.

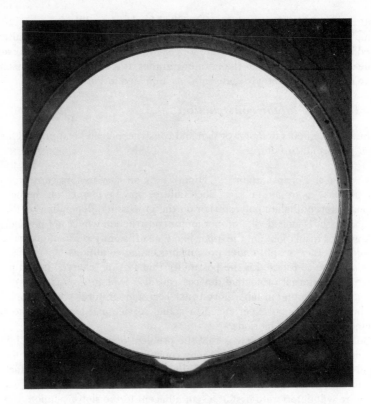

Fig. 5.4   Magnet wire cross section showing presence of a copper sliver

Conductor defects   The incidence of failure due to conductor defects should be considered. The possibility that such defects as slivers may be found in copper and that slivers may cause motor failures is suggested. (See Fig. 5.4.)

5.7.3   Hermetic motor operation

The relation between motor design and construction and insulation failure has a corollary. To what extent is motor performance adversely affected by faulty associated equipment, in this case the compressor and its various component parts? There may be a number of such factors, for example: (1) Fan failures, (2) improper design of evaporators, (3) improper setting of controls, (4) faulty relays, (5) malfunctioning capacitors, and the like.

A full examination of these problems would lead this discussion into the area of product design *per se*, which is beyond the particular scope of this work. There is, however, an all-important point to be affirmed. The optimum selection and application of electrical insulation (indeed, of *all* materials) is a two-way enterprise. In one direction, we are concerned with the effects of materials properties on the service performance and life reliability of a given device or apparatus in which the insulation is applied; in the other direction, we are equally concerned with the effects *on* the material by possible defects of product design and construction.

5.8   Summary

The previous discussion in this chapter has dealt with electrical insulation application problems in selected product areas. It would be useful to enumerate the problems that are most commonly encountered in most product categories. The following list is approximately in the order of most general concern:

1. *Thermal problems* (high temperatures; low temperatures; thermal endurance; thermal shock; thermal conductivity)

2. *Structural and mechanical problems* (physical strength; physical stability; resistance to shock, vibration, fatigue, and the like: compatible thermal coefficient of expansion; workability, that is, adaptability to the most effective fabricating methods)

3. *Chemical factors* (resistance to water and humidity; resistance to solvents, to various other chemical agents, and the like)

4. *Electrical factors* (dielectric strength, dielectric constant; dielectric losses; corona effects)

5. *Special environmental conditions* (effects of radiation, ultrahigh vacuum, cryogenic temperatures, and various extreme conditions, such as are encountered in aerospace, hydrospace, or in hazardous terrestial operations)

Two other points can be made:

1. The product designer needs to be concerned not only with specific items of material properties, but equally with the development of rapid and reliable functional life test methods to determine whether given materials are capable of providing long reliable service in some particular application and enviromnent. (See various IEEE standards for functional life-test procedures on power transformers, random-wound machinery, specialty transformers, electronic devices, among others.)

2. In short, what the designer wants and needs is as comprehensive a knowledge as possible of the properties (limitations no less than the capabilities) of the materials with which he can try to eliminate existing problems, or, at least design *around* such problems.

The extent of insulating and dielectric materials available to the design engineer is very broad and diverse. The succeeding chapters of this book will discuss the various categories of these materials in respect to their basic structure and behavior, processing, performance characteristics, and application technology.

Appendix "A" at the end of the book provides a checklist of some design aids for potting and encapsulated applications and also for transformer and motor windings.

REFERENCES

(1) *Communications* from the following: Harry, C. C., IBM (Rochester, Minn.); Kurtz, M., Ontario Hydro; Licari, J. J., Autonetics, North American Rockwell; Lockie, A. M., Westinghouse Electric Corp. (Sharon, Penn.); Pendleton, W. W., Anaconda Wire & Cable Co.; Salzberg, L. F., Air Force Materials Laboratory; Sanvordenker, K., Tecumseh Products Co.; Spitzer, F., McGill University; Walker, H. P., Naval Ship Engineering Center; Wareham, W. W., Phelps Dodge Magnet Wire Corp.

(2) *High-Temperature-Resistant Wire Insulation Material* (734007), Air Force Materials Laboratory Topical Report, Sept. 22, 1970.*
(3) *High-Temperature Electrical Wire Coatings,* Air Force Materials Laboratory Report, AFML-TR 69-111, Part I, May 1969; Part II, Aug. 1970.*
(4) *Single-Crystal Spinel for Electronic Application,* AFML-TR-68-320, Oct. 1968; and TR-69-315, Feb. 1970.*
(5) Walker, H. P., "Moisture Resistance of Varnish Processes for A-C Induction Motors," paper at Electrical Apparatus Service Association Annual Convention, Miami Beach, Florida, May 1968.
(6) Flaherty, R. J., and Tobin, J. F., "Effect of Varnish Treatment Processes on Environmental Resistance of Class F Insulated Induction Motors," Paper No. 68CP166-PWR, IEEE Winter Power Meeting, New York, N.Y., Jan. 28–Feb. 2, 1968.
(7) "Guide for Loading Mineral-Oil-Immersed Overhead-Type Distribution Transformers With $55°$ C or $65°$ C Average Winding Rise," American National Standard (ANS) Appendix C57.91.
(8) Feather, L. E., and Voytik, P., "Application of Polyamide Papers in Liquid-Cooled Transformers," *Proceedings,* Tenth IEEE-NEMA Electrical Insulation Conference, Sept. 1971, Chicago, Ill.
(9) Dummer, G. W. A., *Materials for Conductive and Resistive Functions,* Hayden Book Co., 1970.

BIBLIOGRAPHY

Note: The papers listed below have *all* been presented at various IEEE-NEMA Electrical Insulation Conferences. In each reference, the page number refers to the respective *Proceedings,* and the date refers to the year of the Conference.

Encapsulated Systems (Power Applications)

Bishop, W. H., and Bartos, D. M., "Silicone Rubber Encapsulant for Dry-Type Transformers," 1969, p. 195.
Fischer, W., "Insulation Systems for High-Voltage Encapsulated Transformers," 1968, p. 51.
Flaherty, Jr., R. J., and Walker, H. P., "Thermal Aging Studies of Encapsulated Motorette Insulation Systems," 1969, p. 139.

Encapsulated Systems (Electronic Applications)

Dunaetz, R. A., and Schwider, A. W., "Optimizing Conformal Coating Reliability and Cost for Military Electronics," 1971, p. 157.
Kerlee, C., "Selecting Encapsulants and Potting Compounds for Electronic Modules," 1971, p. 166.
Lee, H. R., and Alekna, A. J., "Environmental Protection of Semiconductor Devices," 1971, p. 107.
Lem, J. L., and Schenk, C. H., "Protection of Electronic Devices Against Hostile Environments," 1971, p. 185.
Wittenberg, A. M., and Malec, L. F., "Mechanical Stresses Induced on Encapsulated Components," 1968, p. 101.

Nuclear Environments

Campbell, F. J., "The Design of Insulation Systems in Steam and Nuclear Radiation Environments," 1968, p. 63.
Ibid, "Radiation Effects Standards for Electrical Insulation," 1968, p. 68.

Noble, M. G., "Selecting Reliable Cable Insulation for Nuclear Generating Stations," 1971, p. 45.

Stiffler, W. G., and Brenko, W. J., "Qualification Test of a Fan and Motor Designed for Service in Nuclear Containment," 1971, p. 190.

Power Transmission Systems

Dey, P., Drinkwater, B. J., and Proud, S. H. R., "Developments in Insulation for High-Voltage Overhead Transmission Systems," 1968, p. 38.

Doepken, Jr., H. C., "Compressed-Gas Insulation for Concentric Power Lines," 1969, p. 202.

McCullough, C. R., "Electrical Stress Distribution in High-Voltage D-C Solid Dielectric Cables, 1969, p. 198.

Minnich, S. H., "Technical Aspects of Cryogenic Cable Design," 1969, p. 209.

CHAPTER 6 | PAPERS, LAMINATES, FILMS, TAPES, AND FABRICS

This chapter deals with several categories of insulating materials that are manufactured and applied in flat sheetlike form or wound to conform to a given configuration. Generally, these materials are flexible or semiflexible—papers, films, fabrics, and so forth—or they may be rigid—like most laminates and paperboard.

In their number and diversity, the materials discussed here comprise the largest group in the insulation field.

6.1 Paper Insulation

6.1.1 Definitions

Paper may be defined as a substance made from various cellulosic fibrous materials, among which are wood, rags, cotton cuttings, and certain natural fibrous mineral materials such as asbestos. Polymeric materials in fibrous form are another source, as are flakes or platelets of mica and glass, glass fibers, and ceramic fibers of various origin. The term "paper" usually applies to thin sheets, and "paperboard" to thick ones, but there is no sharp line of demarcation. The term "pressboard" is sometimes used for the latter.

The characteristics of a paper material depend on the nature of the raw materials and the conversion process. The process can be controlled and varied in many ways so that a paper of desired characteristics can be produced for given requirements.

6.1.2 Application areas

Paper is historically a major electrical insulating material. It is used for many diverse purposes, both as a discrete material and as a constituent of composite insulations. The composites are able to provide improved performance characteristics

116

since the optimum capabilities of each constituent are combined for a maximum functional effect. The combined effect can be additive or synergistic or both.

As a discrete material, paper is used in the following typical applications:

1. Layer-to-layer insulation of windings
2. Supporting and insulating tubes, separators, and covers of windings
3. Insulation to ground and between coils of windings in rotating electrical machines
4. Insulation on magnet wire and in high-voltage cables for power-transmission lines
5. A base for fabrication of grommets, bushings, sleeves, tubes, caps, covers, and the like, and in tapes
6. As dielectric or part of the dielectric in capacitors (for example, as the separator between the turns or plates of conducting films that comprise the electrodes of dry-type capacitors or as a medium for the absorption and distribution of the liquid dielectric in oil-filled capacitors).

In composites, paper is used in the fabrication of laminates and boards in combination with various resins. Similar composites can be fabricated in the form of rods and tubes (as discussed later). Paper may be combined with polymeric films to form still another category of composites. In fact, the scope for composite materials in which paper is used, or can be used, as a constituent, is extensive. Boards and laminates are applied widely in support, separation, and insulating functions. Copper-clad paper-reinforced plastic laminates are standard in printed circuitry.

To obtain an understanding of the performance capability of papers, it is desirable to know the kind of fiber and the manufacturing process employed. This information is given in subsequent sections.

6.1.3 Basic paper-making processes

The classical "sulfate" process for making paper from wood-pulp provides the basic techniques for most paper-making methods, even where the precursor materials are not cellulosic.

In essence, the sulfate process begins with the logs stripped free of bark, then cut into chips of convenient size. The chips are "cooked" in a mixture of caustic soda, sodium sulfide, and sodium sulfate until all parts of the wood are removed except for the cellulose. The latter is reduced to a pulp in which the fibers are suspended in water. The pulp is fed into the wet end of a papermaking machine where it is formed into a layer on a rotating cylinder (which is covered by a fine wire screen), thence moved to a belt of felt, much of the water being removed during the process. The moving sheet or web of paper deposited on the felt belt then passes through the machine where it is successively dried and then run through rollers for a smoothing operation. This smoothing is technically known as "calendering" and may be extended to "super-calendering" to increase the density of the paper and to impart a gloss to the surface. More than one cylinder may be employed in a machine, thus permitting thick papers to be made in a series of plies.

The widely used Fourdrinier machine differs from the cylinder type only in its use of a fine-wire traveling belt on which the pulp is deposited. Water may be drained off or removed by suction devices as the belt moves forward to deposit the paper on a felt mat. From this point the paper passes through the machine to complete the other operations, as described above. In the Fourdrinier machine there is only one screen, so there are no plies. Thickness is controlled by the quantity and concentration of the pulp and the speed of the moving screen..

6.2   Cellulosic papers

The most extensively used papers are cellulosic in structure, of which "kraft" paper is the most important category. Originally produced in Germany, whence its name (the German word *kraft* means "strong"), kraft paper is made from the wood-pulp of coniferous trees by the sulfate process previously described. Other cellulosic papers are made from raw materials such as rags, cotton cuttings, and hemp. The physical form of papers may be altered to modify their properties (as in crepe paper). Various chemical treatments are also used to improve or change performance characteristics, as in the case of high-temperature layer-to-layer transformer insulation. Vulcanized fibre[1] (discussed in some detail later) is an example of complete chemical alteration.

6.2.1   Properties of cellulosic papers

Cellulosic papers generally exhibit decreasing dielectric constant with increasing frequency. The dielectric constant of dry kraft papers varies between 1.5 and 2.8 at 60 Hz. Dielectric losses increase with increased frequency and temperature. In most apparatus operating at power frequencies (for example 60 Hz) the change in dielectric constant and dissipation factor with frequency and temperature is of little practical importance over the range of normal use. However, when a combination of higher temperatures and higher frequencies is present, the cellulosic papers are not acceptable.

Dissipation factor is important in tissue paper of capacitor grade. There are six or more types of this paper which exhibit DF (at 60 Hz) of 0.11 percent at 40° C to 0.21 percent at 120° C, depending on the type of tissue, each differing in apparent density. The denser papers exhibit the higher dissipation factors. Great care is required in the manufacture, selection, and application of cellulose tissue, including the kraft grades, in capacitor dielectrics.

Similar care is needed in the use of cellulose tissue in high-voltage cable. (This subject is discussed further in later chapters dealing with wire and cable insulations.)

Cellulosic papers used in an insulation system exert a major influence on the life (or thermal endurance) of the apparatus or equipment where the system is applied. This influence is related to two inherent limitations in the characteristics of these papers: (1) their instability when heated either in the presence or absence of oxygen (air), and (2) their highly hygroscopic characteristics. In the first case, excessive heating will break down the chemical structure of the paper and cause embrittlement. In the second case, as moisture absorption increases, the dielectric constant and dielectric loss also increase, with a consequent decline in insulation resistance.

For any given application, it is therefore necessary to give careful consideration and devote much effort to the drying processes that should be employed for any paper used in the insulation system. The operating conditions—ambient temperatures, humidity, hermetic sealing, among others—set up the guidelines for the

---

[1]   Traditionally, this spelling is the industry style for "fiber."

decision. Thus, the maker of high-voltage paper-insulated cable must use a paper drying process that would differ greatly from that needed by the maker of fractional-horsepower motors operating in air in household appliances. If paper is used in hermetically sealed motors, as in refrigerators, removal of moisture is imperative. Usually, paper is dried by heating in an oven or in a vacuum dryer. Regardless of the method selected, the problem is to select a process that will lower the moisture level of the paper not only to a level that will meet performance requirements but that will also be acceptable to the economics of the application.

The presence of some moisture in paper at the time of its application is unavoidable, the acceptable level usually being 6 to 8 percent by weight. If paper is subjected to excessive drying, its chemical structure is impaired and physical properties are lost. Such changes are irreversible. Generally, drying temperatures above 120° C (248° F) are not recommended, although some processes use forced drying at higher temperatures.

Untreated cellulosic papers may be classified as Class 90° C insulation (Class "O"). When treated with suitable organic materials, of which many are available for this purpose, the thermal classification advances to 105° C (Class "A").

6.2.2   Kraft paper and paperboard

ASTM Standard Method D 202, "Sampling and Testing Untreated Paper Used for Electrical Insulation," covers 30 procedures for methods of sampling and testing. ASTM Specification D 1305 covers "Electrical Insulating Paper and Paperboard—Sulfate or Kraft Layer Type." The specification is concerned with products for use in coils, transformers, and similar apparatus. ASTM Specification D 1930 deals with "Kraft Dielectric Tissue, Capacitor Grade." ASTM Specification D 1080 applies to cellulose papers used in the manufacture of impregnated and laminated sheets. There are four types, of which three are of wood pulp and one of cotton fiber.

ASTM Specification D 1305 gives data on four types of untreated kraft insulating tissue, paper, and paper board. Table 6.1 summarizes pertinent data. Chemical requirements are given in Table 6.2, also from the same source.

Table 6.1   Types of Untreated Kraft Insulating Tissue, Paper, and Paperboard

Type*	Thickness, in.†	Density, g/cm³	Minimum tensile strength, machine direction, lb/in. width	Minimum average breakdown voltage per sheet	Maximum conducting paths per sq ft**
I	Min., 0.001	< 0.77	8	350	1
	Max., 0.030		180	3,000	0.005
II	Min., 0.0005	0.77 to	4.6	225	4
	Max., 0.030	0.86	200	3,000	0.005
III	Min., 0.0004	0.86 to	4.1	340	3
	Max., 0.030	0.95	225	3,750	0.005
IV	Min., 0.0002	> 0.95	1.5	250	9
	Max., 0.030		250	3,750	0.005

* As in ASTM D1305.
† Thickness tolerances: I, ±10% all thicknesses; II, 0.002 and less, ±7%; over 0.002, ±10%; III and IV, 0.002 and less, ±5%; over 0.002, ±10%.
** See ASTM D202 for test methods.

Table 6.2 Chemical Requirements for Untreated Kraft Insulating Tissue, Paper, and Paperboard

Source: ASTM D1305

Property	Min.	Max.
Ash, percent	—	0.5
Alcohol soluble material, percent	—	1.0
Aqueous extract		
Conductivity, micromhos per cm	—	15
Water soluble chlorides, ppm	—	32
Fiber composition of unbleached sulfate, percent	100	—
Moisture, as received, percent	4	7
Moisture, as received, for water finished paper, percent	—	9
Hydrogen ion concentration, pH	6.5	8.0

A broad selection of special kraft papers and boards is commercially available, information on which are given in trade literature. Table 6.3 provides data on a relatively low-priced board made of 100-per cent kraft pulp in standard thickness of $\frac{1}{32}$ in. to $\frac{5}{8}$ in. Finish is calendered.

A special grade of kraft capacitor tissue has been developed in Finland to prevent accidental contamination in oil-impregnated capacitors. This tissue contains from 1 to 5 percent of an inorganic material, of submicron size, introduced during the paper-making process, Since these particles are adsorbent, they tend to remove contaminants that may be present in the liquid phase of the capacitor as well as any degradation products that may develop in service.[2]

A thin cellulose-fiber paper resembling fish paper (a variety of vulcanized fibre, but not produced by a chemical process) is used extensively in coil winding.[3] Thickness range is from 0.005 to 0.030 in. Standard finish is supercalendered through thicknesses of 0.020 in.; calender surface is obtainable in 0.021-in. thickness and over. Density is from 1.10 to 1.20 and dielectric strength from 250 to 425 volts per mil. Tensile strength in the machine direction is 10,000 to 13,000 psi and

[2] U.S. Patent 3,090,705, Tervakoski Co., Finland. Tissue manufactured in the United States by Weyerhaeuser.

[3] Kraft paper and Argelec (Case Brothers, Inc., Manchester, Conn.)

Table 6.3 Typical Test Data Kraft Paperboard (Pressboard)

Source: Case Brothers, Inc (see footnote 3). All tests are conducted at 73° ±2° F and 50 ± 1 percent relative humidity in accordance with ASTM and TAPPI methods where applicable.

Property	Thickness, in.			
	$\frac{1}{32}-\frac{1}{16}$	$\frac{3}{32}-\frac{1}{8}$	$\frac{3}{16}-\frac{5}{16}$	$\frac{3}{8}-\frac{5}{8}$
Density, gm/cc	0.95–1.10	0.95–1.10	0.95–1.10	0.95–1.10
Dielectric strength, vpm	250–325	200–300	—	—
Total volts	—	—	Over 30,000	Over 30,000
Tensile, psi				
MD*	20,000–23,000	19,000–22,000	—	—
CMD**	5,000–6,000	4,500–5,500	—	—
pH	6.5–7.5	6.5–7.5	6.5–7.5	
Ash, percent	< 1	< 1	< 1	< 1

* Machine direction.
** Cross-machine direction.

across the machine direction, 3,500 to 4,500 psi.[4] Tear strength in the machine direction of 0.005-in. paper is 125 to 175 grams and across the machine direction, 150 to 200 grams.

6.2.3 Rag papers

The description "rag paper" to identify a certain type of paper used in electrical insulation applications is, in fact, not correct, since the actual raw materials are *new* cotton cuttings. However, the use of the term "rag paper" persists as a matter of industry custom.

These papers are made with the same types of paper-making machines previously described, except that the initial "cooking" process used for kraft paper is not needed since the starting material, cotton, is already cellulose. The highest quality rag paper is made of selected light-colored cotton cuttings. The paper is light grey. No chemical additives, sizing, clay or other fillers are used in the production of rag papers.

Rag papers have the advantage of high tear strength and possess the ability to stretch. These characteristics make it possible to use these papers advantageously in forming such insulating components as slot liners for motors. It is essential that these parts should not crack nor tear during fabrication. A so-called "superstretch" rag paper produced by a proprietary method[5] is compressed in the machine direction to provide reserve material for stretching. The tear strength of this paper is also enhanced through this process.

Table 6.4 gives essential data on rag papers made from light-colored cuttings.

---

[4] Machine direction refers to the direction in which the paper is being processed in the paper-making machine.

[5] Cottrel Paper Co., Rock City Falls, N.Y. (Clupak process: Copaco, Copaco-175, Coparex, and Coparex 125 rag papers).

Table 6.4 Typical Test Values for 100 Percent Rag Paper Made from Selected Light-Color Cotton Cuttings (Data for Both Standard and Mechanically Modified Papers)

Source: Technical data, Cottrell Paper Company (see footnote 5). Mechanically modified papers are "super-stretch" type made by Cottrell's Clupak (tradename) process. Papers made from darker cuttings have sómewhat lower values.

Thickness, in.	Cross-tear resistance		Machine-tear resistance	
	Standard	Modified	Standard	Modified
0.005	220	350	180	200
0.007	380	560	280	320
0.010	700	900	550	600
0.012	900	1,100	675	750
0.015	1,200	1,500	850	950
0.020	1,700	2,000	1,400	1,550
0.025	2,200	2,500	1,600	1,750

Other properties	Standard	Modified
Machine tensile	16,000 psi	11,000 psi
Cross tensile	6,000 psi	6,000 psi
Machine elongation	6–8%	16–20%
Cross elongation	14–16%	14–20%
Dielectric strength	300 vpm	300 vpm
Methanol extractables	0.15–0.20%	0.15–0.20%
Density, 1.25–1.35	pH 7	ph 7

Darker cuttings are used for papers of slightly lower performance values. The table also gives data on the "superstretch" materials.

When paper and paperboard made of 100-percent rag stock are too expensive, lower-priced materials can be made from combinations of kraft and rag stock. The combinations most often found are 50 percent rag and 50 percent kraft and 75 percent rag and 25 percent kraft. These papers are furnished in thicknesses of 0.007 to 0.020 in. The finish may be glazed or supercalendered. Paper with the latter finish has the lower density and dielectric strength. Tensile and tear strength are lower than for the 100-percent rag products. Paperboards made from the combined rag-kraft stock come in standard thicknesses of ⅟₃₂ to ⅝ in. in eleven sizes and in both glazed and calendered finishes.

The rag-kraft papers and paperboards find wide use in transformers, rotating equipment, and many other applications. They make good punched insulating parts.

6.2.4   Glassine

Glassine is a hard-surfaced, glossy, translucent paper, usually bleached, which finds application as layer paper in the "stick" winding process of coil making.[6] Several coils are wound at a time to form a common-layer insulation for all coils on the stick. The stick is cut into individual coils by a saw or a sharp-edged rotating disk. Glassine is available in thicknesses of 0.0007 to 0.003 in. and is put up in rolls of various length to suit the requirements of the user. Density is 1.2 to 1.4 and dielectric strength, 400 to 500 volts per mil.

6.2.5   Rope paper

Rope paper is a very strong, tough paper made of manila (hemp) fibers.[7] These are the strongest known fibers of vegetable origin. The actual raw materials of manila fibers for this purpose are used rope, hawsers, and cordage. One important use for 100-percent rope paper is as insulation on magnet wire. Here, its very high strength permits high-speed application. The covered wire can be bent or formed as required by coil-winding without causing rupture of the paper. Rope paper is made in thicknessess of 0.001 to 0.005 in. It may be slit into ribbons as narrow as ⅟₁₆ in. Rope paper is made on cylinder-type machines. Other fibers of the same family as manila fibers (such as common hemp, sunn, and sisal) are used in making rope paper to suit various requirements. All these fibers have a common origin in plants of the mulberry group. Mixture of wood-pulp and manila pulp serves to reduce costs where lower performance values are acceptable.

6.2.6   Parchment

An interesting paper of unique characteristics, appearance, and surface texture, parchment offers some advantage over glassine and kraft papers in the stick winding of coils.[8] It permits a thinner paper to be employed to support the layers of wire. Its surface texture helps to reduce any tendency of the individual coils to telescope. Its wide range of available thicknesses and surface texture, moreover, provides opportunities for still other uses. Parchment is marked by very low ash (only 0.1 percent). Acidity is low and residual salts appear only as traces.

Originally made by treating rag paper with sulfuric acid and washing, parch-

6   Riegel Papers Corp., N.Y., N.Y.
7   John Manning Paper, Troy, N.Y.
8   Patterson Parchment Paper Co., Bristol, Pa.

ment is now made of high-alpha cellulose pulps to produce "water leaf" of such characteristics that it will react properly when treated in the sulfuric acid bath. This treatment results in the coating of the paper fibers with a gelatinous cellulose. The coating interlocks the paper fibers and fills the interstices. After the acid bath the reacted paper is washed and dried and given such other operations as may be required. Parchment contains no fillers or additives.

6.3   Coated, Saturated, Filled and Custom-Made Papers

Papers, usually krafts, can be treated in many ways, and a number of such papers can be considered as standard industry items. In one type, for example, shellac is applied as a continuous coating on one or both sides of the paper. Or the shellac may be applied in the form of stripes or dots. When used in transformer windings, the paper can be held in place by a hot-melting of the shellac. The use of stripes and dots patterns for the coating is to facilitate penetration of transformer oil. Filled papers include conductive papers made with carbon black fillers. Such papers are used as radio-frequency shielding in cables and electrical apparatus. Varnished papers form still another category used widely in electrical insulation systems. A variety of varnishes is utilized to produce a diversity of papers with many different characteristics. In fact, almost any desired varnish, resin, wax, or any material that will adhere to or saturate paper may be used for making these forms of insulation.

Custom-made papers can be made of most fibrous substances if the fiber size or flake size is such that a pulp may be formed. A proper distribution of particle size is essential. Future developments in the field of custom-made papers seem to be quite promising, since research is active in many potentially useful materials that are still not available in suitable fiber, flake, or particle size.

6.4   Modified Cellulosic Papers

The term "modified papers" is used to identify those papers (mostly krafts) that have undergone some treatment that alters them beyond what is done by simple coating, saturation, filling, or the like.

6.4.1   Mechanically modified papers

Typical of mechanically modified papers is the previously mentioned rag paper (see Par. 6.2.3 and Table 6.4) which features an increased ability to stretch in the machine direction. This is accomplished by a patented process that pushes the fibers back upon each other during the paper-making process.[9] As a result, the extent of available stretch is doubled. Also increased is the tear resistance both in the machine and in the cross-machine direction. The machine direction tensile strength is lowered, but the cross-machine direction tensile strength is not changed. There are also no changes in dielectric strength, thickness, finish, and other properties such as appearance.

A mechanical process known as "creping" crowds the paper into many fine pleats or folds.[10] The presence of these pleats confers on the paper an ability to stretch or elongate to almost any degree from 25 to 300 percent. The creping process can be applied to paper either in the machine or the cross-machine direction, or in

[9]   See footnote 5.
[10]   Denison Mfg. Co. Framingham, Mass.

Table 6.5   Creped Krafts, 50 Percent Elongation

Base paper	Thickness, mils crepe*	Type	Tensile strength, machine direction, lb/in. width
1	5	High density	10
3	15	High density	25
3	6–8	High density, calendered	25
5	22	High density	60
5	10–12	High density, special calendered	60
3	10–12	High density, 2-ply calendered	—

* Thickness is controlled on calendered types only.

both directions. In effect, the paper is capable of either a one-way or a two-way stretch, as desired. The paper can be calendered after being creped, thus making it possible to control paper thickness without loss of stretch.

Creped paper, because of its stretch property, possesses good wrapability, conforms smoothly to irregular surfaces, and has high tear strength. It is very well adapted to taping, manually or by machine. In general, creped paper provides a low-cost extensible tape for many uses. Mostly applied to 100-percent kraft, creping can be also applied to other types, usually on a custom basis.

There are two basic types of kraft crepe paper, differing in density. One is known as the "saturating grade," with an apparent density of 0.65, and the other is known as the "high density" grade, with an apparent density of 0.90.[10] Each grade can be saturated with insulating oils for transformer use, and each can absorb varnishes and other insulating compounds. Choice of type is usually made on the basis of tensile strength, in which the "high density" grade is the stronger.

Table 6.5 enumerates comparative data for a few crepe kraft papers to indicate typical characteristics of the many crepe papers available. The 50-percent grade was selected for this table because of its common use for taping.

6.4.2   Chemically modified papers

Cellulose, in molecular structure, is a string of rings linked together by atoms of oxygen. The ring is the glucose molecule (see below). Nature combines the glucose into cellulose by condensation polymerization. (In 1964, Hassid, Barber, and Elbein duplicated this reaction at the University of California). Thousands of the glucose building blocks may be combined in cellulose. Schulz and Marx suggest that the number of these blocks in wood-pulp is between 1,150 and 3,300.[1]

The OH radicals (shown encircled) appear to be the source of the difficulties encountered in the use of paper as an electrical insulator, especially in oil-filled transformers. In these, water and carbon dioxide are released as a result of operating temperatures and the presence of copper and iron. Since all the elements are contained in a *sealed* case or can, neither the water nor CO_2 can escape but are retained to accelerate the rate of degradation in the insulation system. Chemically modified papers provide an important means of reducing these difficulties by replacing some of the OH groups with nitrogen-containing groups.

Two major processes have been developed: In the *cyanoethylation* process, acrylonitrile is interacted with cellulose pulp so as to increase the nitrogen content of the paper. As a result, the paper is thermally upgraded and is able to retain both mechanical and electrical properties after high-temperature aging. Paper made by this process is known as "cyanoethylated paper." In one research project, a nitrogen content of 2.6 percent was found to provide the best results for long-life paper under elevated temperatures.[11] In the *acetylation process*, the primary purpose is to prevent moisture absorption by blocking the hydroxyl groups that are responsible for chemically linking water to the cellulose molecule. Papers made by this process are known as "acetylated paper."

Special proprietary insulation systems developed by transformer manufacturers are also based on the concept of introducing nitrogen-containing compounds during the paper-making process in order to enhance the thermal capability of the system. These systems are known by their tradenames and, presumably, are not available except under license.[12] Thermally upgraded paper is available from several manufacturers of insulation paper.

6.5   Polymeric-Fiber Papers and Paperboards

The availability of synthetic fibers produced from polymers is not new; both industry and commerce have known and used such fibers in textiles and textile products for many years. Comparatively recent, however, is the development of polymeric-fiber electrical insulation in the form of papers, paperboards, and paper-containing composites.

6.5.1   *Polyamide-fiber paper*

Probably the most important development in polymeric-fiber papers is the one known commercially as Nomex.[13] It is produced from polyamide (the same basic polymer as the familiar nylon, but of quite different thermal characteristics). Unlike the polyamide in the familiar nylon, the high-temperature material used for Nomex will not melt nor support combustion. At 250° C, the melting point of nylon 6-6, Nomex retains 60 percent of its room-temperature strength. It has been recognized by the Underwriters' Laboratories as a component in at least three insulating systems rated as Class 180° C (Class "H") and Class 220° C (Class "C").[14]

---

11 General Electric Company's patent (No. 2,535,690) was reported in AIEE Papers TP-60-58 and CP-59-952.

12 Examples of such proprietary transformer insulation systems include General Electric's "Permalex," McGraw-Edison's "TherMEcel," and Westinghouse Electric's "Insuldur."

13 Tradename, E.I.du Pont de Nemours & Co., Inc.

14 E.I.du Pont, Textile Fibers Dept.

Nomex paper is made on conventional paper-making equipment. Two types of fibers are used: flock fibers, which are relatively short, and fibrids, which are small fibrous binder particles. Mixed in the necessary proportion, these fibers are made into a paper that is furnished, when the process includes calendering, in thicknesses of 2, 3, 5, 7, 10, 15, 20, and 30 mils. The color depends on the thickness, being nearly white in the 2-mil thickness and gradually becoming more off-white until the 30-mil is a light buff. The surface has a "bite" to it so that windings can be made that are tight, firm, and of excellent space utilization. No materials other than the basic polymer fibers are used.

When calendered, the Nomex paper is called Type 410. This type has an apparent density of nearly 1.0. It can be surface-coated with any of the appropriate varnishes or adhesives. It is difficult to impregnate completely, but since it has such excellent inherent characteristics this difficulty presents no drawback to its use. Low-viscosity liquids can be forced into it with alternate vacuum-pressure cycles. Nomex can be cut, punched, and fabricated into the forms and shapes used in the insulation systems of rotating and static equipment. Tables 6.6 and 6.7 contain the important electrical and physical properties.

An uncalendered Nomex paper known as Type 411 is available in a range of thicknesses of 6 to 42 mils. This type has a density of about 0.3 and is relatively porous. It can be completely impregnated and is compressible. Seldom used in its "as-made" condition, its characteristics depend on the nature of resin impregnation and other treatment.

Since the introduction of Nomex in 1961, this paper has met with wide acceptance and can be obtained in laminates, pressure-sensitive tapes and other forms of tape, as well as in creped form and resin-coated form, for application in tubing and standard fabricated parts such as slot liners, wedges, and punched parts. Its use in composites has been particularly noteworthy. Many combinations are possible with other materials, such as glass fabrics, plastic films, and mica.

Paperboard that exhibits the thermal stability of Nomex paper can be made by laminating Types 410 and 411 paper directly to each other, without adhesives, by the application of heat and pressure. Temperatures up to 280° C and pressures starting at 200 psi and increasing to 2,000 psi, with necessary control techniques, are employed in the laminating process. To eliminate the anistropy characteristic of the paper, the plies are cross-stacked as to machine direction.

The growing use of nuclear power and of high-energy devices increases the need for electrical insulation capable of withstanding the deteriorating effects of beta and gamma radiation and of X-rays. The outstanding resistance of Nomex to such radiation is shown in the data given in Table 6.8. The effect of radiation on various fibers including Nomex (under its original designation HT-1 nylon) is summarized in a United States Air Force study.[2]

6.5.2 Other polymeric papers

Papers made from acrylic fibers as well as from polyester fibers find special application in electrical equipment. An acrylic-fiber paper, saturated with an acrylic resin, is used in sealed systems, such as hermetic motors operating in refrigerants.[15] For operation in Class 155° C (Class "F") temperatures, several papers have been developed from polyester fibers with an epoxy resin binder. These papers vary in

---

[15]   Rogers Corp., Rogers, Conn.

Table 6.6   Typical Electrical Properties of Polyamide (Nomex Type 410) Papers

Source: E. I. du Pont, Textile Fibers Dept. (Nomex, a high-temperature nylon, is a Du Pont Company tradename.)

Property	Thickness, mils							
	2	3	5	7	10	15	20	30
Dielectric strength, v/mil*								
a-c rapid rise	450	500	600	750	750	800	700	600
a-c 1-min. hold	—	—	—	—	600	—	—	—
a-c 1-hr hold	—	—	—	—	500	—	—	—
d-c rapid rise	—	—	—	—	1200	—	—	—
Dielectric constant, 10^3 Hz**	2.0	2.1	2.3	2.5	2.6	3.0	3.3	3.4
Dissipation factor, 10^3 Hz**	0.007	0.008	0.010	0.012	0.014	0.016	0.018	0.020

* ASTM D-149, using 2-in.-diameter electrodes.
** ASTM D-150, using 1-in.-diameter electrodes under 20 psi pressure.

Table 6.7   Typical Physical Properties of Polyamide (Nomex Type 410) Papers

Source: E. I. du Pont, Textile Fibers Dept.

Property		Thickness, mils								ASTM test number
		2	3	5	7	10	15	20	30	
Tensile strength, lb/in.	MD*	22	37	70	120	170	270	370	550	D-828-60
	XD**	13	23	45	68	100	180	250	380	
Elongation, percent	MD	10	12	16	19	24	24	25	25	
	XD	8	10	14	16	20	21	21	22	
Elmendorf tear, gm	MD	95	150	260	380	550	850	1,100	2,000	D-689
	XD	140	250	450	620	900	1,400	2,100	2,700	
Finch edge tear, lb/in.	MD	20	40	70	110	130	150	160	220	D-827-47
	XD	10	15	25	30	60	60	60	100	
Shrinkage at 300°C, percent	MD	1.6	1.2	0.9	1.1	1.5	1.4	1.4	1.4	
	XD	1.9	1.6	1.3	1.0	1.0	0.8	0.8	0.8	
Thermal conductivity, BTU/in./hr/ft^2/°F		—	—	—	—	0.76	—	0.78	—	
Basis weight, oz/yd^2		1.2	1.9	3.2	5.1	7.3	11	15	23	

* MD = Machine direction.
** XD = Cross machine direction.

Table 6.8   Radiation Effects on Various Fiber Insulating Materials

Radiation test conditions	Nomex* nylon	Dacron* Polyester	Du Pont nylon 6-6
	Percent breaking strength retained after exposure to radiation		
Beta radiation—Van de Graaf			
200 mega reps (1.9×10^8 rads)	81	57	29
600 mega reps (5.6×10^8 rads)	76	29	0
Gamma radiation—Brookhaven Pile (50° C)			
200 mega reps (1.9×10^8 rads)	70	45	32
1,000 mega reps (9.3×10^8 rads)	55	Radioactive	Crumbled
2,000 mega reps (1.9×10^8 rads)	45	Radioactive	Crumbled
X-rays (50 kV)			
50 hours	85	22	—
100 hours	73	0	—
250 hours	49	0	—

* Du Pont tradenames.

flexibility from a limp, highly flexible grade to one of pronounced stiffness. Typical characteristics of the polyester papers are given in Table 6.9. Data for other commercial materials are included for comparison.[15]

6.6   Asbestos Paper and Boards

There are six minerals of fibrous crystalline structure to which the general name of asbestos is applied. Only one of these is of use as an electrical insulation. This is chrysotile, a hydrated magnesium silicate that is considered to have the empirical formula, $Mg_3 [(OH)_4Si_2O_5]_n$. All the water appears as hydroxyl (OH^-) ions.[3] When free of magnetite, water-soluble electrolytes, and other conducting impurities, chrysotile can be fabricated into a variety of forms used as electrical insulation.[16]

The most generally used asbestos paper is made of selected 100-percent chrysotile fibers, free of iron and other impurities. The basic sheet will maintain its dielectric strength of 200 volts per mil up to 300° C. It should be noted, however, that asbestos paper without binders or saturants is too weak for most uses. For this reason, asbestos paper is made commercially available with several types of saturants or impregnants to improve mechanical characteristics. The specific saturant or impregnant used will also determine the thermal endurance of the paper. Even though the saturant is driven off by operating temperatures, the remaining asbestos structure will still afford adequate insulation protection in most applications.

In order to supply asbestos papers that are still stronger and have increased dielectric strength, that can be wound into place, and that also exhibit certain other desired characteristics, special binders are incorporated into the asbestos fibers during the paper-making process. In addition, the finished paper can be saturated or impregnated with various resins or mineral compounds. In one type of paper, the asbestos fibers are deposited on an interior web of glass cloth during manufacture. In other types, nitrile rubber, starch and wood pulp, and triazine are added as bind-

[16]   Johns-Manville, N.Y., N.Y.

Table 6.9   Typical Properties of Epoxy-Impregnated Polyester Papers Compared to Other Materials

Type	Duroids*		Other Materials		
Reinforcement	Polyester paper	Polyester mat on polyester film	Asbestos paper-glass cloth	Glass cloth	Mica-glass cloth
Resin	Epoxy	Epoxy	Epoxy	Epoxy	Silicone
IEEE Class	155° C(F)	105° C(A)	155° C(F)	155° C(F)	180° C(H)
Physical properties					
Thickness, in.	0.015	0.010	0.011	0.005–0.010	0.015
Specific gravity	1.50	1.00	1.25	1.30–1.45	1.75
Tensile strength, $10^3 \times$ psi					
Longitudinal	6.7	14	13	16–36	12
Transverse	4.3	9	13	9.6–19	12
Ultimate elongation, percent					
Longitudinal	24	34	3	2–3	3
Transverse	5–24	55	3	2–4	3
Finch tear strength, lb/mil					
Longitudinal	1.5–4	18	1	2–3	1.5
Transverse	2–6	21	1	1.5–5	1.5
Elmendorf tear strength, gm/mil					
Longitudinal	40	58	60	27–80	46
Transverse	45–55	168	85	50–90	42
Electrical properties					
Dielectric strength					
Short time, vpm	1,000†	1,000	1,100	900–1,200	1,000
Wrapped electrode, kV	13	9	5	3–11	12
Dielectric half life, hr					
At 410°F (210°C)	—	460	—	10–675	—
At 440°F (227°C)	1,130	—	—	—	—
At 500°F (260°C)	110	35	180	Nil-50	>1,000
Weight loss, hr					
To lose 10% organic					
At 410°F (210°C)	450	460	15	20–30	1,200
At 455°F (235°C)	175	120	10	10–15	500
At 500°F (260°C)	15	20	5	2–5	85
To loss 50% organic					
At 500°F (260°C)	275	420	90	20–25	—

* Duroid is tradename for Rogers Corp. materials (see footnote 15). Values given here are for grades 2305, 2310, and 2315. Except where ranges are shown, values are same for all three grades.

** Ranges shown are for two grades of indicated glass cloth-epoxy.

† Electrical properties are for Duroid 2305, but all three grades are similar.

ers. Saturants include polyvinyl acetate resins, silicones, and polyester resins, among others.

The electrical properties—dielectric constant, dissipation factor, and volume resistivity—will depend on the specific type of paper, the conditioning to which the sample has been submitted, and the test procedures. If an asbestos paper not *specifically* manufactured for use in electrical insulation should be considered for such applications, it is essential that the electrical properties be determined in advance of use.

Asbestos boards are made in several types and in various thicknesses, usually in $\frac{1}{32}$-in., $\frac{1}{16}$-in., $\frac{1}{8}$-in., $\frac{3}{16}$-in., and $\frac{1}{4}$-in. sizes. A 100-percent fiber is the starting point in manufacture. One type, made in an untreated form, is not intended to be a finished insulation. It is, of course, available for further treatment as desired. Another type is silicone-resin-treated with about 20- to 35-percent resin content. Since the resin is uncured, the board can be easily formed and placed as desired, and then cured by heating. At 250° C, the resin will cure in 30 minutes. It is best to heat the board to 100° C, or even to 160° C, to soften the resin content for easy working. As made, the board is tackfree.

Still a third type is a class 130° C insulation made of 93-percent asbestos and 3-percent polyvinyl acetate binder. It is intended as a base sheet to be further treated or processed, but in certain applications it may be used in an as-is condition. Another type of board for 130° C service contains a high-temperature rubber binder. Depending on the application, it may need additional treatment or be left as is.

6.7   Glass-Fiber Paper

A low-density nonwoven paper is made of borosilicate micro-glass fibers.[17] It is available in thicknesses of approximately 0.001 to 0.030 in. and widths up to 38 in. and even 42 in. This paper has a high softening point (over 1,200° F), a decided advantage. Other advantages include high thermal conductivity, low moisture absorption, and good resistance to chemical attack. Since the micro-glass fiber is actually more like a smooth, very fine rod, the paper cannot be pressed like conventional fiber (or otherwise mechanically treated) without fiber damage. Because of the many interstices in the paper, it can be easily impregnated with resin. Fluorocarbon and silicone resins have been used. When epoxy-resin impregnated, glass-fiber paper may be bonded to copper foil for printed-circuit use.

6.8   Glass-Flake Paper

Another glass-type paper is made from glass flakes combined with cellulosic or resinous pulps.[18] The flakes are made by drawing molten glass at high speed into an extremely thin sheet. As the sheet is drawn, it is cooled and broken into random-size flakes, less than 0.0002 in. thick. The paper is available in 0.004-in. and 0.005-in. thicknesses. The electrical, mechanical, and handling characteristics are determined by the nature and quantity of bonding materials employed. Untreated, the paper is fragile, with a tendency to friability when handled. Bonding processes with suitable materials are therefore necessary. Its dielectric properties and resistance to chemicals are similar to those of micro-glass fiber paper.

6.9   Ceramic-Fiber Papers

A ceramic paper is made from washed alumina-silica fibers on a Fourdinier machine.[19] The fibers are made by blasting a steam jet through a molten-mixture stream of 50.9-percent Al_2O_3, 46.8-percent SiO_2, and small amounts of B_2O_3 and Na_2O, with traces of organics. There is no water of combination. Two forms are available, one with no organic binder and one with up to 5-percent organic

[17]   Paliflex Products, Putnam, N.Y.
[18]   Electro Insulation Co., Pittsburgh, Pa.
[19]   The Carborundum Co., Niagara Falls, N.Y.

binder. The paper is white and of low density (12 lb/cu ft). Its melting point is above 3,200°F, and it may be used continuously at 2,300°F. Approximate thicknesses, uncompressed, are $\frac{1}{32}$, $\frac{1}{16}$, and $\frac{1}{8}$ in., but a special paper is $\frac{1}{100}$-in. thick. These papers find their largest field of use in mechanical and thermal applications. Not only are they good heat insulators, they can also be treated and fabricated into many forms. Cements and rigidizers are available for making laminates and composites. Both as papers and as laminates, alumina-silica fibers can be worked, shaped, fastened in place, and the like, with ease.

The alumina-silica papers exhibit excellent resistance to chemicals. They are attacked by very few acids, although they are affected by strong alkalies. They will dry without any change in properties after being wet by water. Although papers without a binder lack strength and must be handled with care, only a small quantity of binder is needed to improve mechanical strength sufficiently to make handling permissible.

Other ceramic fibers have been used in making paper materials; among them a silica paper made of micro-quartz fibers can be noted. This paper, reported to withstand temperatures up to 3,000°F, is available in thicknesses of 0.003 to 0.030 in. Generally, it exhibits characteristics similar to that of other silica fibrous materials. Reinforcement with resin binder is needed.

6.10   Mica Paper

Papers are made from micaceous platelets and flakes by modified paper-making techniques. These papers have assumed an important place in the family of mica products, both in various types of fabricated mica insulation and as important constituents in many composite insulating materials. A general discussion on mica materials appears in Chap. 11.

6.11   Vulcanized Fibre and Fish Paper

Vulcanized fibre is a dense strong material made in four primary forms —sheets, rolls, tubes, and rods. When the thickness of the sheet material is 0.015 in. or less, it is often called fish paper. Many other names have been applied to the thin sheet form, such as leatheroid, leather paper, fiberoid, tarpon paper, and the like.

NEMA Standards Publication VU 1-1963 (or year of latest issue) gives the following general description:

Vulcanized fibre   Vulcanized fibre is made by combining layers of chemically gelled paper. The chemical compound used in gelling the paper is subsequently removed by leaching, and the resulting product, after being dried and finished by calendering, is a dense material of partially regenerated cellulose in which the fibrous structure is retained in varying degrees, depending upon the grade of fibre.

Sheets and rolls   Sheets and rolls are the most commonly used primary forms. After calendering, thin fibre may be run into parent rolls or cut into sheets. Rolls are subsequently cut into coils and sheets into strips for punching, forming, or swaging operations.

Round tubes   Round tubes are made by winding the chemically gelled paper on mandrels of a size suitable for the desired inside diameter. After leaching out the chemical, the tubes are dried and calendered. They are finished by grinding and sanding to the desired outside diameter.

Round rods   Round rods are ground from strips cut from sheets, so that the grain runs lengthwise. The plies are parallel chords of a circular cross section.

Laminated (built-up) fibre   Laminated (built-up) fibre is composed of a number of plies of vulcanized fibre bonded together with an adhesive. It retains all the basic properties of solid fibre, including high arc resistance, and in addition has better dimensional stability and less warpage. Laminated (built-up) fibre can be supplied in heavier thicknesses than solid fibre.

Thirteen grades are described in NEMA Standard VU1, Section 1.01, of which the following five have use as electrical insulation:

Commercial grade (sheets, rolls and rods)   This grade, considered as the general purpose grade, is sometimes referred to as the Mechanical and Electrical Grade. It possesses good physical and electrical properties and can be fabricated satisfactorily by punching, turning, and forming operations. It is made through the entire range of thicknesses from 0.010 to 2 in. Rods are turned from Commercial Grade sheets.

Electrical insulation grade (sheets and rolls)   This grade is primarily intended for electrical applications and others involving difficult bending or forming operations. It is made in all standard thicknesses from 0.004 to $\frac{1}{8}$ in. Thin material of this grade is sometimes referred to as "fish paper."

Bone grade (sheets, tubes and rods)   This grade is characterized by the greater hardness and stiffness associated with higher density. It machines smoother and with less tendency to separate the plies in difficult machining operations. It is made in the limited thickness range of $\frac{1}{32}$ to $\frac{1}{2}$ in. Rods are turned from Bone Grade sheets.

Hermetic grade (sheets and rolls)   This grade is used as electric motor insulation in hermetically sealed refrigeration units. High purity and low methanol extractibles are essential because it is immersed in the refrigerant.

Railroad grade (sheets)   This grade is for use in railroad track joints, switch rods, and other insulation applications for track circuits. It conforms to Parts 58 and 178 of the AAR (Association of American Railroads) Signal Section, "Specifications for Hard Fibre." Its toughness and high compressive strength enable it to withstand the repeated shock loads occurring in track joints.

Although the remaining eight grades are not indicated for use as electrical insulation, some of these grades do find applications as mechanical or structural parts in electrical equipment.[4]

The ASTM Standard Specification for vulcanized fibre is D 710-(year of issue). This covers three grades: bone, commercial, and electrical insulation. Standard methods of testing vulcanized fibre used for electrical insulation are given in ASTM D 619-.

NEMA has Publication No. VU 2-1962 entitled, "Recommended Practices,

Fabricating Vulcanized Fibre." This contains excellent suggestions for cutting, punching, shaving and broaching. Bending and forming operations are described, as are operations performed by drill presses, lathes, screw machines, lathes, milling machines, and the like.

The three most commonly used grades of vulcanized fibre are identified by the following color code:

Form	Grade	Color
Sheet and rolls	Commercial	Gray, red, black
	Electrical	Gray
	Bone	Gray
Round tubes	Bone	Gray, red, black
Round rods	Commercial	Gray, red, black

The white grades are used for nameplates, labels, and the like.

Vulcanized fibre is marked by three important characteristics:

1. Vulcanized fibre has a high strength-to-weight ratio. In fact, it is one of the strongest insulating materials available. Although on a weight-to-volume basis it weighs only one-half the weight of aluminum, it is able to withstand repeated shock and impact forces of considerable magntitude,

2. Vulcanized fibre adjusts its moisture content to the relative humidity of its environment. This characteristic is demonstrated in the data plotted in Fig. 6.1. Any increase in moisture content is accompanied by changes in density and in dimensions. As moisture content increases, so does each dimension of the material. A rough rule is that for each percent of moisture change, the thickness change is 1 percent, the width changes by 0.25 percent, and the length by 0.1 percent. For each 1 percent of moisture increase (over a range of approximately 4 to 32 percent), density drops about 0.3 lb per cu ft.

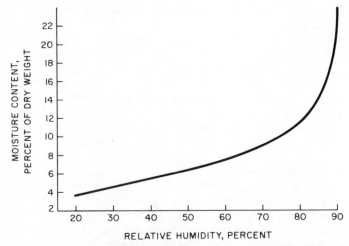

Fig. 6.1   Equilibrium moisture content of vulcanized fiber under various humidities at 73°F (*Courtesy*, NEMA, Authorized Engineering Information 9-25-1963)

The behavior of vulcanized fibre in respect to moisture absorption has both advantages and disadvantages, depending on the application at hand. For example, if impact strength is a vital characteristic, the increase in moisture content is welcome, since it is accompanied by an increase in impact strength. However, if dimensional tolerances have to be closely watched, and ambient humidity conditions are changeable, the tendency of vulcanized fibre to change its moisture content in accordance with ambient conditions would require suitable design precautions.

Subjected to direct contact with water, vulcanized fibre will swell. If soaked for a long period, the material will be completely ruined, but if the damage is not excessive, the material may regain its original form and dimensions on drying.

3. Vulcanized fibre is also marked by its resistance to arcing—an important advantage when it is used in such devices as tubes for fuses, lighting arrester parts, and arc barriers in circuit breakers. In the presence of arcing, vulcanized fibre releases a gas (consisting principally of carbon oxides and water vapor) which quenches the arc.

Detailed properties covering vulcanized fibre materials are available in publications such as NEMA Standard VU-1 and ASTM Standard D 710.

A summary of selected characteristics shows that vulcanized fibre has fair dielectric strength, high dielectric losses, relatively poor insulation resistance, and excellent arc resistance (as noted above). It has excellent compressive and impact strength and good tensile and flexural strengths. Heat resistance is only moderate— it is a Class 105° C (Class "A") material. Acids and alkalis attack vulcanized fibre disastrously, but organic solvents, such as those used in electrical work, do not attack it. Fabricating properties are excellent.

Special grades are made in which properties are modified through various manufacturing techniques. For example, resin-impregnation may be employed to confer improved dimensional stability on vulcanized fibre. Modified grades are also made to provide improved moisture resistance, fire retardance, and resistance to fungus attack. One resin-impregnated grade has Underwriters' Laboratories approval for use as support members of wiring devices.[20] An aniline-type resin is used for this material.

Vulcanized fibre can also be bonded to a variety of other materials to produce composites for different applications.

6.12   Laminated Thermosetting Products (Sheets)

The NEMA Standards Publication LI-1-1965 (with subsequent changes and additions) defines laminated thermosetting products as: "laminated thermosetting products that consist essentially of fibrous sheet materials, such as cellulose paper, cotton fabric or mat, asbestos fabric or mat, wood veneer, nylon fabric, glass fabric, and the like, which are impregnated or coated with a thermosetting resin binder and consolidated under high temperature and pressure into hard solid products of high mechanical strength."

This standard covers industrial products, which includes those intended for use in electrical as well as mechanical applications. An example of an electrical use

---

[20] Anilite Impregnated Vulcanized Fibre (NVF Corp., Yorklyn, Del.)

is as the substrate of copper-clad laminates. For mechanical use, the fabrication of gears is an important example.

The industrial products are made in the form of sheets, rods, and tubes. The word "grade" is used to denote the construction, materials used, and the intended field of application.

There are seven basic classes or divisions of grades. These are paper base, fabric base, asbestos base, glass base, nylon base, polyester, flame-resistant, and post-forming. Each basic grade includes several subgrades although the prefix "sub" is not used. A letter or combination of letters is assigned to designate each grade. The letters are chosen to provide a nomenclature that is not too complex yet enables the engineer to recognize the characteristics of the grade when its letter or letters are known.

NEMA Standard LI-1-1965 includes copper-clad laminates, a category that has grown greatly in the variety of types and in the volume of use. At present, these laminates comprise a most important product group in the electrical industry and give every evidence of maintaining their growth.

In addition to the standard products, many other grades are offered by manufacturers of thermosetting laminates. Far too numerous to be described fully in the space permitted here, a sufficient number will be discussed to provide an availability guide for the reader.

The history of laminated thermosetting products may have begun circa 1910 when the classical studies and work of Dr. L. H. Baekeland resulted in a workable and reproducible process for the production of C-stage phenol-formaldehyde resins. The first NEMA Standard was issued in October, 1927. This grew out of the original work of the Associated Manufacturers of Electrical Supplies. This association became a part of NEMA in 1926. Dr. Baekeland's resin or plastic became known by the trade name "Bakelite," and for many years the name was used almost in the generic sense for this class of resins. Contemporary usage, however, is to describe the broad variety of condensation products of phenols and cresols with formaldehyde as "phenolics."

Resins used   In addition to phenolics, NEMA Standard LI-1-1965 lists the following resins used for thermosetting laminates: melamines, silicones, polyesters, and epoxies. Other resins are used in the manufacture of nonstandard, or special-purpose, laminates. These resins include the acrylics and diphenyl resins. For laminates used in metal-clad form for printed circuitry, TFE-fluorocarbon resin finds application.

Fibrous sheet materials used   The sheet materials used for laminates include cellulose paper, asbestos paper, cotton fabric or mat, asbestos fabric or mat, nylon fabric, glass fiber cloth or mat, and any other suitable fibrous paper or mat.

Manufacturing method   The resins are dissolved in alcohol, alcohol water, or other suitable solvents to form varnish solutions for impregnation and coating of the sheet materials. After being impregnated, the sheets are dried and cut to the desired size. The sheets are then stacked with metal pressing plates and subjected to high temperatures and pressures to form the fully cured thermoset laminates. Temperatures and pressures depend on the resin and the product, with tempera-

Table 6.10   Standard Forms of Laminated Thermosetting Products

Form	NEMA grade designations
Sheets	All grades
Rolled tubes	X, XX, XXX, C, L, LE, A, AA, G3, G5, G7, G9, G10, G11, N1, FR3, FR4, FR5
Molded tubes	XX, XXX, C, CE, L, LE, A, AA, G7, N1, FR3
Molded rods	XX, XXX, C, CE, L, LE, G3, G5, G7, G9, G10, G11, N1, FR3

tures of 130°C to 260°C and pressures as high as 2,500 psi being used. When the press is cooled down, the finished laminates are removed.

Rods and tubes (to be discussed later) are manufactured from sheet materials. Round tubes are made by rolling sheets upon mandrels between heated pressure rolls and curing in an oven. The tubes may also be cured in a mold, depending on the final product desired. Rods are manufactured by curing rolls of sheet material in a mold and then grinding to size. The rods can also be machined from strips cut from cured sheet stock. The laminations are parallel chords of the circular cross section. In addition to round tubes and rods, square and rectangular tubes are molded from sheet material.

Standard forms of laminated thermosetting products   Table 6.10 taken from the NEMA Standard LI-1-1965 lists the various standard grades and the forms in which products of the grades are available. Table 6.11 shows the standard colors for the various forms and grades. Grades GPO-1, 2 and 3 are polyester-glass mat sheet laminates. Grade CF, post-forming, and copper-clad laminates are not listed in the tables but are covered by the NEMA Standard. Other products are available that will, no doubt, be recognized ultimately by the Standard.

Table 6.12 shows the grade letters and the fibrous materials and the resin used in them. A column headed "Applications" briefly gives the outstanding characteristics and the field of use.

Table 6.11   Standard Colors of Sheets, Tubes and Rods

Form	NEMA grade designations	Colors
Sheets	X, XX, XXP, XXX, C, CE, L, LE	Natural or black
	XP, XPC	Natural, black, or chocolate
	XXXP, XXXPC, AA, G2, G3, G5, G7, G9, G10, G11, N1, FR2, FR3, FR4, FR5	Natural
	A	Natural or gray-black
Tubes and rods	X, XX, XXX, C, CE, L, LE	Natural or black
	AA, G3, G5, G7, G9, G10, G11, N1, FR4, FR5	Natural
	A	Natural or gray-black

**Table 6.12   Checklist of NEMA Grades of Laminates,
Composition, and Applications**

Grade*	Fibrous component	Resin component	Applications
X	Paper	Phenolic	Mechanical; high-humidity conditions affect it markedly
XX	Paper	Phenolic	Machines well; suitable for usual electrical uses
XXX	Paper	Phenolic	Suitable for radio frequency work; good in high humidity
XP	Paper	Phenolic	An X grade for punching
XXP	Paper	Phenolic	An XX grade for punching; better electricals and moisture resistance than XX
XXXP**	Paper	Phenolic	Better electricals still and good for hot punching; high insulation resistance and low dielectric losses at high humidity
XXXPC**	Paper	Phenolic	Low temperature punching grade of XXXP; as good electricals as XXXP
ES1	Paper	Melamine	Engraving stock; black or gray surfaces, white core
ES2	Paper	Phenolic	Similar to ES1; white sub core, black core
ES3	Paper	Melamine	Like ES1 but white or gray surface, black core
C	Cotton fabric, 4 oz/yd²	Phenolic	Mechanical, in gears; no control of electricals
CE	Cotton fabric, 4 oz/yd²	Phenolic	Mechanical; better moisture resistance; controlled electricals; not recommended for primary insulation above 600 volts.†
L	Fine-weave cotton fabric, 4 oz/yd² or less	Phenolic	Mechanical; fine machine applications; surface threads 75 × 152 or more as means of identification of grade; no control of electricals
LE	Fine-weave cotton fabric, 4 oz/yd² or less	Phenolic	Mechanical; better machining and appearance and moisture resistance; not recommended for primary insulation above 600 volts†
A	Asbestos paper	Phenolic	Slightly more resistant to heat than cellulosic grades; high organic content; not recommended for primary insulation above 250 volts; small dimensional changes when exposed to moisture
AA	Asbestos fabric	Phenolic	Better resistance to heat, stronger and tougher than Grade A; not recommended for primary insulation at any voltage; lower dimensional changes with moisture than Grade A
G2	Staple-fibre glass cloth	Phenolic	Good electricals; mechanically weakest of glass-base grades; good dimensional stability; lower in dielectric losses than other glass-base grades (except silicones)
G3	Continuous-filament glass cloth	Phenolic	Good electricals under dry conditions; high impact and flexural strength; bonding strength poorest of glass-base grades; good dimensional stability
G5	Continuous-filament glass cloth	Melamine	Highest mechanical strength and hardest; good flame resistance; excellent insulation under dry conditions; low insulation resistance under high humidity; good dimensional stability

6.12.1   Criteria for selection of laminate sheets

The selection of a suitable grade for a given application is primarily an engineering function. Although the 32 grades listed in NEMA Standard LI-1 (see Table 6.12) can be used as electrical insulation, many are not made under conditions where the electrical characteristics are controlled. Of these 32 grades only 12 are specifically made for electrical use. They are identified in Table 6.12 by italic grade designations. Several non-NEMA types are described in Table 6.12A Table 6.13

Table 6.12   Checklist of NEMA Grades of Laminates, Composition, and Applications (Cont'd)

Grade*	Fibrous component	Resin component	Applications
G7	Continuous-filament glass cloth	Silicone	Low-dielectric loss, high insulation resistance and good electricals under dry conditions; percent change from dry to humid conditions is high; excellent heat and arc resistance; meets IEEE requirements for 180° C
G9	Continuous-filament glass cloth	Melamine improved for heat resistance	Excellent electricals under wet conditions; high mechanical strength; hard; good flame; heat, and arc resistance; good dimensional stability
G10**	Continuous-filament glass cloth	Epoxy	Extremely good flexural, impact, and bonding strength; good electricals under both dry and humid conditions
G11**	Continuous-filament glass cloth	Epoxy	Similar to G10 but retains at least 50 percent of room-temperature strength at 150° C after 1 hr at 150° C
L1	Nylon cloth base	Phenolic	Excellent electricals under high humidity; can flow or creep at temperatures higher than normal; good impact strength
FR2**	Paper	Phenolic modified	Similar to XXXPC** but self-extinguishing after source of ignition is removed
FR3**	Paper	Modified epoxy	Self extinguishing after source of ignition is removed; compares well with XXXPC**
FR4**	Continuous-filament glass cloth	Modified epoxy	Self-extinguishing after source of ignition is removed; otherwise similar to grade G10**
FR5**	Continuous-filament glass cloth	Modified epoxy	Self extinguishing after source of ignition is removed; otherwise similar to grade G11**
CF	Cotton fabric, 4 oz/yd²	Phenolic curable polymer	Can be fully cured by heat and pressure after post-forming; Part 12 of NEMA Standard LI-1 covers the special requirements of this grade
GPO-1 and GPO-2	Random-laid glass fibers	Polyester	GPO-1 for general purposes; GPO-2 for self-extinguishing properties; see Part 11 of NEMA Standard LI-1
GPO-3	Random-laid glass-fiber mat	Polyester with fillers	Recommended for mechanical and electrical applications; exhibits resistance to carbon tracking and has self-extinguishing properties; see Part 11a of NEMA Standard LI-1

 * Grades in *italics* are indicated for electrical uses.
 ** Available in copper-clad form (further details on copper-clad laminates are given in the text).
 † "Primary insulation" is that insulation which is in *direct contact* with current-carrying parts; laminated insulation used where terminals have separate insulation is not considered to be "primary insulation."

gives NEMA standard values for the eight most important and widely used characteristics of laminate sheets.

Military or Federal specifications require that vendors qualify their products by submitting what are called results of preproduction tests of samples of the grades designated in the specification. The characteristics chosen for Table 6.13 include the principal properties selected for preproduction tests. The values given are from the NEMA Standard, which agrees (in most cases) with those required by the military or Federal specifications. Once made with qualification approval given, the pre-

Table 6.12A   Selected Commercial Laminates Not Covered by NEMA Standards

Designation	Fibrous component	Resin component	Applications
HHT*	Random-laid glass-fiber mat	"Poly nuclidic ester"	Offered as Class 180° C (H) laminate, good electrical and mechanical characteristics; low weight loss, high dielectric strength, and high flexural strength after 240 hours at 180°C
MG1365** (copper-clad)	Non-woven glass base	Epoxy	Good mechanical and electrical properties; green color
MG1365-FR** (copper-clad)	Non-woven glass base	Epoxy	Same properties as MG1365, but flame-retardant; red color
Phenolic laminated wood†	Wood veneers	Phenolic	For use where parts are made for switch gears, "Stay Wire" insulation, insulating links, bolts, rods, nuts, spacers, wedges, beams, handles and levers, and the like

* Haysite Corp. Bulletin 228, Erie, Pa.
** Norplex Corp., LaCrosse, Wis.
† Westinghouse Electric Corp., Hampton, S.C.

production tests need not be repeated. Testing for acceptance of a given lot is much simpler.

Table 6.14 provides data on the range of thicknesses of standard sheets. Data on tolerances for dimensions and related values can be obtained from NEMA Standard LI-1 or from vendors' catalogs. ASTM standards applicable to laminates and referred to in the NEMA standards are listed in Table 6.15. Military specifications are in substantial agreement with NEMA.

Table 6.16 gives temperature classifications for laminates. Values for grade designations identified by italics have been derived from a research project conducted jointly by the Underwriters' Laboratories in Chicago and NEMA at the University of Delaware. These are tentative values to be fixed later on the basis of continuing experience and as additional data are obtained. The remaining values in the table are derived from other authoritative sources.

6.13   Laminated Thermosetting Products (Rods and Tubes)

Rods and tubes are made only from a portion of the sheet laminates listed in Table 6.12, as indicated below. Only those rods and tubes with appropriate electrical characteristics, and with adequate processing controls of these characteristics, are suitable for electrical use. However, other grades may also be used for electrical purposes, provided the user can determine to his satisfaction that proper performance can be expected. Advantages and disadvantages for the suitable grades are noted in the following enumeration:

Grade XXX—Molded rods and tubes having characteristics similar to those for sheets except as limited by differences in construction and shape.

Grade XXX—Rolled tubes (best electrical properties for rolled paper-base tubes under humid conditions)

Grade G-10—Molded rods (mold seams are weak points mechanically and electrically)

Grade G-10—Rolled tubes only, no molded tubes (characteristics similar to those for sheets)

Table 6.13 Principal Properties of Laminate Sheets for Preproduction Tests (Derived Mostly from NEMA Standards)

Notes:

1. Thicknesses: All values in table apply only to thicknesses of $\frac{1}{16}$ in. and $\frac{1}{4}$ in., unless otherwise stated.
2. Dielectric strength (minimum kilovolts): All values are parallel to laminations, obtained by step-by-step to failure test method (ASTM D229).
3. Dielectric constant and dissipation factor (maximum): All values at frequency of 1 MHz (ASTM D150 and D229).
4. Surface Resistance (megohms minimum): Values given are based on applicable military specifications and test methods; test methods also appear in ASTM D257.
5. Impact strength (minimum): Values are in ft-lb/one-in. of notch for sheets from $\frac{1}{8}$ in. to 2 in., but not exceeding maximum thickness for specified grade; ASTM D256 gives standard test methods.
6. Flexural strength (minimum): These values are for specimens measured flatwise, cut lengthwise and crosswise; ASTM D790 (referred to in ASTM D229) is applicable; all values in psi $\times$ 10^3.
7. Bonding strength (minimum): In pounds per sheets $\frac{1}{2}$ to 2 in. thick, but not exceeding maximum thickness for the grade; see ASTM D229.
8. Maximum water absorption (percent): ASTM D570 and D229 are applicable.
9. Condition specifications: Condition A = tested as received. Condition C = tested after 96 hr at 35° C and 90 percent RH. Condition D24/23 = tested after immersion in distilled water for 24 hr at 23° C. Condition D48/50 = immersion in distilled water for 48 hr at 50° C. Condition E48/50 = tested after 48 hr at 50° C.

Grade*	Minimum dielectric strength		Maximum dielectric constant (Condition D24/23)		Maximum dissipation factor (Condition D24/23)		Minimum surface resistance (Condition C96/35/90)†
	(Condition A)	(Condition D48-50)	$\frac{1}{16}$ in.	$\frac{1}{2}$ in.	$\frac{1}{16}$ in.	$\frac{1}{2}$ in.	
XXX	50	6	5.90	5.70	0.047	0.043	10
XXXP	60	15	4.80	**	0.035	**	3
XXXPC	60	15	4.80	**	0.035	—	3
G10	45	40	5.40	5.40	0.035	0.035	1,000
G11	45	40	5.40	5.40	0.035	0.035	1,000
FR2	60	15	4.80	4.80	0.035	—	
FR3	60	30	4.80	4.80	0.040	—	1,000
FR4	45	40	5.40	5.40	0.035	0.035	1,000
FR5	45	40	5.40	5.40	0.035	0.035	
GPO-1	40	15	4.30‡	4.40‡‡	0.03‡	0.06‡‡	—
GPO-2	40	15	—	—	—	—	—
GPO-3	40	15	4.50¶	—	0.05¶	—	—

Table 6.13 Principal Properties of Laminate Sheets for Preproduction Tests (Cont'd)

Grade*	Minimum impact strength (Condition E48/50)		Minimum flexural strength (Condition A)				Minimum bonding strength		Maximum water absorption	
	Lengthwise	Crosswise	Lengthwise		Crosswise		(Condition A)	(Condition D48/50)		
			1/16 in.	1/2 in.	1/16 in.	1/2 in.			1/16 in.	1/2 in.
XXX	0.40	0.35	13.5	11.8	13.5	11.8	—	—	1.4	0.60††
XXXP	—	—	12.0	10.5	12.0	10.5	—	—	1.0††	0.040††
XXXPC	—	—	12.0	10.5	12.0	10.5	—	—	0.75††	0.10
G10	7.0	5.5	60.0	50.0	45.0	35.0	2000	1600	0.25	0.10
G11	7.0	5.5	60.0	50.0	45.0	35.0	1600	1500	0.25	0.40††
FR2	—	—	12.0	10.5	**	**	—	—	0.65††	0.25††
FR3	—	—	20.0	16.0	**	**	—	—	0.60††	0.10
FR4	7.0	5.5	60.0	50.0	45.0	35.0	2000	1600	0.20	0.10
FR5	7.0	5.5	60.0	50.0	45.0	35.0	1600	1500	0.20	0.35
GPO-1	8.0	8.0	18.0	18.0	18.0	18.0	850	800	1.00	0.25
GPO-2	8.0	8.0	18.0	18.0	18.0	18.0	850	800	0.80	0.35
GPO-3	8.0	8.0	18.0	18.0	18.0	18.0	800	800	0.60	

* All grades tabulated are specially suitable for electrical use.
† All figures are for sheets 1/8 in. thick.
** Maximum thickness is 1/4 in.
†† For sheets 1/4 in. thick.

‡ For sheets 1/8 in. thick, Condition A.
‡‡ For sheets 1/8 in. thick, Condition D24/23.
¶ For Condition A.

Table 6.14   Standard Thicknesses for NEMA Laminate Sheets*

| | Thickness range, in. | |
Grade	Minimum	Maximum
X	0.010	2
XP	0.010	¼
XPC	1/32	¼
XX	0.010	2
XXP	0.015	¼
XXX	0.015	2
XXXP	0.015	¼
XXXPC	1/32	¼
ES-1	3/64	¼
ES-2	0.085	¼
ES-3	3/64	¼
C	1/32	10
CE	1/32	2
L	0.010	2
LE	0.015	2
A	0.025	2
AA	1/16	2
G-2	1/32	2
G-3 and G-7	0.010	2
G-5, G-9	0.010	3½
G-10, G-11, and N-1	0.010	1
FR-2 and FR-3	1/32	¼
FR-4 and FR-5	0.010	1

* NEMA Standard LI-1

Table 6.15   ASTM Standards Applicable to Plastics Laminates*

Pub. No	Title
D150	"Methods of Test for A-C Capacitance, Dielectric Constant, and Loss Characteristics of Electrical Insulating Materials"
D229	"Methods of Testing Rigid Sheet and Plate Materials Used for Electrical Insulation"
D256	"Methods of Test for Impact Resistance of Plastics and Electrical Insulating Materials"
D257	"Methods of Test for Electrical Resistance of Insulating Materials"
D348	"Methods of Testing Laminated Tubes Used for Electrical Insulation"
D349	"Methods of Testing Laminated Round Rods Used for Electrical Insulation"
D495	"Method of Test for High-Voltage, Low-Current Arc Resistance of Solid Electrical Insulating Materials"
D570	"Method of Test for Water Absorption of Plastics"
D621	"Methods of Test for Deformation of Plastics Under Load"
D635	"Method of Test for Flammability of Rigid Plastics Over 0.127 cm (0.050 in.) in Thickness"
D709	"Specifications for Laminated Thermosetting Materials"
D790	"Method of Test for Flexural Properties of Plastics"
E53	"Method for Chemical Analysis of Copper (Electrolytic Determination of Copper)"

* American Society for Testing and Materials, Philadelphia, Penn.

Table 6.16   Temperature Classification for Laminates Based on Values Derived from Underwriters' Laboratories Research*

Grade†	Provisional temperature rating,** °C	
	$\frac{1}{32}$ in.	$\frac{1}{16}$ in.
X, XP	110 (120)	130 (145)
XPC, XX, *XXX*	100 (105)	130 (140)
XXXP, XXXPC	125 (125)	125 (135)
FR-2	95 (100)	130 (145)
FR-3	115 (125)	130 (135)
C, CE, L, LE	110 (120)	130 (145)
FR-4, *G-10*	130 (150)	130 (150)
A, AA	135 (150)	135 (155)
GPO-1, GPO-2	110	120 (130)
G-5, G-9	140 (150)	140 (155)
G-7	170 (190)	180 (200)
G-11, FR-5	170 (180)	170 (190)

* The classifications proposed by UL are tentative and provisional and will be firmly established when further data are available. They have been developed to be applied to systems as well as materials: (1) For insulating systems—Class A (105°C), B (130°C), F (155°C), H (180°C), 200°C, C (220°C), 240°C, and over 240°C; (2) For insulating materials—5°C increments up to 130°C; 10°C increments from 130°C to 180°C (except that 155°C is included as an exception); 20°C increments over 180°C.

† The grades in *italics* are those indicated as being particularly suitable for electrical purposes.

** The temperature values given in parentheses are ratings which appear to be possible, but insufficient data have been accumulated at the time of publication to establish them as certain.

Grade G-11—Same as for Grade G-10.

Grade FR-3—Rods and tubes (can be fabricated, but no standards exist)

Grades FR-4 and FR-5—Molded rods and rolled tubes, no molded tubes (molded rod seams are weak points mechanically and electrically; rolled tubes have characteristics similar to those of sheet materials except as limited by inherent differences in construction and shape)

In addition to specific Standards, NEMA Standard LI-1- includes a considerable amount of data on mechanical, electrical, and thermal properties, also on solvent resistance. These data, intended to serve as general information on sheets, tubes, and rods, are identified by NEMA as "Authorized Engineering Information."

Laminate sheets are also made in square and rectangular molded tubes and NEMA's LI-1- provides some data, but in a rather limited degree, primarily tolerances for warp or twist and both inside and outer dimensions.

Of the eight grades enumerated above made in tubes or rods, Grade XXX is considered to be the prime electrical grade. Its general properties for tubes are summarized in Table 6.17.

6.14   Copper-Clad Laminates for Printed Circuitry

In this discussion of available materials for various types of printed circuitry, the emphasis will be on the materials and their characteristics. Methods (etching or other) for converting the copper-clad laminate (rigid or flexible) into a circuit board will be discussed only briefly, and only where the information will aid in understanding materials properties.

The functions of a circuit board incorporated as a component in a given elec-

Table 6.17   Typical Properties for NEMA Grade XXX Tubes

ROUND ROLLED TUBES: Inside diameters are from ¼ to 8 in.; outside diameters, from 5/16 to 10 in.; wall thicknesses, from 1/32 to 1 in.

Water absorption, per cent, maximum, condition D 24/23 following condition E 1/105:*

Wall thickness, in.	Percent absorption
1/32 up to 1/16	3.5
1/16 up to 3/32	1.5
3/32 up to 1/8	1.3
1/8 up to 3/16	1.0
3/16 up to 1/4	0.8
1/4 up to 3/8	—
3/8 up to 1/2	—
1/2 up to 1	—
1	

Dielectric breakdown, minimum, condition D 48/50:*

(a) Parallel to laminations; short-time test; all sizes and wall thicknesses; 18,000 V.

(b) Perpendicular to laminations; short-time test; dielectric strength as follows:

Wall thickness, in.	Volts/mil
1/16	225
1/8	250
1/4	250

ROUND MOLDED TUBES: Inside diameters are from ¼ to 8 in.; outside diameters, from 5/16 to 10 in.; wall thicknesses, from 1/16 in. to 1 in.

Water absorption, per cent, maximum, condition D 24/23 following condition E 1/105:*

Wall thickness, in.	Percent absorption
1/16	1.4
3/32	1.2
1/8	1.1
3/16	1.0
1/4	0.9
1/2	0.8
1	0.7

Minimum density: 1.22 gm/cu cm

Minimum compressive strength (axial direction): 20,000 psi

Dielectric strength, minimum, perpendicular to laminations, short-term test, condition A:*

Wall thickness, in.	Volts/mil
1/16	300
1/16 to 1/8	220
1/8 to 1/4	150
Over 1/4 to 1/2	110

* Condition D 24/23 = immersion in distilled water for 24 hr and test at 23°C. Condition D 48/50 = immersion for 48 hr and test at 50°C. Condition E 1/105 = test after 1 hr at 105°C. Condition A = as received.

tronic circuit are dual: (1) It acts as a support for various components, such as resistors and capacitors, and for the interconnections as well, and (2) it provides insulation for this assembly. Obviously, the laminate circuit board carries a heavy responsibility for the reliable performance of the entire circuit. The selection of the circuit board requires careful evaluation of materials and their properties and involves important engineering decisions.

There are three types of copper-clad laminates or film:

1. Rigid copper-clad laminate sheet for conventional printed circuits.

2. Thin rigid and/or flexible laminates used mostly for building multi-layer printed circuits.

3. Copper-clad film and thin flexible sheets for printed circuits where flexibility of the completed product is essential.

There are two standards or specifications of prime importance. The NEMA Standard Publication No. LI-1-, "Industrial Laminated Thermosetting Products," covers a selection of eight grades of laminates selected as worthy of use in copper-clad composites. Military Specification MIL-P-13949D with Amendment 3 covers a selection of nine types. The term "grade" in the NEMA document and the term "type" in the MIL document are interchangeable in meaning.

In addition to the NEMA and MIL documents, The Institute of Printed Circuits issues publications of importance in the field, and ASTM Standards, such as D 229, D 669, D 1867, among others, contain much useful information. Finally, manufacturers of laminates issue technical literature that contains detailed information of design value. This literature provides information not only on laminates but also, in some cases, on so-called "prepreg" (pre-impregnated, semi-cured materials) used primarily in multi-layer printed-circuit fabrication.

The grades covered by the NEMA Standard are XXXP, XXXPC, FR-2, FR-3, FR-4, FR-5, G-10, and G-11 (see Table 6.18). One or both surfaces may be copper-clad. Since most applications involve the use of soldered connections, copper with its high conductivity and ease of soldering, is the metal of choice. Where welding is

Table 6.18   NEMA and Comparable MIL Standards for Laminates

Source: NEMA "Authorized Engineering Information"

NEMA designation	Comparable military designations from MIL-P-13949C	Type
XXXP	—	Paper base, phenolic resin
XXXPC	—	Paper base, phenolic resin
FR-2	—	Paper base, phenolic resin, flame-resistant
FR-3	Type PX	Paper base, epoxy resin, flame-resistant
FR-4	Type GF	Glass-fabric base, epoxy resin, general-purpose, flame-resistant
FR-5	Type GH	Glass-fabric base, epoxy resin, temperature- and flame-resistant
G-10	Type GE	Glass-fabric base, epoxy resin, general-purpose
G-11	Type GB	Glass-fabric base, epoxy resin, temperature-resistant

necessary, nickel foil may be the cladding metal. Other metals or alloys are used in special applications.

The nine types covered by MIL-P-13949D (Paragraph 1.3.1, Plastic Sheet, Laminated, Copper-Clad) are described as follows:

PH — Paper base, epoxy resin, hot punch, flame retardant
PX — Paper base, epoxy resin, flame retardant
GB — Glass (woven-fabric) base, epoxy resin, temperature resistant
GC — Glass (combination woven and nonwoven fabric) base, polyester resin, flame retardant
GE — Glass (woven-fabric) base, epoxy resin, general purpose
GF — Glass (woven-fabric) base, epoxy resin, flame retardant
GH — Glass (woven-fabric) base, temperature resistant and flame retardant
GP — Glass (nonwoven-fiber) base, polytetrafluoroethylene resin
GT — Glass (woven-fabric) base, polytetrafluoroethylene resin

Of these, four types appear to be the most important and are used in the greatest volume. Their designations and the NEMA Grade equivalents are:

Type MIL-P-13949	Grade NEMA LI-1
PX	FR-3
GB	G-11
GE	G-10
GF	FR-4

6.14.1   Copper foil properties

Pure aluminum can be rolled into very thin foils with comparative ease and at a low cost compared with this same operation with pure copper. Thin pure copper foils can also be made by rolling, but their cost is high. Copper foil so produced and suitably annealed could exhibit a purity of 99.9 + percent and a conductivity of 100 percent IACS (International Annealed Copper Standard) equal to a resistivity of 0.15328 ohm-gram/meter^2 at 20 ° C which is the accepted method of expressing the IACS of mass resistivity. [21]

A fast and economical electroplating method for producing copper foil was introduced some thirty years ago by Anaconda American Brass Company. [22] Marked by excellent bonding characteristics, this foil, trademarked "Electro Sheet," was used as a roofing material and as an element in composites of copper foil and building papers. However, since the electrodeposited foil was less dense than the rolled variety, its mass resistivity was higher. The resistivity decreased with an increase in thickness, as shown below:

Nominal weight of copper sheet, ounces per square foot	Nominal thickness, in.	Percent conductivity IACS
1	0.0014	96.3
2	0.0028	98.0
3	0.0042	98.5
5	0.0070	99.5

[21] The term "meter^2" denotes not an *area* but the product, (ohm/meter) $\times$ (grams/meter).

[22] In this electroplating process, a wide-surfaced, large-diameter wheel forms the cathode of the plating couple. The wheel is dipped into, and revolves within, a bath of copper in solution. The deposited copper foil is then stripped off the wheel as it revolves. One surface of the foil reflects that of the wheel surface; the other surface is formed at the interface with the electrolytic bath.

With an emerging market for copper foil in printed or etched circuits, a special grade of electrodeposited foil has been developed that is known as "Printed Circuit Grade" and that meets the requirements of both NEMA and MIL specifications. NEMA simply requires that the copper foil shall have a minimum purity of 99.5 percent (silver counted as copper). The MIL specification requires the same degree of purity and also calls for a resistivity of 0.15940 ohm-gram/meter^2 at 20° C. Compared to a resistivity of 0.15328 for a conductivity of 100 percent IACS, the MIL requirement is equivalent to a conductivity of 96.16 percent IACS or $(0.15328/0.15940) \times 100$.

Both NEMA and MIL specifications include requirements for surface finish of the copper foil and define limits for processing imperfections such as dents, pinholes, and the like. The imperfections are defined as follows:

Pits—Small holes that do not penetrate entirely through the foil.

Dents—Depressions that do not significantly decrease the foil thickness.

Pinholes—Small holes not exceeding the area of a 0.005-in. circle that penetrate through the foil.

Inclusions—Any foreign matter enclosed partly or entirely in the foil (usually metallic lead).

Wrinkles—Small ridges or furrows or creases in the smooth surface.

Scratches—Any number of marks less than 140 microinches deep.

For surface finish, the requirements are as follows: "The surface finish is permitted to not exceed an arithmetical average exceeding 20 microinches as determined by a profilometer (0.5 gram-0.030 in. cutoff)." (see Tables 6.19 and 6.20.)

The electrolytic deposition process for producing copper foil has been greatly refined (by at least two suppliers) so that now it is possible to produce foil meeting all the requirements. The plating process used is apparently much like that of the original development, but the quality of the foil surfaces is better and the purity of the copper as high as 99.8 percent.

Although *rolled* copper foil is economically out of line with electroplated foil, as previously noted, the higher cost may still be justified in some applications. In such cases, laminates with rolled foil can be obtained from some manufacturers.

6.14.2   *Laminate base criteria*

It is obvious that the completed circuit board must be a combination of optimum-quality copper-foil cladding and optimum-quality thermosetting laminate base. The edges of the printed or etched foil conductors may be separated by only 0.016

Table 6.19   Tolerances for Copper-Clad Laminates after Etching

Nominal thickness excluding copper,* in.	Thickness tolerances of base laminate after etching, ± in.
0.002 up to 0.0045	0.001
Over 0.0045 up to 0.006	0.0015
Over 0.006 up to 0.012	0.002
Over 0.012 up to but not including ⅟₃₂	0.003

* Copper surfaces must be free from defects which may affect serviceability, such as blisters, wrinkles, cracks, holes, dents, and scratches, as covered by military specifications.

Table 6.20   Standard Requirements for Copper-Foil Surfaces

Minimum purity of the copper foil must be 99.50, and the minimum purity analysis must be made in accordance with ASTM E53-48.

Nominal weight, oz per sq ft	Nominal thicknesses, in.	Nominal tolerances, in.	
		Plus	Minus
1	0.0014	0.0004	0.0002
2	0.0028	0.0007	0.0003
3	0.0042	0.0006	0.0006
4	0.0056	0.0006	0.0006
5	0.0070	0.0007	0.0007

in. There are numerous live connections that pass completely through the laminate. There are also many points where the applied voltages are perpendicular to the laminate. Creepage (current leakage) across the surface may be caused by the voltages parallel to the laminations and voltages perpendicular to the laminations. The laminate must exhibit sufficient insulation strength to resist such hazards. Moreover, the laminate must maintain this resistance after it has passed through the processes of circuit-board fabrication.

Apparently, the G-10 grade (epoxy-resin/glass cloth laminate) with 2-oz copper on one or both sides is used in the largest volume. (Weight of cladding is understood to be in oz/sq ft.) The most common thickness is 1/16 in. The thickness of copper-clad laminates is measured overall; that is, it includes the thickness of laminate and that of the cladding.)

To maintain high quality in the production of copper-clad laminates, utmost care in cleanliness is mandatory during all steps of manufacture, including handling, packing, and other necessary operations. The use of "white rooms" is a necessity. It is also essential that all manufacturing steps be carried out under closely controlled conditions.

Standard tests are provided for the adhesives used in bonding copper foils to the base laminate. When the adhesive is left exposed after the foil has been etched, the surface resistance of the adhesive in effect constitutes the surface resistance to the base. Precautions are taken, therefore, not to destroy the adhesive in the test specimen so that resistance tests can be made. (In epoxy-resin laminates, this precaution is unnecessary because epoxies have very strong adhesive properties and no adhesives are required.)

In the process of fabrication into circuit boards, copper-clad laminates must withstand various mechanical and chemical actions. The laminate is subjected to cutting to shape and size, punching of holes, application of resist coating, etching in a bath of ferric chloride (or other etchant), removal of excess etchant, drying, mounting of components, and soldering by "floating" in a bath of liquid solder at 500° F. When finished, the copper must adhere at all points and the electrical and mechanical characteristics of the clad laminate must not have been impaired.

Plating of copper conductors with some other metals (gold, for example) is sometimes desired. Other requirements sometimes call for hot-pressing the etched board so that the copper is very nearly flush with the surface. Important features of the complete clad laminate are:

1. Easy machinability, for sawing, punching, drilling with ordinary machine tools

2. High resistance to the etching process

3. High resistance in the solder bath

4. Stability against changes of temperature, presence of moisture, and the effects of solvents, alkalies, and oils.

Within the space available here, it is not possible to describe in detail all the requirements for the production of circuit boards able to withstand all rigors of fabrication and subsequently able to perform satisfactorily under designated operational and environmental conditions. In the addition to the information given here and the related tables, the previously cited NEMA and MIL standards and specifications should be consulted for fully detailed characteristics and requirements.

The Underwriters' Laboratories give brand recognition to laminate products submitted to test. A reference file number is assigned to approved vendors. Military specifications require vendors' products to be given recognition on a "Qualified Products List." The vendor must present proof of quality maintenance at stated intervals. These specifications also require that flame-retardant grades contain an internal red trademark, whereas non-flame-retardant grades must also be identified.

6.15   Thin Rigid and Flexible Copper-Clad Laminates

NEMA Standard and MIL Specifications cover laminates copper-clad in a range of thicknesses from 1/32 in. to 1/4 in., inclusive. The thickness, as noted previously, include both the base laminate and the copper (or other metal) cladding. These laminates may be considered as "rigid," although the 1/32-in. material does not require much force to bend it. Copper-clad laminates thinner than 1/32 in., however, are not covered by the NEMA and MIL documents. For practical purposes, though, the NEMA and MIL values will be found to be in general agreement with those for the thinner materials. There is an exception in the manner of identifying the thickness: For the thicker laminates, the given thickness is understood to include both the base laminate *and* the cladding, whereas for the thinner laminates, the thickness is given only for the base material. The thinner laminates may also be color-coded by some producers or manufactured in translucent or natural colors.

The available thin copper-clad laminates range from 0.002 in. to 0.030 in., *excluding* the copper foil. The grades used in greatest volume are G-10 and FR-4, which are made with 1- or 2-oz cladding on one or both sides. Some producers specialize in these two grades; others make a wider line. The following copper-clad cloths and papers are available in addition to G-10 and FR-4 (thicknesses are for the base laminate only):

1. Class F (155° C) varnished glass cloth (0.0035 to 0.012 in.)

2. Epoxy-varnished glass cloth (0.0035 to 0.012 in.)

3. Self-extinguishing insulation (0.0035 to 0.012 in.)

4. Moisture-resistant paper (0.005 in., 0.007 in., and 0.010 in.)

5. Resin-impregnated paper (0.005 in., 0.007 in., and 0.010 in.)

As will be noted, these grades offer a good choice to meet special selection requirements, ranging through high-temperature solderability, dimensional stability, self-extinguishing ability, moisture resistance, and low cost.

Two special laminates have been developed to meet exacting requirements.

The first is a flexible epoxy copper-clad laminate 0.007 in. thick and made for service up to $130°$ C.[23] It can be bonded by heat and pressure without the use of adhesives or prepregs. Among its outstanding characteristics is its ability to be soldered by conventional means, high bond strength, high tolerance to severe creasing and flexing, and resistance to etchants. The second is a Type GT material (MIL-P-13949D) with nominal thicknesses of 0.010, 0.015, and 0.020 in. When clad, this laminate is available in nine sizes from 1/32 in. to 1/2 in. Both of these laminates are produced under exacting controls of dielectric constant, thickness, and low thermal expansion. Within these parameters it is possible to use the laminates in such applications as microwave striplines. Tables 6.21 and 6.22 give typical values for the epoxy copper-clad laminate.

Prepreg materials for making multilayer circuits are usually made as B-stage epoxy-impregnated glass cloth, in two grades, G-10 and F-4.[24] Prepregs are usually specified in thicknesses *after pressing*, although some producers also give "as is"

---

[23] Flexible Epoxy, CuClad (3M Company, St. Paul, Minn.).
[24] The B-stage of a resin is the condition of being insoluble and fusible.

Table 6.21   Typical Electrical Properties for Special-Purpose Epoxy Copper-Clad Laminates

All data are based on laminates of $\frac{1}{16}$ in. thickness. Test conditions are as follows: Condition A = tested as received. Condition D24/23 = tested after immersion in distilled water at 23° C for 24 hr. Condition D48/50 = tested after immersion in distilled water at 50° C for 48 hr. Condition C96/35/90 = tested after conditioning at 35° C., 90% RH for 96 hr. Condition E-1/150 = 1 hr at 150° C.

	Test method	Condition	6098*	K6098*	MIL P-13949 requirements
Dielectric strength, parallel, stepwise, kV	D-149	A	50	50	—
		D-48/50	35	35	20 min
Volume resistivity, megohm cm	D-257	C-96/35/90	10^7	10^7	1×10^6 min
Surface resistance, megohms	D-257	C-96/35/90	$> 10^4$	$> 10^4$	1×10^4 min
Dielectric constant *(K)*					
at 1 MHz	D-150	A	2.6 ± 2%	2.5 ± 2%	—
		D 24/23	2.6	2.55	2.8 max
at 10 GHz†	3M stripline method	A	2.485 ± 1.6% (6098-11, $\frac{1}{32}$ in.)	2.40 ± 1.8%	None
Dissipation factor					
at 1 MHz	D-150	A	0.00085	0.00085	—
		D 24/23	0.002	0.002	0.005 max
at 10 GHz††	3M stripline method	A	0.0018	0.0018	None
Arc Resistance	D-495	A	> 180	> 180	180 min

* Epoxy CuClad 6098 and K6098, 3M Company, St. Paul. Minn.
† By controlling K to ± 2% (3M stripline method at X-band frequency), resonant frequency of resonators, phase delay, and stripline wavelength are reproducible to ± 1%. On some constructions, the tolerance on K will be less than ± 2%.
†† Dissipation factor includes losses due to copper treatment which do not show up in usual cavity measurements made on unclad materials.

Table 6.22   Typical Physical Properties for Special-Purpose Epoxy Copper-Clad Laminates*

	Test Method	Condition*	Copper	6098**	K6098**	MIL-P-13949 requirements
Copper peel strength, min. avg., lb/in.	MIL P-13949	A	1 oz	10	10	8
			2 oz	12	12	10
		E-1/150	1 oz	10	10	8
			2 oz	12	12	10
Flexural strength, psi	D-790	A				
lengthwise				17,000	16,500	15,000
crosswise				14,000	13,500	10,000
Tensile strength, psi	D-229	A				
lengthwise				26,000	26,000	—
crosswise				20,500	20,500	—
Warp and twist	MIL P-13949	A		Class A	Class A	Class A
Water absorption, percent	D-570	D24/23		0.024	0.024	0.10 max
Linear coefficient of thermal expansion, $\times 10^{-6}$/°C		(25°–150°C)				
lengthwise				9.10	9.10	—
crosswise				9.26	9.26	—
Thermal conductivity, perpendicular to lamination, cal/sec/°C/cm²/cm		(23°–100°C)		2.6×10^{-4}	2.6×10^{-4}	—
Heat capacity, BTU/lb/°F		(−40°F)		0.16	0.16	—
		(100°F)		0.21	0.21	—
		(200°F)		0.24	0.24	—
		(400°F)		0.30	0.30	—
Solder float	MIL P-13949	(260°C for 20 sec)		Passes	Passes	No blisters or delamination
Shrinkage after etching				Negligible under all circumstances		
Cold flow				Negligible under all circumstances		
Chemical resistance				Inert to virtually all printed-circuit processing chemicals and solvents		
Flammability				Nonflammable under all circumstances		

* All data for 1/16-in. thickness. Definitions for test conditions are given in Table 6.21.

** 3M Company epoxy CuClad laminates.

Table 6.23   Data for FEP-Coated Polyimide Film

Data is for DuPont's Kapton polyimide film, 3 mils thick with 1 mil coating of FEP on each side ready for bonding.

Tensile strength, psi	16,000
Ultimate elongation, percent	85
Dimensional stability, percent at 250°C	0.3
Moisture absorption, percent at 100% RH	1.5
Dielectric strength, vpm	3,600
Dielectric constant, 1 kHz	3.0
Dissipation factor, 1 kHz	0.0009
Volume resistivity, ohm-cm	5×10^{17}

thicknesses. Producers of prepregs make available technical information that includes specifications covering properties and directions for care, handling, storage testing, and use of the materials.

Polymeric films (discussed in detail in Par. 6.16) find use as base materials for flexible copper-clad laminates. For temperatures up to 260°C, polyimide film with an FEP (fluorinated ethylene propylene) coating on one or both sides is available in thicknesses of 0.001 in. to 0.005 in. with a 1-oz or 2-oz cladding. Polyester films are also used for copper-clad laminates. Table 6.23 gives data for FEP-coated polyimide film and Table 6.24 for unclad polyester film.

No moisture absorption data appear in various property tabulations for thin laminates. Experience has shown that the experimental errors in moisture absorption tests for these laminates is large enough to negate the results. In multilayer circuit assemblies, in fact, there is no exposure to environments where water absorption is an important factor.

The wide diversity and multiplicity of copper-clad laminates available for electrical and electronic applications have made possible a rapid growth in printed-circuit design. At present, these materials represent a very substantial volume of use. They have been particularly effective where production quantities are large and the cost of necessary production equipment can be easily absorbed, as in TV and radio sets.

Table 6.24   Typical Properties for Unclad Flexible Polyester Film

Grades	0.001, 0.003, 0.005, 1, 2, 3-oz copper foil on 1 or 2 sides may be bonded thereto.
Tensile strength	125 lb per inch of width for 0.005 in.
Dissipation factor at 1 MHz	0.016
Dielectric constant at 1 MHz	3.0
Dielectric strength (short time, Condition A),*vpm	5,000
Surface resistivity (etched), ohms/sq in.	1×10^{12}
Fold endurance (both machine directions, 1-kg load), cycles 5 mil	14,000

* Condition A = tested as received.

6.15.1   Deposited-copper flexible printed circuits

A departure from the use of copper foil is a method described as "photo-selective metal deposition," developed by Western Electric's Engineering Research Center. This method is intended specifically for flexible printed circuits used in the company's telephone equipment. In the new process, photosensitized bare plastic film is exposed to ultraviolet light that forms an invisible circuit pattern on the surface. The image is developed in a bath of palladium chloride, and several plating techniques are then used to selectively deposit copper on the desired circuit pattern.

A versatile type of machinery has been designed for putting this process into commercial production. The machinery is described as being able to fabricate flexible circuits bent to a variety of shapes. Saving of copper is said to be substantial in comparison to the quantities required in conventional cladded laminates and etching techniques.

6.16   Films and Sheeting (Unsupported)

Films may be described as thin, transparent, or translucent extrusions or castings in thicknesses of 10 mils (0.010 in.) or less, available in a wide variety of chemical origin or composition. Films of concern to this discussion are, for the most part, plastics.[25] In thicknesses greater than 10 mils, the film materials are usually described as sheets or sheeting. In this category, we find elastomeric materials as well as plastics.

Films, in their various aspects, represent one of the most important developments in contemporary materials science and technology and find many applications beyond those in the field of electrical insulation and dielectrics. Film technology, for example, has grown to be of major importance in magnetic and solid-state devices.

The versatility of film materials is strikingly evident in the field of insulation and dielectrics. Specific applications are literally legion and cannot be enumerated within the space available here. The films are used as wire and cable insulation, as the basic structure of adhesive tapes, as layer insulation in coils, as insulation to ground and as covers for coils, as the dielectric in capacitors, as motor and generator insulations, and as the material from which to fabricate many parts or components. The last-named may include bobbins, spools, core tubes, spacers, end washers, grommets, spirally wound tubing, and the like.

As constituent in composite structure, films have made an especially valuable contribution to the field of materials. Combined with papers, cloths, mats and many other materials, films have expanded the areas of application open to films alone. Composite design is not simple, however, but imposes many problems of materials compatibility between individual elements, including adhesives (if any is used). The main thrust of a composite design is to develop a product that not only reflects the capabilities of each constituent but that also meets *combined* functional requirements that the individual materials cannot meet.

---

[25] Most of the polymers from which films or sheets are extruded or cast are used also in molded parts, encapsulating compounds, and the like. Additional discussion on these polymers will be found in Chap. 8, which is devoted to insulation in bulk form.

The characteristics of various films (and sheets) that are used extensively in electrical and electronic design will be described in the following sections. Some attention will also be given to those films that are primarily used in other areas, such as packaging, but that have inherent characteristics of possible electrical use.

6.17   Cellulosic Films

Celluloid and cellophane may be considered as the forerunners of modern cellulosic films. Celluloid, a product of cellulose nitrate, camphor, and alcohol, has long ago been ruled out for use in electrical apparatus owing to its high degree of flammability. Cellophane is essentially made up of regenerated cellulose, a softening agent, water, and for many uses, some sort of coating. There are some 100 types of cellophane, one of which (Du Pont's PUD-0) is described as the purest form of regenerated cellulose commercially available. In the electrical field it is used in batteries as a membrane for dialysis. This type of cellophane contains water but no softener.

There are three cellulosic films of importance in electrical insulation at present, all acid esters of cellulose. These films are cellulose triacetate, cellulose acetate, and cellulose acetate butyrate (a mixed ester). Ethyl cellulose also has been made into films, but at the time this is being written it is not available, since the basic compound is pre-empted for military requirements.

6.17.1   Cellulose triacetate films

Chemically, the triacetate is the starting point for the production of cellulose acetate. Triacetate carries the process of acetylation to completion. To make cellulose acetate of desired characteristics, the acetyl level is lowered by hydrolysis. The triacetate has few solvents, is hard, and has high tensile strength, great folding resistance, and is self-extinguishing. It is also characterized by high short-time dielectric strength and heat resistance. It is often used as a tape in wire insulation because the high tensile strength permits high-speed application as a wire serving.

6.17.2   Cellulose acetate films

Two types of acetate films are produced, one by casting and the other by extrusion. A wide range of grades is manufactured in both types, differing in plasticizer content, clarity, and surface quality, to meet the demands of various markets, including packaging and other nonelectrical areas.

For electrical use, the solvent cast films are most often selected. They have good mechanical and dielectric strength. A most important property is an ability to resist corrosion. In direct contact with current-carrying copper conductors they do not release any corrosive acids. Consequently, acetate film can be used in fine-wire winding since the wire will not be destroyed by corrosion even when a considerable d-c voltage is imposed. To facilitate winding operations one surface of the film may be given a matte finish. This finish does not substantially affect the electrical characteristics.

Acetate films are used as backing in the production of pressure-sensitive adhesive tapes. The unsupported film may be combined with acetate cloth, glass filaments, and rayon filaments to provide additional strength. This application is an example of the use of films in composites, as discussed above. (Additional information is given in Par. 6.33.)

Cellulose acetate films are rated as "slow burning" by Underwriters' Laboratories. Characteristics of the films are given in Tables 6.25 and 6.26.

6.17.3   Cellulose acetate butyrate films

This mixed ester of cellulose is made by the use of acetic and butyric acids. The characteristics of the resulting ester is dependent upon the relative amounts used of the two acids. A wide selection of the mixed esters is used in many applications other than the production of unsupported films. The film, as regularly supplied, is much like cast cellulose acetate film. It can be slit into narrow ribbons for application to wires. A variety with one coated side is used where heat-sealing is desired. The film is not rated as self-extinguishing.

6.18   Polyester Films

Polyester films are probably better known by their trademark names such as Mylar (Du Pont), Celanar (Celanese Corporation), and Scotchpak (3M Company). These versatile films have found many applications as insulations and dielectrics. They are produced in a wide variety of types so that a wide selection of properties is available to the design engineer.

Essentially these polyesters are the result of reacting a dibasic acid, such as terephthalic, with a polyhydric alcohol such as glycol. The reaction is of the condensation type where water is produced with the ester. The reaction is reversible; if the ester is heated in a closed vessel with water the result is a mixture of acid and alcohol. Consequently, the presence of water cannot be tolerated if these polyester films are used in any hermetically sealed unit.

Table 6.25   Physical Properties for Cast Cellulose Acetate Films

All tests were made after the samples had been conditioned for 24 hr at 50% relative humidity and 23° C (73° F), and the test were made under the same conditions. Data are given on films made by Celanese Plastics Co. Newark, N.J.

Formula	High plasticizer P-903	Low plasticizer content P-904
Thickness range, in.	0.00088–0.002	0.00088–0.002
Specific gravity	1.27	1.30
Tensile strength, psi	7000–9000	9000–12,000
Elongation, percent	25–35	15–25
Softening temperatures*		
°C	125	175
°F	257	345
Heat shrinkage, percent		
48 hr, 60°C (140°F)	1.5–2.5	0.05–0.2
48 hr, 116°C (240°F)	4.0–6.0	0.9–1.5
Moisture absorption, percent		
24 hr, 0% RH plus		
48 hr, 90% RH	4.0	8.5

* Softening temperature = temperature at which sample becomes limp when heated on a polished block at rate of 15 °C per min. Samples conditioned before testing for at least 24 hr at 50% RH and 23 °C.

Table 6.26   Electrical Properties of Acetate Films

Property*	ASTM method	High plasticizer, P-903†	Low plasticizer, P-904†
Dielectric strength, volts/mil	D149-4		
0% RH		2800	3600
50% RH		2300	3000
95% RH		1300	1800
Volume resistivity, ohm-cm	D257-49T		
0% RH		10^{15} (range)	
50% RH		10^{13} (range)	
95% RH		10^{11} (range)	
Dielectric constant, at 1 kHz	D150-47T		
0% RH		4.3	3.8
50% RH		4.8	4.8
95% RH		6.8	7.5
Dielectric constant, at 1 MHz	D150-47T		
0% RH		3.7	3.5
50% RH		4.1	4.4
95% RH		6.1	7.0
Power factor, at 1 kHz	D150-47T		
0% RH		0.019	0.015
50% RH		0.019	0.016
95% RH		0.031	0.037
Power factor, at 1 MHz	D150-47T		
0% RH		0.039	0.023
50% RH		0.043	0.031
95% RH		0.044	0.037

* All at 25°C
† Celanese Plastics Co. data

Under the usual conditions found in nonhermetically sealed units these films exhibit so many excellent characteristics and perform so well that they are used in countless applications. To list all of them would take much space.

The polyester films are used as *barriers*, both electrical and thermal—for example, as electrical insulation in motors and transformers and as thermal barriers in wires and cables.

They can be used as supporters or covers in such applications as bobbins, tubes, and related parts or components. As carriers they are used in magnetic tape and printed circuits. In capacitors, thin grades are used as the dielectric. It is apparent that their field of use covers much of the field of electrical and electronic apparatus and devices.

The films are available in a thickness range of 0.00025 to 0.014 in. (1/4 to 14 mils), in widths up to 60 to 72 in., depending on thickness and type. The useful temperature range is from −60° to +150°C.

For convenience the various types of Mylar will be described as typical of the range in types available. (It is to be understood that comparable types are available in most cases from other makers.) The types are as follows:

1. *Type A* is the general-purpose material, a clear, transparent, brilliant surface of excellent insulating and dielectric functions.

2. *Type CS* (also Scotchpak by 3M Company) is Type A made heat-sealable by being coated on one or both sides with selected polyolefins. It may be colored.

Flame-retardant and self-extinguishing types are included in this category (Scotch-pak offers two types.)

3. *Type HS* type shrinks 50 percent in both width and length when heated to 100° C. The film is used as a cover or jacket in capacitors and other components.

4. *Type T* has a high-tensile strength in the longitudinal direction that makes it particularly useful as a backing in magnetic tapes and pressure-sensitive tapes.

The films are also available in a corrugated form to facilitate their application in wire and cable. The corrugated surface provides a stronger bond and therefore may be preferred for certain applications and fabricating conditions.

Polyester films can be laminated, embossed, metallized, punched, dyed, coated, and will accept painting. Heavy gages can be vacuum-formed. There are many adhesives that are compatible with these films in making laminates with other materials.

Detailed property and application data are available from the producers of these films.[26,27,28] Selected data appear in Table 6.27.

6.19   Fluorocarbon Films (FEP and TFE)

There are two basic types of completely fluorinated resins. FEP is a copolymer of tetrafluorethylene and hexafluoropropylene monomers. TFE is the fully saturated fluorinated polymer. They differ in molecular construction but in many ways they are alike. They differ in ease of fabrication into films where FEP has the advantage. The temperature range of TFE is $-450°$ to $+500°$F whereas that of FEP is $-450°$ to $+400°$F. The coefficient of friction, static and dynamic, of TFE is lower than that of FEP by about one to two. They are alike in having excellent electrical and chemical properties. Their strength is "good" but not comparable with the polyesters. Their moisture absorption is zero.

FEP film (Du Pont) is transparent, thermoplastic, can be heat-sealed, thermoformed, laminated, metallized and, with the help of a special surface. bonded by use of conventional adhesives. It is inert to most solvents and chemicals. (Fluorines and related compounds and molten alkali metals attack it.) It does not "stick" and is capable of continuous use at 250° C. Dielectric properties and insulating characteristics remain constant over wide ranges of temperature and frequency.

FEP film is used in coil winding, capacitors, rigid and flexible printed circuits, and as insulation for wire and cable. The film is available in thicknesses of 1/2, 1, 2, 5, 10, and 20 mil. Surfaces may be treated to accept adhesives on one or both sides.

TFE film is available in cast form in thicknesses from 1/8 to 3/4 mil. Two or more layers of film (if of the same thickness) can be combined for increased ease in application. The film can be metallized with vacuum-applied aluminum. Such film finds use as electrostatic shielding material.

In mill-run quantities, TFE film can be pigmented. Since normally, the film is smooth and waxy, it is sodium-etched for the reception of adhesives. It can be specially coated so that epoxy resins will adhere directly to it. TFE film is used in pressure-sensitive tapes. Conductive coatings may be applied to one or both sides.

[26]   Du Pont Film Dept., Wilmington, Del.
[27]   3M Company, St. Paul, Minn. (polyester film)
[28]   Celanese Plastics Co., Newark, N.J. (acetate and polyester films)

Table 6.27   Selected Properties of Polyester Films

Values are for DuPont's Mylar polyester film.

Property	Typical Value	Electrical		
		Test Condition	Test Method
Dielectric strength (short term for 1 mil film), vpm	14,000	25°C—d-c	500 volts/sec
	7,500	25°C—60 Hz	ASTM D149-64 and D2305-68
	5,000	150°C—60 Hz	—
Dielectric constant	3.30	25°C—60 Hz	ASTM D150-65T
	3.25	25°C—1 kHz	—
	3.0	25°C—1 MHz	—
	2.8	25°C—1 GHz	—
	3.7	150°C—60 Hz	—
Dissipation factor	0.0025	25°C—60 Hz	ASTM D150-65T
	0.0050	25°C—1 kHz	—
	0.016	25°C—1 MHz	—
	0.003	25°C—1 GHz	—
	0.0040	150°C—60 Hz	—
Volume resistivity, ohm-cm	10^{13}	25°C	ASTM D-257-66 and D2305-68
	10^{13}	150°C	—
Surface resistivity, ohms/sq.	10^{16}	23°C—30% RH	ASTM D257-66
	10^{12}	23°C—80% RH	—
Insulation resistance, ohms	10^{12}	35°C—90% RH	ASTM D257-66 and D2305-68
Corona resistance (3 mil Type A), hr per sheet	30	25°C—300 V a-c, 60 Hz	Modified ASTM D2275-64T

Table 6.27   Selected Properties of Polyester Films (Cont'd)

Property	Physical			Test Method
	Typical Value		Test Condition	
	1 MIL TYPE A	1 MIL TYPE T		
Ultimate tensile strength (MD)*	25,000 psi	45,000 psi	25°C	ASTM D882-64T Method A—100%/min
Stress to produce 3% elongation (MD)	13,000 psi	18,000 psi	25°C	ASTM D882-64T Method A—100%/min
Stress to produce 5% elongation (MD)	15,000 psi	23,000 psi	25°C	ASTM D882-64T Method A—100%/min
Ultimate elongation (MD)	120%	40%	25°C	ASTM D882-64T Method A—100%/min
Tensile modulus (MD)	550,000 psi	800,000 psi	25°C	ASTM D882-64T Method A—100%/min
Impact strength	6.0 kg-cm/mil	6.0 kg-cm/mil	25°C	Du Pont Pneumatic Impact
Folding endurance (MIT)	14,000 cycles	—	25°C	ASTM D2176-63T (1 Kg loading)
Tear strength, propagating (Elmendorf)	20 gm/mil	20 gm/mil	25°C	ASTM D1922-61T
Tear strength, initial (Graves)	800 gm/mil	450 gm/mil	25°C	ASTM D1004-61
Tear strength, initial (Graves)	1800 lb/in.	1000 lb/in.	25°C	—
Bursting strength (Mullen)	66 psi	55 psi	25°C	ASTM D774-63T
Density	1.395	1.377	25°C	ASTM D1505-63T
Coefficient of friction (kinetic, film-to-film)	0.45	0.38	—	ASTM D1894-63
Refractive index (Abbé)	1.64	—	25°C	ASTM D542-50
Area factor (sq in./lb/mil)	20,000	20,500	25°C	Calculation
Melting point	250°C (480°F)		—	Fisher-Johns
Service temperature	−70°C to 150°C (−100°F to 300°F)		—	—
Coefficient of thermal expansion	1.7×10^{-5} in./in./°C		30° to 50°C	Modified ASTM D696-44

* Machine Direction.

The uses of both TFE and FEP films are many and reflect the unique characteristics of their chemical structures. The dielectric constant is 2.0, and power factor is very low: < 0.0003 to 60 to 10^8 Hz. (Dissipation factor at this level is numerically the same.) Since the characteristics of both films are so nearly alike the data in Table 6.28 are given as typical of each.

6.19.1 Vinylidene fluoride

Vinylidene fluoride is a high-molecular-weight thermoplastic polymer (CH_2-$CF_2)_n$ containing over 59 percent of fluorine by weight. It is a chemical relative of TFE and is resistant to most strong acids and bases. Organic compounds, both aliphatic and aromatic, alcohols, acids, and chlorinated solvents have virtually no effects on it. Strongly polar solvents, such as ketones and esters, act as partial solvents at elevated temperatures.

Its film form, known particularly by the tradename Kynar (Pennwalt Corp.), finds its best electrical use as an insulation for hookup wire. Heat-shrinkage tubing can also be made from this polymer (as discussed in Chap. 7). The film is commercially available in thicknesses of 1 to 10 mils. Up to 3 mils thickness, the films are clear, but thicker films are slightly hazy.

The film exhibits very low permeability to O_2, N_2, CO_2, and the halogens. Temperature range for useful service is $-80°$ to $+300°$ F. Melting point is $340°$ F. The material is self-extinguishing. No thermal degradation occurs after exposure for one year at $300°$ F.

Vinylidene fluoride outgasses at $100°$ C and a pressure of 5×10^{-6} mm Hg. Under these conditions, a stationary rate of weight loss is shown of 1.3×10^{-11} g/cm^2/sec.

Electrically this polymer is marked by a very high dielectric constant which decreases, however, as frequency is increased. This behavior is also true of the dissipation factor. The dielectric strength is moderately high. A very low water absorption rate is a featured property.

The particular combination of properties exhibited by vinylidene fluoride appears to indicate that it may offer a good field for investigation in many electrical applications.

6.19.2 Polyvinyl fluoride

Still another member of the TFE family of fluorine-containing polymers, polyvinyl fluoride films (known as Du Pont's Tedlar, or simply as PVF) are outstanding for their weathering ability, which has led to increasing use as a permanent finish in architectural applications. In addition, the excellent aging properties, chemical inertness, high dielectric strength and dielectric constant of these films offer functional opportunities in electrical insulation and dielectrics. The film is available in 1/2, 1, and 2-mil thicknesses, and in widths up to 72 in. There are four types of film, supplied in three different surface characteristics.

Properties of PVF films are detailed in Table 6.29. In summary, these films are chemically inert, resist most known solvents, and have very low moisture absorption. Service temperature is $-70°$ to $+105°$ C. The films melt with decomposition at $> 300°$ C. Shrinkage rate varies with type: it is 7 percent for Type 20 after 30 min at $300°$ C, drops to 4 percent for Type 30, and is less than 1 percent for Type 40.

Table 6.28   Selected Properties of FEP-Fluorocarbon Films

Values are for DuPont's fluorinated ethylene propylene (FEP-Teflon).

Property	Typical Value (for 1 mil film)	Test Condition	Test Method
		Electrical	
Dielectric Strength (1 mil)	6500 volts/mil	25°C, 60 Hz	ASTM D-149-61†
Dielectric Strength (20 mil)	1800 volts/mil	25°C, 60 Hz	ASTM D-149-61
Dielectric Constant	2.0	25°C, 100 Hz to 100 MHz	ASTM D-150-59T
Dielectric Constant	2.02–1.93	1000 Hz, −40°C to 225°C	ASTM D-150-59T
Dissipation Factor	0.0002–0.0007	25°C, 100 Hz to 100 MHz	ASTM D-150-59T
Dissipation Factor	0.0002	1000 Hz, −40°C to 225°C	ASTM D-150-59T
Dissipation Factor	0.0005	100 MHz, −40°C to 240°C	ASTM D-150-59T
Volume Resistivity	$> 10^{17}$ ohm-cm	−40°C to 240°C	ASTM D-257-61
Surface Resistivity	$> 10^{14}$ ohm/sq	−40°C to 240°C	ASTM D-257-61
Surface Arc Resistance*	> 165 sec	—	ASTM D-495-58T
		Physical	
Ultimate Tensile Strength (MD)††	3000 psi	25°C	ASTM D-882-61T
Yield Point (MD)	1700 psi at 3%	25°C	ASTM D-882-61T
Stress at 5% Elongation (MD)	1900 psi	25°C	ASTM D-882-61T
Ultimate Elongation (MD)	300%	25°C	ASTM D-882-61T
Tensile Modulus (MD)	70,000 psi	25°C	ASTM D-882-61T
Impact Strength	2 kg-cm/mil	25°C	DuPont Pneumatic Impact
Folding Endurance (MIT)	4000 cycles	25°C	ASTM D-643-43
Tear Strength—propagating (Elmendorf)	125 gm/mil	25°C	ASTM D-1922-61T

Table 6.28   Selected Properties of FEP-Fluorocarbon Films (Cont'd)

Property	Physical		
	Typical Value (for 1 mil film)	Test Condition	Test Method
Tear Strength—initial (Graves)	270 gm/mil	25°C	ASTM D-1004-61T
Tear Strength—initial (Graves)	600 lb/in.	25°C	ASTM D-1004-61T
Bursting Strength (Mullen) (1 mil)	11 psi	25°C	ASTM D-774-46
Density	2.15	25°C	ASTM D-1505-60T
Coefficient of Friction (Kinetic)			
(Film-to-Film)	0.57	25°C	ASTM D-1894-61T
Refractive Index (Abbé)	1.341-1.347	—	ASTM D-542-50
Area Factor	12,900 sq in./lb/mil	—	Calculation
Melting Point (°F)	500°-535°F	—	Hot stage microscope with crossed
(°C)	260°-280°C	—	polaroids
Service Temperature—continuous			
(°F)	−425° to +400°F	—	—
(°C)	−255° to +200°C	—	—
intermittent			
(°F)	−425° to +525°F	—	—
(°C)	−255° to +275°C	—	—
Coefficient of Linear Expansion	4.61×10^{-5} in./in./°F	−100°F	—
	5.85×10^{-5} in./in./°F	+160°F	—
	9.0×10^{-5} in./in./°F	+212°F	—

* Samples melted in arc; did not track.
† Short-time test in air at 23°C, 60 Hz, ¼-in. electrode, 3/32-in. radius, 500 V/sec rate of rise.
†† Machine direction.

Table 6.29 Selected Properties of Polyvinyl Fluoride (PVF) Films*

Values are for DuPont's Tedlar (tradename) PVF film. Electrical data apply to clear film, Types 20 and 30. Physical data apply to clean film, Type 20, except for thermal properties, which apply to all types.

Electrical	
Property	Typical value
Dielectric strength (1 mil)	3,500 volts/mil
Dielectric constant	8.5
Dissipation factor	0.02
Dissipation factor	0.09
Volume resistivity	10^{13} ohm-cm
Surface resistivity	10^{15} ohms/sq

Physical	
Ultimate tensile strength (MD)*	19,000 psi
Yield point (MD)	6,000 psi at 2%
Stress to produce 5% elongation (MD)	6,000 psi
Ultimate elongation (MD)	100%
Tensile modulus (MD)	280,000 psi
Impact strength	5–6 kg-cm/mil
Folding endurance (MIT)	47,000 cycles
Tear strength	
propagating (Elmendorf)	12 gm/mil
initial (Graves)	452 gm/mil
initial (Graves)	997 lb/in.
Bursting strength (Mullen)	70 psi/mil
Density	1.38 gm/cc
Coefficient of friction (kinetic, film to film)	0.47
Refractive index (Abbé)**	1,467
Area factor	40,000 sq in./lb/½ mil
Melting point (°F)	> 570° F with decomposition
Melting point (°C)	> 300° C with decomposition
Service temperature (°F)	−100°F to + 225°F
Service temperature (°C)	−70°C to + 105°C
Coefficient of thermal expansion	2.8×10^{-5} in./in./°F
Flammability	Slow burning to self-extinguishing

* Machine direction.
** ASTM D542-50

The PVF films have high dielectric strength (3,500 volts per mil in the 1-mil thickness). Dielectric constant is very high (8.5 at 23°C and 1 kHz), and increases with temperature and frequency. Dissipation factor increases with frequency at 23°C and 50°C but falls slowly at 100°C and rapidly at 140°C. Volume resistivity is moderately high.

Type 20 has high tensile strength and high resistance to folding and creasing. Type 30 is a medium-tensile-strength film available in colors. It finds use as a base for pressure-sensitive tapes that are often employed as release sheets in printed circuitry.

6.20 Polyethylene Films

Polyethylene was the first of the polyolefins used as electrical insulation and as dielectrics. The simple ethylene molecules $CH_2 \!=\!\!=\! CH_2$ or $C_2H_4(C_2H_2n + 2)$

are joined end to end to form polyethylene. Unfortunately all is not quite as simple since the final products, in the form of giant molecules, are mixtures of molecules of varying length, some of which may have branches of varying composition.

These large molecules are nearly nonpolar, are highly moisture resistant, and have a power factor of 0.0002 and a dielectric constant of 2.27 over a very wide range of frequencies.

Since the first polyethylene was introduced different processes have resulted in a need to type the various products as recognized by ASTM Specification D1248, as follows:

> *Type 1* Density range 0.910 to 0.925 g/cc, called low density (branched).
>
> *Type 2* Density range 0.926 to 0.940 g/cc, called intermediate density.
>
> *Type 3* Density range 0.941 to 0.965 g/cc, called high density or linear.

The density and molecular structure of the different types are determined by the particular process used (Ziegler, Phillips, Hercules, Standard Oil of Indiana, or other), which in turn vary in the type of catalyst, applied temperature and applied pressure. The types are sometimes identified as "high pressure" or "low pressure," but it is preferable and simpler to stick to the standard ASTM designations.

As a film, polyethylene has excellent electrical insulating characteristics and is outstanding in high-frequency applications. It is chemically inert and forms a good moisture barrier. Polyethylene is also used widely as extruded wire and cable insulation.

An extensive line of commerical polyethylene film is available in various thicknesses and formulations. Detailed data are offered by manufacturers.[29]

6.20.1   Irradiated polyethylene film

Irradiated polyethylene film (and sheet) is produced by bombardment of polyethylene with high-energy electrons. A Van de Graff accelerator is the usual source of electrons. The electrons induce a cross-linking action that radically alters the molecular structure of the polymer and also alters certain properties. Specifically, the irradiated film (compared to the standard film) exhibits a much higher resistance to environmental stress-cracking, is nonmelting, and (above the melting point) becomes elastomeric. Passing through the transition point, the irradiated film acquires shrinking force. In the elastomeric form, the irradiated film, therefore, can be used for encapsulating purposes. It can be wrapped around any given configuration (including irregular shapes) and heat-shrunk to fit, thus providing a moisture-proof, dust-proof seal.

Irradiated film is available in two general-purpose grades (low density and high density) and an encapsulating grade. The general-purpose grades come in thicknesses of 2.5 to 10 mil; the encapsulating grade is available in thicknesses of 4 and 8 mil. There is also a semiconducting grade used for shielding and like purposes in cables, among other applications.

In sealed systems where oxidation is not a problem, clear films are used. Clear films that contain an antioxidant and black films that also contain an antioxidant are used at 105° C and have the ability to withstand short-time overloads at 200° C. General-purpose films exhibit lengthwise shrinkages of 30, 25, and 15

---

[29] See Footnote 26.

percent, depending on grade, at 150° C. The encapsulating grade exhibits shrinkage of 40 to 50 percent.

Since most of the property characteristics of standard polyethylene are not markedly changed by the irradiation, except as noted above, the usual published data for electrical and physical properties will suffice in most cases for the irradiated material. Where design considerations demand it, it is recommended that some experimental trials be made to assure suitability of the irradiated film.

The combination of the inherent excellent electrical properties of polyethylene with the special properties of the irradiated material indicate advantageous applications in wire and cable insulation, microwave assemblies, printed circuitry, and also where moisture-proof barriers are needed.

Irradiated polyethylene is frequently identified by the tradename Irrathene (General Electric Company). Irradiated polyethylene fabricated insulation is made by several suppliers.

6.21   Polypropylene Film

Propylene $CH_2 = CH - CH_3$ is polymerized to form polypropylene, which, like polyethylene, is a member of the polyolefin family. The polymer is cast into brilliantly clear sheets and exhibits a much higher melting point than that of polyethylene. Like polyethylene, polypropylene has excellent electrical properties. It has high dielectric strength, a low dissipation factor, and a dielectric constant of 2.2 to 2.3. Its mechanical properties are good, and it is chemically inert.

Polypropylene films are biaxially oriented (that is, they are stretched in two directions) to increase the density and improve mechanical properties. The degree of improvement is related to the degree of stretching. The increase in tensile strength attributed to biaxial orientation is noteworthy—from 7,000 to 25,000 psi in the "machine direction." The electrical properties remain unchanged. Selected property data are given in Table 6.30.

6.22   Polyimide Films

Polyimide is the product of a condensation reaction between an aromatic tetrabasic acid and an aromatic diamine. The source of the acid radicals may be pyromellitic dianhydride (PMDA). Made into film, by Du Pont under the tradename, Kapton, it possesses a unique combination of properties. It does not melt. It maintains its electrical and mechanical properties over a wide temperature range. It has been used from −269° to +400° C. It is flame-resistant, char beginning above 800° C. No organic solvents are known. At room temperature the properties of this film and polyester films are similar. As the temperature is increased or decreased, the polyimide film characteristics are less affected than those of polyester films.

Adhesives are available for bonding polyimide film to itself, to metals, papers, and other films. The film can be laminated, metallized, punched, and formed. It cannot be heat-sealed except by suitable pre-coating with fluorocarbon resins, as discussed later.

The number of application fields is extensive. Polyimide film is an important material in wire and cable insulation. Rotating apparatus, transformers, chokes,

Table 6.30   Typical Properties of Polypropylene Films

Source: Avisun Corp., Philadelphia, Pa.

Property	Unoriented	Biaxially oriented
Density	0.89	0.89
Tensile strength for 1-mil film, psi	7,000 MD	25,000
	4,000 cross MD	23,000
Elongation, percent	850 MD	90
	800 cross MD	90
Dielectric strength, vpm		5000
Volume resistivity, ohm-cm	3×10^{15}	5×10^{14}
Dielectric constant	2.2	2.2
Dissipation factor at 1 kHz	0.0007	0.0005
Temperature range	$-40°$ to $105°$C	$-40°$ to $105°$C
Flexibility	Fair	Fair
Resistance to cold flow	Good	Good
Effect of solvents		
boiling water	Excellent	Excellent
aliphatic hydrocarbons	Good	Good
aromatic hydrocarbons	Good	Good
trichloroethylene	Good	Good
resistance to acids	Good to Excellent	
resistance to alkalis	Good to Excellent	

capacitors, magnetic tapes, pressure-sensitive tapes all can benefit by its use since it maintains its properties over a wide range of temperature and frequency.

The tensile strength of polyimide is high, comparing well with that of polypropylene at 25°C, at which temperature both exhibit 25,000 psi. Moreover, at $-195°$C it has a tensile strength of 35,000 psi and at 200°C, 17,000 psi. Its zero strength temperature is 815°C.

Like that of all condensation-reaction polymers, the properties of polyimide are affected by water. If exposed to high temperature and pressure in a closed system with water, it will be affected markedly. However, after 2,400 hr in boiling water the tensile strength levels out at 18,000 psi from an initial 26,000 psi at 0°C. The ultimate elongation after 2,400 hr in boiling water levels out at 20 percent from 70 percent at 0°C.

Dimensional stability depends upon its normal coefficient of expansion and the residual stresses of film manufacture. Polyimide shrinks on its first exposure to elevated temperature, but thereafter the normal values of thermal coefficient of expansion are observed.

Unlike most other organic films polyimide exhibits very low tendency to cold flow and has very high cut-through temperature. It is subject to oxidation degradation although this does not appear to be serious for practicable use. After exposure to 300°C for 1,000 hr, the tensile strength, in air, drops to 18,000 psi from an intial 25,000 psi. In helium the drop is to 19,000 psi.

In air the weight loss of polyimide above 500°C is large. Comparable results are not shown at even higher temperatures upon aging in helium. Although there are no known organic solvents, it is destroyed by strong bases such as 10-percent sodium hydroxide.

Kapton Type H is available in ½-, 1-, 2-, 3-, and 5-mil thickness, Standard width is up to 12 in.

As noted above, for heat-sealing purposes polyimide is coated with FEP resin. A film of this type is designated as Kapton Type F and comprises Type H coated on one or two sides with the FEP resin. The heat-sealable surface also serves to improve resistance to bases and strong acids. The effects of oxygen are reduced as is the rate of moisture permeability.

Type F is available in nine standard combinations. The first is 1-mil Type H with ½-mil FEP on one side for a total nominal thickness of 1½ mils. The ninth member of the series of standard constructions is 6 mils thick consisting of 5-mil Type H with 1 mil of FEP on one side. Table 6.31 provides a summary of data of three constructions.

6.23   Miscellaneous Films

Many polymeric films can be produced that do not find use in the electrical field. On the other hand, some of the films described in this chapter might not have found use in electrical applications if their research and development had not been first supported by nonelectrical markets. An interesting example is the emerging use of polyvinyl fluoride for electrical insulation (Par. 6.19.2) although its first, and probably still its major market, is the architectural field.

An even more striking example has taken place during the preparation of this work. Most of the polypropylene film has been used as a packaging material, since its characteristics were found ideal for food packages. In fact, one large film producer has done no research on it for possible electrical use for the past five years or so. But by now the two largest electrical apparatus makers have announced capacitors with polypropylene film of the biaxially oriented form as the solid dielectric. The results claimed are lower loses, longer life expectancy, smaller size for equal performance, and more stable capacitance.

These paragraphs therefore review briefly the characteristics of some films that have one or more property of particular interest in electrical/electronic design. Their commercial availability, though not necessarily in the electrical market, gives the designer an opportunity for experimental work that otherwise might not have been possible at reasonable cost and effort. It might be noted that most of the films described here may be laminated to other films or sheets, papers, glass-fiber cloth, and the like to form composites.

Polycarbonate film   This film is produced from a unique class of resins known as "Lexan" (General Electric Company) described as the first in which the polymer is joined by a carbonate radical. Although the actual reaction is not revealed, a possible method may involve a reaction of bisphenol A and carbonyl chloride to give a linear, thermoplastic polymer. (Bisphenol A is also used in making epoxy resins.) The polycarbonate film is extruded without the use of plastizers. It is available in thicknesses of 1 to 20 mils, in widths of up to 36 in.

Although the tensile strength and elongation properties of polycarbonate film are not outstanding, they are retained to a marked degree despite increases in temperature. Moisture does not bring about shrinkage after several weeks at 50 percent relative humidity followed by heating at 100°C. Heating at 150°C

Table 6.31 Selected Data for Heat-Sealable FEP-Coated Polyimide Film

Values are for DuPont's FEP-coated Kapton (Type F) polyimide film.

Property	Film Construction*		
	019	929	051
Ultimate tensile strength (MD), psi			
25°C	17,000	17,000	21,000
200°C	11,000	11,000	14,000
Yield point (MD) at 3 percent, psi			
25°C	7,300	7,300	8,600
200°C	4,000	4,000	5,000
Stress at 5 percent elongation (MD), psi			
25°C	9,000	9,000	11,000
200°C	5,500	5,500	7,000
Ultimate elongation (MD), percent			
25°C	75	> 80	100
200°C	85	—	—
Tensile modulus (MD), psi			
25°C	320,000	320,000	310,000
200°C	173,000	173,000	179,000
Impact strength at 25°C			
kg-cm	7	14	32
kg-cm/mil	4.6	4.6	5.3
Tear strength, propagating (Elmendorf)			
gm	20	—	120
gm/mil	13.5	—	20
Tear strength, initial (Graves)			
gm	650	—	—
gm/mil	435	—	—
Weight percent, polyimide	57	57	77
Weight percent, FEP	43	43	23
Density	1.67	1.67	1.54
Dielectric strength			
total volts	6,300	11,400	18,500
volts/mil	4,200	3,800	3,100
Dielectric constant	3.0	3.1	3.4
Dissipation factor	0.0014	0.0011	0.0013
Volume resistivity, ohm-cm			
25°C	10^{18}	8×10^{17}	10^{17}
200°C	10^{14}	9×10^{13}	10^{13}
Moisture absorption at 25°C, percent			
50% RH	0.8	0.8	1.0
98% RH	1.7	1.7	2.2

* Construction: 019 = 1 mil polyimide + 0.5 mil FEP on one side; 929 = 2 mil polyimide + 0.5 mil FEP on both sides; 051 = 5 mil polyimide + 1 mil FEP on one side.

produces only 2 percent shrinkage. Boiling water rapidly affects the elongation but continued immersion does not change the tensile properties. Resistance to heat aging at 75°C is good but at 125°C aging the final elongation is lowered from 97 to 9 percent.

Chemical resistance is selective. Many food materials are of no effect. Methylene chloride is a good solvent. The hydrocarbon refrigerants attack it as do aromatic hydrocarbons, ketones, and esters. Outgassing in vacuums as high as 10^{-7} mm Hg at $100°$ C is not exhibited.

The dielectric strength, dielectric constant, and volume resistivity are relatively constant with change in temperature and frequency. Power factor at 60 Hz exhibits a flat curve from $25°$ C to $142°$ C where a sharp upward break occurs.

Polyphenylene oxide (PPO)   One of the newer "engineering thermoplastics," PPO has the outstanding characteristic of maintaining its dielectric constant over a frequency range of 60 to 3 GHz. The polymer has found use in structural parts in electrical equipment, owing to a combination of properties such as dimensional stability, impact strength, self-extinguishing ability, and good formability. In addition, the electrical and mechanical characteristics have led to the development of a film material made from a modified PPO resin and indicated particularly for application in strip transmission lines and other microwave-frequency devices.[30] Dielectric constant of this material is given as 2.72. PPO was commercially introduced by General Electric Company.

Ionomer film   Ionomers are defined as polymers linked by ionic forces as well as by covalent bonds. An ionomer known commercially as Surlyn A (Du Pont) can be made in a wide range of molecular weight and degree of crystalline order. Although the "ion" is not identified by the producers, it is probably a metal, for example, zinc. The unusual characteristic of this ionomer is its exceptional resistance to corona effects—about 100 times that of polyethylene. Because of its range of molecular properties and its crystal structure, Surlyn A may be "tailored" to fit a variety of uses, including, possibly, electrical insulating film. At present, however, the indicated applications seem to be packaging.

Nylon film   Nylon films have been available in a large variety. The very high rate of moisture absorption hinders the use of this film for electrical insulation.

Polystyrene film   Biaxially oriented polystyrene film of high purity provides stable dielectric loss properties that are useful in such applications as coaxial cable. The film is also used as capacitor dielectric. The generally good resistance to chemicals is an asset.

Polysulfone films   Another of the engineering thermoplastics, the polysulfones are notable for their excellent mechanical properties and their use in structural and mechanical parts. Creep resistance, for example, is particularly good. The polymer can be extruded in film form and there is therefore a possibility that the basic properties, which include high dielectric strength, may lead to electrical insulating applications.

Vinyl films   The widely used vinyl films and sheet (polyvinyl chloride and polyvinyl chloride-acetate copylmer) also merit special electrical insulating uses where they can provide good electrical properties with excellent resistance

---

[30]   The Polymer Corp., Reading, Pa.

to chemicals and moisture. The vinyls are important tape materials which will be discussed later in the section dealing with this category of materials.

Elastomeric sheets Silicone rubber and other elastomeric sheets will be discussed in Chap. 8 along with other elastomeric insulations.

6.24 Insulating Fabrics and Cloth

This section will deal with fabric and cloth treated or coated with diverse resins or elastomers so that the finished product is capable of performing an electrical function. In essence, the base material provides the mechanical strength and other required physical properties, whereas the saturant, impregnant, or coating carries the primary responsibility for electrical performance.

Fabrics for electrical insulations span a very wide historical period. Their use in insulation goes back to the early 1800's in pioneer investigation in the phenomenon of electromagnetism. The American physicist, Joseph Henry (1797–1878), for whom the unit of induction (the henry) has been named, adoped silk fabric for wire insulation in order to solve a pressing problem of finances and working space for his studies in this field. The thin fabric made it possible for him to wind many turns and even layers on soft-iron cores. As a result, a small battery sufficed to produce a large magnetic effect. Henry said, "The principal object in these experiments was to produce the greatest magnetic force with the small quantity of galvanism [electricity]." In effect, Henry was pursuing a modern engineering objective—the highest possible performance at the lowest cost.

Insulating fabrics are made from several different categories of fibers. Cotton, silk, and asbestos have been used for many years. Among synthetic fibers, conventional polyamide (the familiar nylon) has been used for some years, whereas newer fibers, such as the high-temperature Nomex polyamide and polyester, are emerging in the fabric form. Looming over the entire field are glass-fiber materials in their ability to combine with a variety of coatings and treatment, as well as composites with other fabrics.

Some discussion of fiber definitions should be useful at this point. In the textile industry, fibers are defined as filaments having a length of at least 100 times the diameter with a minimum length of 5 mm. The fibers used in fabrics for electrical insulation, however, far exceed this requirement. Cotton has a length-diameter ratio of 500 to over 1,000. Glass, nylon, and silk fibers are of continuous length. The cross sections of synthetic fibers are circular, cotton has a flat tubular cross section, and silk is generally found to have a triangular cross section. Density of nylon fiber in grams per cu cm is 1.14, of silk, 1.37, cotton, 1.55, glass, 2.57. Asbestos (which will be discussed with other asbestos materials in Chap. 11) has a fiber density as high as 2.80 gm/cm^3. ·

The "fineness" of a fiber is defined by the relation between weights for a given length. The common measure of linear density is the "denier," which is the weight in grams of a fiber 9,000 meters long. By assuming a circular cross-sectional area and by knowing the denier density, various engineering data for a given fiber material can be calculated.

Detailed discussion of textile fibers is, of course, beyond the scope of this book, but the information given above should be helpful in following the description of various *finished* fabrics used in electrical insulation as given in succeeding paragraphs.

In one respect, all woven cloths as they come from the loom have a common characteristic. The dielectric strength is a function of the air spaces in the weave and is roughly equal to the air gap length, or may be even less. The actual dielectric strength of finished cloths depends, for the most part, on the saturant with which the fabric has been impregnated, or on the coating.

A multiplicity of insulating fabrics and cloths is available for design purposes, representing a diversity of combinations of fabrics, resins, coatings, treatments and other factors. A catalog of such materials, however, is beyond the purpose of this book. Such information is readily available in the commercial literature and trade directories. The following paragraphs will concentrate on the most important materials and their salient characteristics.

6.25   Glass-Fiber Cloth

Since the design trend is toward high thermal endurance at temperatures of 130°C and above, woven glass cloth has captured a large market. It has many advantages such as dimensional stability, high heat resistance, good thermal conductivity, and low moisture absorption. It is resistant to attack by most chemical solutions and when saturated the resulting electrical properties can be controlled over a very wide range. Since glass fibers possess high tensile strength, the fabrics made from them are very strong. The glass cloth itself will not burn although in the finished form the saturants are usually flammable.

The uses of glass cloth are wide-spread and diverse. They are used in the form of varnished flexible sheets (coated with fluorocarbon resins, silicone rubber, or silicone varnish, among other coatings), made into laminates and composities, and are used as foundation for printed circuits. Glass cloths may be obtained as sheets cut to size or in continuous rolls.

For electrical work, a borosilicate glass composition, designated as E-glass, contains silicon dioxide 52 to 56 percent, calcium oxide 16 to 25 percent, aluminum oxide 12 to 16 percent, boron oxide 8 to 13 percent, sodium and potassium oxides from 0 to 1 percent, and magnesium oxide from 0 to 6 percent. The following nominal values apply: Softening temperature, 1,400°F; moisture absorption (as a solid "patty"), 0.5 percent; dielectric constant, 6.3; power factor, in percent, 0.20 max at 60 to 37 MHz.

NEMA GF 1 provides applicable standards and specifications for continuous-filament glass yarns. NEMA VF 40 does the same for continuous-filament woven glass fabric. The specifications for the woven glass fabric is for the material as it comes from the loom, known in the textile industry as "greige" (rhymes with beige). An applicable ASTM specification for glass fabric is D 2518, but it excludes tapes and fabrics intended primarily for reinforcements in plastics laminates. Thicknesses for fabrics range from 0.0010 in. to 0.050 in.

Most glass cloths are made as "plain" or straight weave. Here the warp and the filling threads cross alternately. This weave contains the greatest density of yarns. Since it can be tightly woven or loosened, it permits a more "open" weave. Since it has good porosity and takes up saturants, the finished cloth is of the best possible quality. (The lengthwise warp threads or yarns and the filler or woof threads across the fabric are put in place by the shuttle.) There are other weaves such as basket, twill, satin, and leno, but these are seldom used for electrical purposes. Greige has to be cleaned of sizings, oils, and starches present during the weaving or preweaving processes.

After cleaning, a "finish" or "treatment" is applied in order to bond the glass to the saturant. The maker of the finished cloth must select the proper treatment or finish depending upon the saturant. The most widely used finish is tradenamed Volan (Du Pont), a methacrylate chromium chloride. It is used when polyester, epoxy, and phenolic saturants are employed.

NEMA Standard GF 1 contains a nomenclature for glass filament designation. The first letter indicates the glass composition. The letter "E" as already noted designates the electrical (borosilicate) designation. The second letter indicates continuous filament by "C" and staple by "S." The third letter indicates the diameter of the individual fiber in decimal fractions of an inch in accordance with the following code:

$$D = 0.00021 – 0.00025$$
$$E = 0.00026 – 0.00030$$
$$F = 0.00031 – 0.00035$$
$$G = 0.00036 – 0.00040$$

The strand number follows the letters. The number is 1/100th of the yardage of raw glass fiber per pound. The first digit of the strand number shows the number of original strands that are twisted and the second digit the number of these yarns that are plied. The total number of strands is the product of the two numbers. (Should either be zero, replace with 1.) To calculate yards per pound, multiply the first digit by 100 and divide by the number of strands.

For example, ECD, 450-4/3 is an electrical-grade (E), continuous-filament (C) yarn, with individual fibers in the diameter range of 0.00021 to 0.00025 in. (D). It has a nominal yardage of 45,000 yd per pound of single-strand fiber (450 × 100) and is made up of 12 strands (three plies, each consisting of four strands twisted together before plying). The approximate yardage is 3,750 yd per lb, a value derived by dividing the nominal yardage (45,000) by the total number of strands (12). Table 1 of NEMA Standard VF 40 provides the style number of the greige cloth and all details of thread count, nominal thickness, yarns types, and the like. This table is of primary use to converters.

6.25.1   Varnished glass fabrics

NEMA Standards Publication VF-7-1958 covers *yellow varnished glass fabric,* formulated and designed for use in Class 130 ° C (Class "B") insulation systems. Thicknesses covered are 3 to 30 mils of finished cloth. Minimum dielectric strength in volts per mil is 1,200 for the 3-mil cloth to 300 for the 30-mil cloth after 24 hr at 23 ° C ± 2 ° C and 50 percent relative humidity. After aging for seven days at 130 ° C ± 3 ° C (as provided in ASTM D 902), these values drop to 1,100 and 100 vpm, respectively. This fabric is available in sheets, rolls, and tapes.

Black varnished glass fabric is covered by NEMA Standards Publication VF-6-1954. The stated requirements are substantially the same as for the yellow products.

The oil-resistance of oleoresinous varnish-coated glass fabric determines its

Table 6.32   Typical Product Values for Yellow Varnished Glass Cloth

Property	Value
Finished thickness, in.	0.010
Approximate weight (#127 Fabric), lb/sq yd	0.76
Tensile strength, PIW*	250
Elmendorf tear strength, gm	$> 1,600$
Water absorption (ASTM D570), 24 hr, percent	1.25
Dielectric strength	
(short time, room temperature), vpm	1,200
1,000 hr at 125°C	1,400
Power factor, 25°C, 60 Hz	3.5
Thermal classification	130°C (Class B)

* PIW = pounds per inch of width.

suitability for use in oil-immersed apparatus. Here the yellow is superior as it softens and swells less than the black. Table 6.32 provides typical product values for yellow varnished glass cloth, as given in ASTM D 902.

Resin-varnished glass cloth for continuous service at 155°C (Class "F") is generally available with three kinds of saturating coatings. A straight-weave glass fabric is used for each of the three saturants or varnishes. A brief description of each type follows:

Polyester-coated glass cloth with good resistance to solvents and moisture is used in motors, transformers, and electronic components.

Epoxy-coated glass cloth is used in liquid-insulated equipment such as oil-immersed transformers and as a substrate for flexible printed circuits.

Polyurethane-coated glass cloth may be used where solvent resistance, conformability, and resistance to physical abuse is required.

Table 6.33   Typical Product Values for Coated Glass Cloth

Resin	Finished thickness	Base cloth	Weight, lb/sq yd	Tensile, PIW	Elmendorf tear, gm
Polyester*	0.010	112	0.59	90	$> 1,600$
Epoxy	0.003	1070	0.195	95	—
Polyurethane	0.003	108	0.20	70	—
Polyurethane	0.010	127	0.68	235	625 MD††

	Dielectric strength, vpm	Percent power factor, 60 Hz		Dielectric constant, 60 Hz
		25°C	80°C	
Polyester*	1,500	2.3	2.8	3.6 (C96/23/0)**
Epoxy	2,000	0.035 (DF)†	—	3.5 (C96/23/0)
Polyurethane	2,000	2.20 (DF)†	4.00 (DF)†	—
Polyurethane	1,500	4.90	6.50	—

* For polyester-coated glass cloth, thermal endurance data are available—50,000 hr at 155°C (ASTM D1830), also moisture absorption—0.6 percent after 24 hr (ASTM D520).
** Condition C = 96 hr at 23°C, zero humidity (dry).
† DF = Dissipation Factor
†† Machine direction.

Typical product values for these materials are summarized in Table 6.33, also based on ASTM D 902.

For continuous use as Class 180° C (Class H) two kinds of coated glass cloth are available. Both are based on the use of straight-woven glass fabrics:

Silicone-varnished glass fabric is covered by NEMA Standards Publication VF5-1956.

Silicone-elastomer (rubber)-coated glass fabric is covered by NEMA Standards Publication VF8-1961.

The silicone-varnished product has high resistance to moisture, oils, ozone, and is used for layer and ground insulation in dry-type transformers and motors. Rotational vibration in enclosed motors, however, may cause release of silicon dust from silicone-treated insulation. Deposited on commutators, the dust may cause brush wear through abrasion.

Silicone-rubber-coated glass cloth is used as layer insulation and as cable tape for power cable. The elastomer does not soften or flow at high temperature and retains its flexibility. The use of this cloth in cables has the advantage that there is no conducting residue should the cable be destroyed by fire.

Typical performance values for both types of silicone-treated glass fabrics are given in Table 6.34. The values are for 7-mil to 10-mil thicknesses.

6.25.2 *Adhesive-backed varnished glass cloth*

For use as Class 130° C insulation, yellow varnished glass cloth can also be obtained with pressure-sensitive adhesive on one side. The material is used for wrapping coils, windings, bus bars, and in related applications. It is available in rolls 1/2 in. to 36 in. wide and in sheets cut to size. Four types are commercially made:

1. A straight-woven glass fabric with a flexible yellow oleoresinous varnish, coated one side with non-thermosetting adhesive.

2. The same as (1) above, but in an open weave and of a lower tear strength.

3. A straight-woven glass fabric with a flexible yellow oleoresinous varnish, coated on one side with a *thermosetting* adhesive.

4. The same as (3) above but in an open weave and of lower tear strength.

All types are made in thickness of 0.007 in. and 0.010 in.

Table 6.34   Typical Product Values for Silicone Varnish and Silicone Rubber-Coated Glass Cloths

	Silicone varnish	*Silicone rubber*
Finished thickness	7 and 10 mils	10 mils
Base cloth	#116, 4 mils	#116, 4 mils
Approximate weight, lb/sq yd	0.48	0.68
Tensile strength, PIW*	150	125
Dielectric strength		
24/23° C/50% RH**	1,400	1,000
aging at 250° C	1,100	1,050
Dissipation factor, 180° C	0.04	—
Dielectric constant, 180° C	2.7	—

* PIW = pounds per inch of width.
** Conditioning 24 hr at 23° C and 50 percent relative humidity.

6.26   Cotton Fabrics

Varnished cotton fabrics have been made for many years for use at Class 105 °C (Class A) temperatures. The following NEMA Standard Publications apply:

VF 2-1957   Yellow straight-cut varnished cotton fabric
VF 1-1957   Black straight-cut varnished cotton fabric
VF 4-1957   Yellow bias-cut varnished cotton fabric tape
VF 3-1957   Black bias-cut varnished cotton fabric tape

ASTM D 373-64T Specification covers all four of the above. ASTM D 295 covers standard methods of testing these products.

The black varnished fabrics are made in three different surfaces: greasy, dry, and tacky. The yellow fabrics come only in the dry and tacky finishes. In tape form, the fabrics are packed in dry rolls, or dipped in wax, or oil-immersed in cans.

Typical performance data, based on ASTM D 373, are available in Table 6.35.

6.27   Nylon Fabrics

NEMA Standard No. 48-139 covers *yellow varnished nylon fabric*, a Class 105° C (Class A) material used where a thin, high-dielectric strength insulation is needed. The dielectric strength is given as 1,800 vpm for the 2-mil thickness and 1,500 for the 8-mil fabric. Tensile strength is 14 lb per in. of width for the 2-mil thickness and 16 lb for the 8-mil thickness. The NEMA Standard includes data for nominal thicknesses of 1, 3, 5, 7, and 8 mil.

6.28   Silk Fabrics

Yellow varnished silk fabric is covered by NEMA Standard No. 48-140. This standard and the one for nylon fabric are very much alike.

6.29   Prepreg Materials

Various materials—such as glass-fiber cloth, combined weaves of glass and polyester fibers, combined glass and polyamide fibers, glass fiber mats, polyester mats, among others—are preimpregnated or coated with semi-cured B-stage resins or elastomers. These materials are known as "prepregs" and are particularly indicated for use where it is desirable to mold an insulation to a given configuration and then cure to final form under application of heat and pressure, or heat alone. Coil wraps, structural insulating parts, layer insulation, banding for armatures, transformers, and the like, are among the applications frequently made for the prepregs.

A generous choice of base materials, saturants, coatings, and processing methods provides a wide selection range for design application. The base materials include a choice not only by type of precursor fiber, but also by type of weave construction, such as unidirectional, multidirectional, and specially oriented fibers.

Table 6.35   Typical Product Values for Varnished Cotton Cloths

Nominal thickness, mils	Cut	Minimum breaking strength, PIW* Black and yellow		Load, PIW	Minimum elongation Percentage		Average minimum dielectric breakdown, kV					
							Black			Yellow		
		Average	Individual		Black	Yellow	Unstressed	Elongation	Hot oil	Unstressed	Elongation	Hot oil
7	Bias	34	30	10	6.0	4.0	7.7	4.2	7.7	7.0	3.2	7.0
7	Straight	38	32	20	1.5	1.5	7.7	—	—	6.3	—	—
10	Bias	36	32	10	7.5	5.0	11.0	9.0	11.0	11.0	7.0	11.0
10	Straight	39	34	25	1.5	1.5	11.0	—	—	10.0	—	—
12	Bias	38	34	10	7.5	6.0	13.8	11.0	13.8	13.2	10.0	13.2
12	Straight	40	34	25	1.5	1.5	13.8	—	—	12.0	—	—

* PIW = pounds per inch of width.

6.30   Composite Fabrics

As indicated above, composite weaves are made in which one type of fiber is the warp and a second type provides the filler. For example, a fabric of this type is made of polyamide fibers for the warp and glass fibers for the filler. It has the special property of high temperature resistance and is designed for service at temperatures over 180° C. Similar composites have been made of glass and polyester fibers.

6.31   Mats and Nonwoven Materials

There have been references to "mats" in preceding paragraphs. These materials are of nonwoven construction and comprise a random distribution of fibers. They are frequently used as reinforcements in reinforced plastics compounds, and, in flexible sheet form, can be combined with suitable saturants and coatings to function as electrical insulation.

6.32   Sheet Materials in Plastics Laminates

A considerable amount of information on glass-fiber cloth and other base materials, including prepregs, has been given in Par. 6.14 and those thereafter.

6.33   Tapes

A long narrow flat strip of any material or combination of materials may be called a *tape*. For the purposes of this section, tapes are considered in their use as electrical insulation in such applications where their form is an advantage as a permanent component of a given electrical device. Or the tape may be used to facilitate the manual assembly of one or more parts into the desired position. In this sense the tape may be thought of as "an extra or third hand."

Tapes are used for two principal reasons, which are physical (or mechanical) and electrical (or insulating). In most applications both reasons for the use of tapes are present. The tape may hold something or some part in place and at the same time function as electrical insulation. Or the tape may impart some desired electrical characteristic to the device of which it is a part.[31]

Tapes used essentially as aids in device fabrication or assembly are not left permanently in the device. Here the tape is removed when it has served its purpose and is not essential to the proper operation of the device. For example, tape may be used as a mask in solder-dip of painted-circuit boards.

Tapes find their principal use in coils or windings of magnet wire (also magnet strip). They may be in direct contact with the magnet wire, on the insulation or conductor, or both. It is essential that the conductor, its insulation, and the tapes be compatible. Examples of use in windings are many. A few of them are (1) to hold

---

[31] The data in Table 6.36, as well as much of the description in the text, was taken from the sales literature of the following firms: Dielectric Materials and Systems Div., 3M Company, St. Paul, Minn.; Kendall, Polyken Div., Chicago, Ill; Permacel, New Brunswick, N.J.; Mystik Tape Div., Borden, Inc., Northfield, Ill.; Tapecon Inc., Rochester, N.Y.; Dielectrix Corp., Farmingdale, N.Y.; Chemplast, Inc., Wayne, N.J.; Acme Chemicals & Insulation Co., New Haven, Conn.

coil ends and coil leads in place, (2) to cover all or part of the coil, (3) to separate the core or bobbin from the wire, (4) to separate windings of a multisection coil, and (5) to hold terminal boards in place on the coil. In rotating apparatus the entire area or partial area of armatures or field coils may be covered by tape.

Tapes may also be used to bundle a number of single insulated wires into cable form by wrapping (serving) the wires with tape. The serving may cover the entire surface of the bundled wires or may be applied as separate straps spaced along the length of the bundle.

Still another use for tapes is as insulation of joints in many types of conductors. Application may range from insulation of joints in fine magnet wires to that in high-voltage power cable.

Capacitors and many other electrical devices may be covered by tape or tape may be so placed as to form part of the dielectric.

Just as there are many uses for tapes so are there many kinds and types. The major categories of tapes will be described and data given to aid in selecting the tape most suitable for the desired use. Some kinds of tapes may require coverage or mention in other chapters (for example, asbestos tapes, woven glass tapes, mica-backed or mica-filled tapes, and other inorganic-filled tapes).

6.34   Pressure-Sensitive Adhesive-Coated Tapes (PSA Tapes)

Pressure-sensitive adhesive-coated tapes have captured a large market because they are quick and easy to apply, requiring only a light pressure to secure adhesion. They offer a wide selection of suitable thin-sheet insulation, and they can form a permanent part of finished devices able to withstand operational hazards. The ability of these tapes to adhere strongly and instantly to most materials encountered in manufacture of electrical devices and apparatus facilitates many operations that would otherwise be very difficult if not commercially impossible.

In its simplest form a PSA tape consists of a backing (substrate) to which is applied an adhesive coating. A primer may be used between backing and adhesive. The tape is usually packaged in the form of a roll. Suitable release agents may be employed to prevent transference of adhesive to backing surfaces within the roll.

The tape manufacturer must elect to use primers and release agents as needed to permit production of a usable roll of tape. When tape is unrolled *all* of the adhesive must remain on one side of the backing and none upon the other side. When the PSA tape is primarily intended for use as electrical insulation, it may be of the comparatively simple form (four, three, or two elements). For other uses more complex constructions are required. They will be mentioned briefly where their use may be in close connection with the manufacture and performance of an electrical device.

Treatment of some kind of the backing or substrate is common practice. This is done to impart some essential characteristic to the tape such as a higher voltage breakdown resistance than the air-space equivalent. There are many other reasons for treatment of backings. A few are mentioned in following paragraphs. The treatment may eliminate the need for a primer layer in some tape construction.

Paper tapes may be creped to provide stretch. A saturant may be used to reduce moisture effects and to reduce corrosion. Color is introduced to aid in identification.

Cloth and mat tapes virtually demand some sort of treatment of the cloth.

In common use are various varnishes, silicone rubbers and varnishes, epoxy compounds, and fluorocarbons.

Film tapes made of polyester may be treated to provide a bondable surface for impregnating and encapsulating compounds. This treatment prevents the formation of voids in which corona may form that may lead to failure.

Composite tapes. Treatment of the components or the entire composite structure may follow the practice commonly used for the *individual* constitutent elements.

6.34.1   *Mechanical and electrical properties of PSA tapes*

The mechanical and electrical properties of PSA tapes reside in the backing or substrate. There is a wealth of backing materials from which to select those having the desired characteristics. Since the tape must function for long periods of time in an electrical device (if it is a permanent part), the selection of a list of usable backings is restricted. Such backings may be grouped into four categories or classes. These are paper, cloths or mats, plastic or rubber films, and composites or combinations. The outstanding characteristics are as follows:

1. *Paper*   Lowest in cost, excellent for use where the mechanical and electrical requirements are not demanding or the tape is a temporary component.

2. *Cloths and mats*   Greatly increased mechanical strength over papers or films is obtained. Mats may be strong if reinforced but are used for other reasons such as porosity which aid impregnation with resins.

3. *Films*   Films are thin, substantially continuous, relatively free of pinholes, and are good-to-excellent insulations. There are so many tapes in this category that they differ greatly in performance.

4. *Composites or combinations*   This class of tapes offers performance capabilities in a range that a tape made of one backing material cannot exhibit.

The adhesives may also be grouped in four categories or classes. These are natural rubber, synthetic rubber, acrylics, and silicone rubber.

The function of the adhesive is to bond the tape backing to the material on which it is placed so that no relative motion can occur. At the same time the tape must not exhibit any damaging effects when enduring the rated operating conditions of the electrical device where it is applied.

Most PSA tapes have a thermosetting adhesive. When the in-place-tape undergoes heating (for example, in drying or curing the impregnant such as varnish), the adhesive polymerizes or cross-links or sets so it will not again heat-flow nor be solvent in the environment. For example, when in transformer oil the cured tape must not be dissolved or damaged. A few nonthermosetting tapes are available, but these are primarily used for temporary holding operations during assembly or where service conditions would not affect stability of the tape.

When considering the choice of a PSA tape, the following check list of characteristics may be helpful:

Thickness	Flagging
Width	Oil resistance
Breaking strength	Temperature classification
Elongation	Accelerated aging
Dielectric breakdown	Unwind force
Adhesion strength	Moisture effects

Resistance to penetration	Processing needs
Resistance to corrosion	Fungus resistance
Bond separation	Tear resistance
Shear strength after solvent immersion	Cost

All of these seem self-evident with the possible exception of the term "flagging" which refers to the lifting of a terminating end of a tape from the surface to which it should adhere, or to the loosening of part of the turns of a spiral wrapping of tape.

Under processing it is well to consider the effects of storage at temperatures that might be encountered and the length of time to be stored. In general, storage for 1 year at 74°F and 50-percent relative humidity can be tolerated. Cool storage at 35°F will prolong storage life. Difficulty in unwinding, decreased adhesion, and changed appearance of the roll may be noted as storage problems.

The curing or polymerization of the adhesive, if thermosetting, usually requires an hour at 300°F for a complete cure. Exceptions are the silicone adhesives which may require up to 24 hours at 480°F.

Operating temperature limits and the intensity of stress during elevated temperatures should be considered. Although tapes may operate satisfactorily for limited periods at temperatures above their temperature ratings, such thermal exposure tends to embrittle the tapes and reduce their holding strength.[5] Low temperature operation below 32°F may affect the choice of a tape. In general, the low temperature problem is loss of adhesion. The more extreme the operating temperature and the greater the stress, the more essential is the need to know the ability of the tape to endure and hence the need to have good performance data to aid selection.

Pressure-sensitive adhesive tapes are the subject of specifications or standards issued by ASTM, NEMA, and Department of Defense agencies. The Underwriters' Laboratories also have guides for tape selection.

ASTM D-1000 is an omnibus specification covering a large number of PSA tapes for many uses. MIL-I-15126 is the military counterpart. Other specifications for PSA tapes include:

MIL-I-23594	For high-temperature use
MIL-I-595	For low-temperature application
ASTM D-2754	For high-temperature glass cloth
ASTM D-2484	For polyester film
ASTM D-2301	For vinyl chloride plastic

Table 6.36 shows typical characteristics of a selected group of PSA tapes. These data are intended as a guide to the wide variety of tapes available for selection to fill a given need. There are many other kinds of tapes and many other makers. It is suggested that the makers of PSA tapes be asked for data relating to their particular products. Makers can also offer suggestions and help in the use of their tapes. Some of them offer dispensers and applicators which assist greatly in production uses of PSA tapes.

Among the many kinds of PSA tapes not included in Table 6.36 nor in preceding paragraphs are the following:

Friction tape   Probably more people have used or are familiar with friction tape than any other kind. It is made of cotton cloth with a rubber or resin

Table Number 6.36 Typical Characteristics of Some Pressure-sensitive Adhesive Tapes

Source: See firms in footnote 31.

Backing, adhesive*	Mfr†	Thickness over adhesive, mils	Tensile strength, lb/in. width	Elongation at break, percent	Adhesion steel, oz/in. width	Dielectric strength (single-layer), volts to breakdown	Electrolytic corrosion,†† megohms	Thermal classification¶	Military specification
Black crepe paper (RT)	3M	10	25	12	40	1400	0.88	105°C	MIL-I-15126
Yellow flat back paper (T)	M	5.5	48	4	80	2300	0.90	105°C	MIL-I-15126
Tan flat rope paper (T)	P	17	45	8	60	2500	0.85	105°C	MIL-I-15126
Nomex flat paper,** silicone (T)	M	4.5	25	5	27	3500	1.0	180°C	—
White cotton cloth (T)	3M	7	55	5	60	3000	0.85	105°C	MIL-I-15126
Glass cloth (RT)	3M	7	150	5	50	2000	0.97	130°C	MIL-I-15126
Epoxy resin, glass cloth (T)	3M	5	100	3	50	6000	1.0	155°C	—
Glass cloth, silicone (ST)	3M	7.5	150	5	40	2500	1.0	180°C	MIL-I-19166
White acetate cloth (RT)	3M	8	50	20	50	2500	1.0	105°C	MIL-I-15126
Acetate film, white (RT)	3M	3.5	25	40	70	6000	1.0	105°C	MIL-I-15126
Acetate fiber (RT)	3M	7	55	4	60	4500	1.0	105°C	—
Polyester film (RT)	3M	2.5	25	100	45	5000	1.0	130°C	MIL-I-15126
Kapton film** (S)	P	3	25	60	18	6500	1.0	180°C	—
Vinyl film (N)	P	7	20	200	42	2500	1.0	—	MIL-I-7798
Teflon TFE** (S)	C	3	6	275	15	2800	1.0	180°C	MIL-I-23594
Acetate film, glass filaments	3M	8	275	2	45	5500	1.0	105°C	—
Polyester film, rope paper (TS)	M	5.6	57	4	70	5600	1.0	105°C	—

* Key to adhesive descriptions: R = rubber resin; A = acrylic; S = silicone; O = oil proof; T = thermosetting; N = nonthermosetting.

† Manufacturers: 3M = 3M Company; P = Permacel; M = Mystic; C = Connecticut Hard Rubber.

** Registered Trade Marks: Nomex = DuPont; Kapton = DuPont; Teflon = DuPont.

†† As per ASTM 1000 or by manufacturer's wire test.

¶ Manufacturer's data.

adhesive. There is an ASTM Standard D-69, "Friction Tape for General Use for Electrical Purposes." It is published in part No. 28 of the book of ASTM Standards. (Part No. 29 is devoted to Electrical Insulating Materials.) Friction tape should not be used in any device where its electrical behavior is important.

Conductive PSA tape   Copper or aluminum foil with a conductive pressure-sensitive adhesive are used to form a composite conductive tape. These tapes are used as effective radio-frequency shielding. Lead foils are supplied with a rubber-adhesive for use as a nuclear radiation shield. Flat copper foils or wires laid parallel are bonded to PSA tapes of polyester, acetate, paper, glass cloth, and other backing. Other metals may be applied in like manner. Conductive PSA paper tapes are also available for use as shields.

Films   It appears that any film or sheet material may be made into a PSA tape. Films not previously mentioned are so made. Some of these films are polyvinyl fluoride, polypropylene, and polyethylene. Among the many TFE tapes of the PSA type some have a bondable backing which will adhere to varnishes and the like with a thermosetting silicone adhesive.

Tapes with PSA on both sides are available. They can be used for laminating copper to a laminated board, among other uses.

6.35   Unsupported and Supported Electrical Tapes

Electrical tapes other than the previously described pressure-sensitive-adhesive (PSA) tapes can be categorized according to structure (unsupported or supported) and according to composition. Chapter 8 has descriptions and data of a selected number of rubbers used as electrical insulation (including tapes). (According to ASTM 1418, the terms *rubber* and *elastomer* are synonymous, with the former used as a generic term for both.)

Unsupported tapes are those in which the rubber is sufficiently strong so that it is alone capable of withstanding the physical and electrical forces that may act upon it.

Supported tapes, often of silicone rubber, are made by calendering or otherwise coating (for example, by dispersion) a fabric or web.

Adhesives are those substances capable of holding materials together by surface attachment. For the purposes of electrical insulation the materials must be held together strongly and without exhibiting any damaging effects. The adhesives used in making PSA tapes are most often rubber-based. For the unsupported or supported tapes of this section, rubber is present in the entire structure of most types. Thus the necessary adhesive ability is present and no separate compound for providing adhesion is required. However, some types of the tapes of this section do require that use be made of an added adhesive layer. Thus some of the tapes can be classified as PSA in the sense in which they are discussed in this chapter.[32]

6.35.1   *Applications for unsupported and supported tapes*

The unsupported and supported tapes dealt with in these paragraphs are employed to provide electrical insulation for joints, splices, or terminations of wires

---

[32]   Physics uses the term *adhesion* to describe the bonding together of unlike substances. Cohesion is used to describe the bonding together of like substances.

and cables. When the joint is properly protected, usually by use of suitable tapes, the joined wires or cables may be used indoors, outdoors, or buried.

An omnipresent, time-honored application is in insulating the joints in building wires and cables where a rubber tape is wrapped around the joint, which is then covered with a wrapping of friction tape. Untold numbers of joints have been so made and when the tapes were of good quality have given many years of trouble-free performance. The rubber tape provides the electrical insulation and the friction tape provides the resistance to abrasion and also prevents the wire from being punctured through the insulation of the joint.

A single, dual-purpose elastomeric tape has been approved by the Underwriters' Laboratories for some uses that formerly required both rubber and friction tapes. The elastomer is usually a vinyl compound. This method of insulating may be used in original equipment.

Joints, splices, and terminations in high-voltage power cables are made using the tapes discussed in these paragraphs as insulation, shields, and protection against the environment. In cables having solid insulation (such as polyethylene) these tapes may be used with operating voltages as high as 35 kV. Here resistance to corona, ozone, and thermal aging plus the essential dielectric, physical, and moisture-resistant properties must be present in the joints, splices or terminations.

Rubber and elastomer tapes may be used in rotating as well as static equipment for insulation of coils, bus bars, and similar components and structures.

The cross-sectional shape of unsupported tapes has an important bearing on

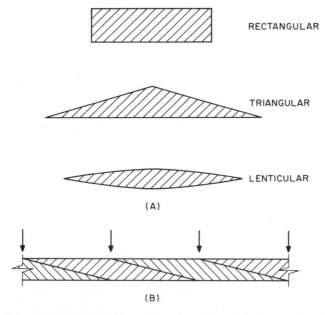

Fig. 6.2   (A) Standard cross sections of unsupported electrical insulating tapes; and (B) precise half lapping provided by a colored line placed at the apex of triangular tapes (indicated by vertical arrows)

applications. The cross section may be rectangular, triangular, or lenticular, as shown in Fig. 6.2A. A colored line is placed at the apex of the triangle that permits precise placement of the tape so that exact half-lapping can be obtained. Taping of form-wound coils with triangular tape provides void-free, smooth, and uniform insulation, as shown in Fig. 6.2B.

6.35.2 Vulcanization and other processing methods

Several processing methods are applied to tape materials. These are vulcanization, cross-linking (by chemical means or by irradiation), and heat-curing. The selection of any particular process, or combination of processes, depends on the composition of the tape, or the use for which it is intended, or on both factors. A brief description of these methods follows:

Vulcanization of rubber is done by mixing sulfur or sulfur compounds with the raw rubber. The mixture is heated and the resulting compound is characterized by substantial improvements in the properties of the original material. In essence, the process of vulcanization changes the linear polymer structure of raw rubber to a branched or cross-linked structure with marked enhancement of such properties as tensile strength, hardness, elasticity, and the like. Compounds other than those of sulfur may be, and are, employed in vulcanization.

Chemical cross-linking is, in fact, a variant of the vulcanization process, except that it usually refers to methods for improvement of organic or organo-inorganic polymers (other than rubber). Silicone rubber, for example, is subjected to cross-linking by reacting it with peroxide compounds, such as 2-4-dichlorobenzoyl peroxide or benzoyl peroxide. Cross-linking can be accomplished by heating for a few seconds to 10 minutes. Chemical cross-linking is also applied to other polymers or resins, such as the olefins.

Irradiation cross-linking is usually accomplished by exposure of a material to high-energy bombardment in an electron beam accelerator. As with chemical cross-linking, improved properties are obtained. Self-adhering silicone-rubber tape, for example, exhibits increased adherence and higher resistance to tracking after irradiation.

Curing is a heat treatment that may be applied *in addition* to vulcanization. Curing serves to drive off volatile materials and to stabilize the rubber or elastomers for high-temperature operation.

Processes such as cross-linking by irradiation and heat-curing can be applied to fabricated tapes, as well as to basic tape materials. Certain tapes are available not only fully cured, but also semicured as well as uncured. In the two latter cases, curing takes place *after* the tape has been applied in place.

6.36   Rubber Unsupported Tapes

There are two ASTM Standard Specifications covering rubber-insulating tapes, D-119 for rubber-insulating tape and D-1373 for ozone-resistant rubber-insulating tape.

The principal requirements of these tapes are given in Table 6.37. The standard nominal dimensions are 0.030 in. thick ± 0.003 in., ¾ in. wide or 2 in. ± 1/32 in.

Table 6.37   ASTM Requirements for Rubber Tapes

Characteristics	D-119 Specification	D-1373 Specification
Tensile strength, min., psi	250	250
Elongation, min., percent	300	300
Dielectric strength, min., volts/mil	350	350
Fusion	Ability of the tape to fuse together when applied with 200-percent elongation and a weight of 4 lb/in. of width	Ability of the tape to fuse together when applied with 200-percent elongation and a weight of 3 lb/in. of width
Tackiness	Ability of the tape to adhere to itself	Ability of the tape to adhere to itself
Free sulfur, max., percent	0.5	—
Ozone resistance	—	No injury after 3-hr test. (ASTM D-412 and D-470 specify ozone concentration between 0.010 and 0.015 percent by volume)

with no individual thickness more than $\pm$ 0.004 in. and no individual width more than $\pm$ 1/16 in. Roll lengths, separator and packing are specified.

The rubber insulating tape of D-119 is also called splicing tape, splicing compound, and insulating tape. It is used for insulating splices in wires and cables and other insulating purposes. When applied under proper tension it forms a moisture-resistant coating. It is standard operating procedure to apply a covering of friction tape (ASTM D-69) over the rubber tape applied to joints in building wires and cables.

For insulating joints in wires and cables having solid insulation and operating at voltages over 2000 V, with conductor temperatures up to 75° C, the "Ozone-Resistant Rubber Insulating Tape" of ASTM D-1323 may be used. The specification covers "unvulcanized or partially vulcanized rubber insulating compound . . . wound in rolls with a separator between layers."

Tape makers offer tapes that exceed the required properties of D-1373 for insulating splices of cables having solid insulation of butyl, rubber, polyethylene, and cross-linked polyethylene. A variety of non-standard widths may be obtained.

Some tapes do not require a layer separator. These tapes form an homogeneous mass which is moistureproof and resistant to ozone, ultraviolet light, and heat. Tapes in the form of "pennants" are available for easy production of stress cones without need for constant measurement as the cone is being formed.[33, 34, 35]

"Pennants" are made of the same sheet material as the desired tape. The maximum width is equal to the length of the cone covering the joint. The minimum width is the final, shortest wrap of the cone. The length of the pennant is sufficient to build a cone covering the joint to the desired thickness.

Joints and splices in cables having laminated insulation with oil impregnation and filling are not insulated with rubber tapes because of the solvent action of the oil. Tapes for use with these cables are described elsewhere in this chapter.

---

[33] Splicing Tape, 3 One 3 (Chase & Sons, Randolph, Mass.)
[34] Splicing Tape, Slipknot No. 25 (Plymouth Rubber Co., Canton, Mass.)
[35] High Voltage Splicing Tape 8380 (Insulating Materials, General Electric, 1 Campbell Road, Schenectady, N.Y.)

6.37 Vinyl Unsupported Tapes

Polyvinyl chloride (PVC) sheeting is made into useful unsupported vinyl plastic tapes.[34,35,36] These tapes are used as sole insulation for splices in electrical wiring, as an outside wrap over cable joints insulated with corona-resistant tapes, as sealing of insulated splices for direct burial in underground installations, and for cable harnesses and binding. Table 6.38 provides average properties for four representative tapes.

To provide quick adhesion when applied, certain vinyl tapes have a backing with a permanently fixed quick-sticking adhesive. In this respect, vinyl tapes may be also classified as pressure-sensitive adhesive (PSA) types. (See discussion on PSA tapes in earlier sections of this chapter.) The usual color for vinyl tapes is black, but a thin (4-mil) pressure-sensitive tape is made in six colors. (ASTM Specification D 2301-65 covers vinyl chloride plastic pressure-sensitive electrical insulating tapes.)

6.38 Silicone-Rubber Unsupported Tapes

Silicone rubber offers the advantage of marked resistance to the effects of corona. In this respect it compares favorably with mica. It has excellent thermal stability, moisture resistance, high thermal conductivity, and resiliency.[35-39] With the development of self-adhering compounds and use of triangular cross-section tape with a colored guide-line at the apex of the triangle, easy application has been secured without the need for the use of pastes or adhesives.[38] A single layer of triangular tape half-lapped results in the formation of a continuous mass whose thickness is the thickness of the tape at the apex of the triangle (as shown in Fig. 6.2B).

Self-adhering silicone rubber tapes are not tacky to the touch, but the layers will bond together at room temperature in two or three days. Heating the tapes for

[36] 3M Company, Electrical Products Div., St. Paul, Minn.
[37] The Budd Company, Polychem Div., Newark, Del.
[38] Dow Corning Corp., Midland, Mich.
[39] Moxness Products, Inc., 1914 Indiana St., Racine, Wis.

Table 6.38 Average Properties of Vinyl Tapes

Source: See footnote 36.

Type*	Thickness, mils	Tensile strength lb/in.	Elongation at break, percent	Adhesion, oz/in.	Breakdown strength, volts
General-purpose†	7	20	200	25	10,000
All-weather**	8.5	20	200	20	11,000
Heavy-duty††	10	30	250	25	12,000
	20	65	450	25	19,000
High-temperature and oil-resistant‡	8.5	20	300	16	11,000

* All tapes resist electrolytic corrosion and do not require friction-tape overlayer.
† Usually ¾ in. width
** For use in all-weather conditions; pliable at 0° F.
†† Same tape as general-purpose but thicker.
‡ For continuous use up to 105° C and oil-resistant to splash or brief immersion; leaves no residue if re-entry to a splice is desirable.

three to five hours at 150° C will produce a uniformly bonded mass that exhibits maximum performance characteristics superior to those of room-temperature wire.

Fully cured silicone rubber tapes with a silicone rubber adhesive offer an outer wrapping for out-door high-voltage terminations in contaminated areas. Cross-linking by irradiation results in a tape that does not require the use of an adhesive coating.

ASTM D-2148-66 covers "Tentative Methods of Testing Bondable Silicone Rubber Tapes Used For Electrical Insulation." It applies to self-adhering silicone rubber tapes having rectangular, triangular, and lenticular cross sections. Tape thickness at the apex of the triangle is given as 0.020, 0.030, 0.040, 0.050, 0.060, 0.070, and 0.080 in. For rectangular tapes the thicknesses are 0.010, 0.015, 0.020, 0.025, and 0.030 in. Widths of tapes which may vary from ¾ to 2 in. are generally subject to a ± $\frac{1}{16}$-in. tolerance.

The usual color of the tapes is red, although this is not mentioned in the test method. The colored guide line for triangular tapes may have different colors assigned to the different tape thicknesses.

The temperature operation range is said to be from well below 0° to 260° C. The usual temperature classification is 180° C (Class H).

Typical electrical properties for self-adhering tapes are as follows:

Dielectric strength:
    200 volts/mil for 0.080 in. thick
    600 volts/mil for 0.012 in. thick
Dissipation factor:
    0.002 at 100 Hz
    0.001 at 10^6 Hz
Dielectric constant:
    3.0 at 100 Hz
    2.9 at 10^6 Hz
Volume resistivity (ASTM D-257):
    3×10^{14} ohm-cm

The tensile strength is 500 to 1,000 psi, depending on cure; elongation is 450 to 950 percent; the durometer hardness (Shore A) is 50. It is recommended that tape makers supply the specific characteristics of their tapes.

6.39   Fluorocarbon Unsupported Tapes

The three basic tapes in the fluorocarbon family are: (1) Skived TFE (tetra-fluoroethylene), (2) extruded TFE, and (3) extruded FEP (fluorinated ethylene propylene).[40] They are similar in most characteristics except FEP, which is heat-sealable, meltable (260° to 280° C), and can be thermoformed. The continuous service temperature of TFE is − 265° to + 285° C and of FEP, − 225° C to + 200° C. (Physical, electrical, and thermal properties of the respective polymers are given in Chap. 8). Both plastics in tape form have good dielectric strength, dielectric constant of 2.0 to 2.1, with TFE showing a wider range of frequency with very

---

[40] Chemplast, Inc., 150 Dey Road, Wayne, N.J.

small change. Both have low dissipation factors: 0.0003 over a very wide range of frequency for TFE and 0.0002 to 0.0007, depending on frequency, for FEP.

Uses for these tapes are many where it is desired to utilize the basic properties of the plastics. They are used as wrap on insulation, for the base of printed circuits, and as barriers to prevent electrolytic corrosion. In rotating equipment, layer insulation between coils and phases, slot liners, and coil wrappings can be made of these tapes.

These tapes are available in rolls in the following forms (further data may be obtained from the makers):

1. *Skived TFE tapes*   Available in thicknesses of 0.002 to 0.125 in. and widths of ½ to 12 in.

2. *Skived TFE bondable tapes*   One or both sides are treated for use with adhesives, usually epoxy. The adhesive is applied by the user of the tape, which comes in thicknesses of 0.002 to 0.125 in. and widths of ½ to 12 in.

3. *Skived TFE with pressure-sensitive adhesive*   The adhesive used is a silicone. The tape is often used as "a third hand" or for masking areas during dip soldering. These tapes leave no residue when removed and are available in thicknesses of 0.003 to 0.010 in. and widths of ½ to 6 in.

4. *Skived TFE tapes in perforated form and as expanded mesh*   Tapes of these constructions are used mostly for mechanical functions. Electrical uses may be indirect, as for example in protective wrap-on jackets for multiconductor cables that contain a number of individual insulated wires.

5. *Extruded TFE tapes in fully sintered form*   Complete sintering has the advantage of developing all the desirable characteristics of the tape. It will sinter into one piece where laps of tape are placed. This self-fusing permits the tape to become continuous insulation over the area of tape operation.

6. *Extruded TFE tapes in unsintered and sintered types*   These tapes are available with or without one or two treated sides. Unsintered tapes can be fused *in place*. They are available in rolls 0.002 to 0.005 in. thick and ¼ to 1 in. wide.

7. *Extruded heat-sealable FEP tapes*   When heated at 285° to 295° C, these tapes are self-bonding. When heated at 345° to 370° C these tapes are bondable to TFE tapes. The FEP tapes can also be bonded under heat and pressure to metals or high-temperature resistant fabrics of glass fibers or asbestos.

8. *Extruded FEP tapes, surface-treated to accept low-temperature adhesives on one or both sides*   This type of FEP tape is used where the substrate cannot tolerate heating to high bonding temperatures.

Both types of the FEP tapes are available in rolls. The thicknesses of the non-treated type range from 0.0005 to 0.020 in., and of the treated type, from 0.001 to 0.020 in. Width of tape is up to 46 in.

Specifications covering fluorocarbon tapes include the following government, military, and commercial issues:

1. *TFE skived*   AMS-3651, AMS-3661, MIL-W-7139, PS3-66, AMS-2491 (bondable).

2. *TFE skived, with pressure-sensitive adhesive*   MIL-T-23594

3. *TFE extruded and sintered*   MIL-T-23594; also for unsintered extruded.

4. *FEP extruded*   L-P-00523

6.40 Miscellaneous Unsupported Tapes

The wide range of rubber and elastomeric materials that are now commercially produced makes possible a diversity of tape products for insulating purposes. In addition to the tapes previously described in some detail, a few more will be described here more briefly.

6.40.1 Irradiated ethylene-propylene tape

This cross-linked insulation in tape form (one of the Irrathene products of General Electric Company) is suitable for insulating splices of power cable (up to 15 kV and above) and control cables. Only a minimum build-up of tape layers is needed. This tape is a self-bonding type that forms an homogeneous mass from its own compressive force. It remains flexible at temperatures below 40°C but will not melt or flow at 125° C. It has a normal cable temperature rating of 90° C and an emergency rating of 130°C.

The irradiated ethylene-propylene tape is compatible with polyethylene, cross-linked polyethylene, butyl, ethylene-propylene rubber (EPR), vinyl, silicone, and other insulations. It has excellent weather resistance.

Typical properties are as follows:

Physical
Tensile strength (ASTM D 1373)	585 psi
Elongation (ASTM D 1373)	940 percent
Ozone resistance (ASTM D 1373)	No effect
Brittle point (ASTM D 473)	55° C
Water absorption (UL 510)	0.19

Electrical
Dielectric strength (ASTM D 1000)	1,165 volts/mil
Dielectric constant (ASTM D 295-58)	2.36
Power factor (ASTM D 295-58)	0.0066
Corona resitance (IPCEA S-19-81)[41]	
at 150 volts/mil	$\gg$ 1,000 hr

6.40.2 Brown natural-rubber splicing tape and friction tape

These tapes are made from smoked sheet natural rubber.[34] They exceed the requirement of ASTM D-119 for rubber insulating tape and ASTM D-69 for friction tape. The brown color stems from the use of the natural rubber. The two tapes are intended to be used together where a superior insulation of their kind is needed. The tapes are available in ¾-in. width rolls.

6.40.3 Neoprene tapes

For use with neoprene-insulated wire and cable and for extreme weather resistance, a self-curing tape of neoprene is available. It has good resistance to oils and alkalis. The dielectric constant is 3.8; power factor, 1.7 percent; dielectric strength, 450 volts/mil. Thickness is 0.030 ± 0.003 in. and width, ¾ in. The tape may also be obtained unvulcanized for extra-tough jacketing requirements such as

---

[41] Insulated Power Cable Engineers' Association

high-resistance to oils, alkalies, and sunlight. The tape is available 0.030 in. thick and 1 in. wide.

6.40.4   Rubber barrier tapes

Vulcanized rubber tape is available for use with vulcanized tape as a barrier to prevent penetration of stranded wire, or as a separator between insulating tapes and sheathing tapes in multiconductor construction. This type of tape is available in 0.030-in. thick, 1-in. wide rolls.

6.40.5   Unvulcanized rubber tapes

Two types of unvulcanized rubber tape are used on joints and splices. A gray-colored tape is used directly on the joint, whereas a black tape is used for covering. Both tapes have high dielectric strength of 500 volts/mil, a dielectric constant of 2.5, and a power factor of 1.0 percent. When heated at 175° C for 15 min, the tapes will vulcanize. These tapes meet U.S. Bureau of Mines requirements for permanent splices.

6.40.6   Double-rubber tape

A self-vulcanizing layer of white rubber fused to a layer of black vulcanized rubber has thickness of 0.045 in. It fuses into a homogeneous mass that may be used under water. It is available in ¾-in. and 2-in. wide rolls.

6.40.7   Semiconducting cable splicing tape

In terminating and splicing shielded power cable, it is necessary to extend the shield up to the apex of the cone of tape used to protect the cable insulation where it ends. The sketch of Fig. 6.3 shows the construction. Note the use of the stress relief cone for a termination and how the shield stops at its apex.

A joint may be represented as two terminations, end-to-end. Here, the shield is continued from cable shield to cable shield, that is, across the entire joint. *All* shields must be grounded.

A conducting rubber tape is available that may be used in place of the metallic mesh usually employed as cable shielding. The tape is self-bonding as it is taped in place. To guard against misapplication as ordinary tape, this tape is clearly marked to identify it as a *conducting* tape. It is available on rolls ¾ in. wide by 0.030 in. thick.

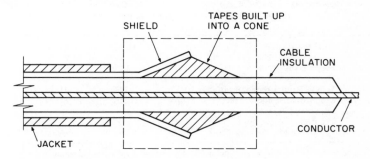

Fig. 6.3   Construction of a stress relief cone for terminations in power cables

6.41   Tape Materials in Sheet Form

Some of the materials described under the various kinds of tapes are available in the form of sheeting of the same properties as given for the tapes. Makers of tapes should be consulted for details of availability.

6.42   Supported Rubber and Elastomer Tapes

The growing availability of magnet wires capable of continuous operation at temperatures of 180° C (Class H) and above has created a demand for compatible members of a complete insulation system. This demand has been met by the commercial development of silicone rubber tapes (and sheets) supported by glass fibers or glass cloth. These materials found many uses as insulation on rotor and stator coils in rotating equipment. Other uses are as cable wrapping and insulation, as insulation of form-wound coils and bus bars, and for wrapping cable harnesses.

The silicone rubber used in these insulations may be uncured, semivulcanized, or vulcanized. If glass cloth is used as the substrate, one or both sides may be covered with the silicone rubber. If glass fibers are used to support the rubber tape, they may be laid parallel with the edge of the tape or as a sinusoidal glass yarn web embedded in the silicone rubber.[39] The parallel glass fibers do not permit stretch of the tape; the sinusoidal glass yarn web does. "Wicking" (moisture absorption by fibers or yarn) is prevented in both tape constructions since in each mode the glass fibers are not exposed at tape edges.

There are two types of the stretch tape, one with 25 percent and the other with 15 percent elongation. Both are of rectangular cross section 0.020 in. thick and ½ or 1 in. wide.[42]

NEMA has Standards Publication VF8-1961 covering "Fully Cured Silicone

---

[42]   New Jersey Wood Finishing Co., Woodbridge, N.J.

Table 6.39   Typical Properties of Silicone Rubber-Coated Glass Cloth, Sheets, and Tapes

Source: See footnote 41. Data are for fully cured material. Values for semicured and fully cured are substantially the same. Semicured will bond together and may be cured by applying a pressure of 50 psi at 135° C. A post-cure of 25° to 30° C above max. use temperature for 2 to 3 hr will insure maximum performance. Color is red.

Total thickness	10 mils	15 mils	25 mils
Fabric thickness	4 mils	7 mils	14 mils
Fabric style*	1,165	1,528	1,523
Thread count	60 × 52	42 × 32	28 × 20
Tensile strength, min, lb/in.	100	160	190
Breakdown strength, volts	6,000	8,000	10,000
Hardness-Shure A			
cured	45	45	45
post-cured	75	75	75

* The number of warp and filling yarns per unit of cloth length or width and the weight of the cloth affect mechanical performance. The dielectric strength and dissipation factor are also controlled. For example, higher thread count means higher dielectric strength.

Table 6.40   Properties of Supported Silicone Rubber Stretch-type Tape

Source: See footnote 39.

Configuration	Flat
Color	Black
Thickness, in.	0.020 ± 0.002
Width, in.	1.000 ± 0.005
Dielectric strength, volts/mil	750
Controlled stretch, percent	25
Dielectric constant, 25° C	3.0
Optimum tension, lb	6 to 10

Elastomer (Rubber) Coated Glass Fabrics" in sheet form and tape. This standard refers to ASTM D-902 for testing methods. There are also several governmental and military specifications which may be applicable, such as MIL-I-22444A and MIL-C-2194.

Table 6.39 gives typical properties of 10-, 15-, and 25-mil thick silicone rubber-coated glass cloth. Table 6.40 gives typical properties of supported silicone rubber, stretch-type tape.[39]

6.43   Supported Fluorocarbon Tapes

Glass fabric is coated with TFE (tetrafluoroethylene) to form supported tapes exhibiting increased strength and resistance to cut-through as compared with the unsupported tape. The essential characteristics of TFE are retained. This tape is available plain, bondable, or pressure-sensitive in a broad range of thicknesses and widths. Typical properties are given in Table 6.41.[39]

6.44   Fabric Tapes, Nontreated (Plain)

Nontreated (plain) fabric tapes used in most applications are woven of cotton or glass fibers. Because of their wide use this discussion will be devoted primarily to these tapes. Other fiber tapes are used in special applications. These tapes include those made of polyester, acetate, nylon, and silica fibers, and combinations such as glass-polyester fibers. Asbestos is also produced in woven tape form and is described in Chap. 11 with other inorganic materials.

Cotton tapes have been made and used since the beginnings of the electrical manufacturing industry. The plain tape is intended to be used in the form as produced. After it is put in place by hand or machine application, the device is usually impregnated with a varnish or other insulating compound.

Glass tapes are a comparatively recent addition to the family of insulating materials. Their great advantage is the increased thermal endurance. Operating temperatures may be met at levels which would destroy cotton tapes. However, cotton tapes are able to serve the requirements of 105° C (Class A) operation where they find a good market.

Both the cotton and glass tapes are used to cover an infinite variety of coils and windings in rotating and static equipment. They are also used to cover bus bars

Table 6.41   Typical Properties of Supported Fluorocarbon (TFE) Tapes

Source: See footnote 39.

Breaking strength versus thickness

Thickness, in.	Breaking strength, lb/in. of width
0.003	70
0.005	120
0.006	120
0.008	160
0.010	350
0.014	350
0.015	350

Dielectric strength versus thickness

Thickness, in.	Dielectric strength, volts/mil
0.003	800
0.005	700
0.006	650
0.008	550
0.010	500
0.014	500
0.015	500

Dissipation factor (60 Hz to 2×10^8 Hz)	< 0.0007
Volume resistivity, ohm-cm	$> 10^{15}$
Surface resistivity, ohms at 100% RH	$> 10^{13}$
Continuous-service temperature rigid applications	$-150°$ to $+290°C$
flexing applications	$-31°$ to $+290°C$
Temperature class	$180°C$ (Class H)

and any similar conductor where they are applicable. Used as ties, bindings, bands, straps, and like applications, they must be nonconductors yet are not part of the active insulation system. In some uses such as in hermetic motors and encapsulated windings, it is necessary to prevent any possible undesirable chemical interactions between the tapes and other materials. When cotton tapes are used in hermetic motors, it is therefore necessary to remove the wax that is normally contained in cotton. Glass tapes need to have the proper surface treatment to insure compatibility with encapsulants.

6.44.1   Specification for woven cotton tapes

The standard specification for woven cotton tapes is ASTM D-335. Three types are described as: Type A, a plain weave; Type B, a herringbone weave (at least single-point twill, two up and two down); and Type C, a nonelastic weave. Standard widths with permissible tolerances are given as ½, ¾, 1, 1¼, 1½, and over 1½ in. Thicknesses of 0.005, 0.007, 0.013, 0.020, and 0.030 in. are listed with the recommended construction and physical requirements.

ASTM D-259 gives "Standard Methods of Testing and Tolerances for Woven Tapes." This standard is applied to cotton and silk tapes.

6.44.2 Specifications for woven glass tape

Woven glass tapes are the subject of ASTM D-580, "Standard Methods of Testing and Tolerances for Woven-Glass Tapes." The military specifications MIL-Y-1140C and MIL-Y-17205 also have requirements and data for these tapes. Although tapes are not included, ASTM D-2518, "Specification for Woven Glass Fabrics for Electrical Insulation," provides useful complementary information.

Medium Weave A Quality tapes are those originally specified as ECC-11A and are made of 225 continuous yarns in thicknesses of 0.005, 0.007, 0.010, and 0.015 in. and widths of ½, ¾, 1, and 1½ in.

Tight Weave B Quality tapes are in greatest demand. The number of yarns used is 450, and the thicknesses are 0.003, 0.005, and 0.007 in. The thinnest tape has made possible a substantial reduction in electric motor sizes. The widths are ¼, ⅜, ½, ¾, 1, 1½, 2, and 2½ in. The first six widths are available in thicknesses of 0.003 and 0.007 in. The entire list of widths is available in 0.005 in. thickness. A ⅝-in. wide tape is also available in a 0.003 in. thickness. Both this and the preceding quality tapes meet MIL-Y-1140 requirements.

Medium Weave C Quality tapes are made of 150 yarns. Their minimum average breaking strength is greater than that of 225-yarn weaves. One-inch wide tape of 0.005 in. thickness has minimum breaking strength of 160 lb for 225-yarn tapes, 335 lb for 450-yarn tapes, and 200 lb for 150-yarn tapes. This quality is lower in cost than the A and B quality tapes but is suitable for many commercial applications, such as motor repairs. Tapes made of 225 and 450 yarns are more resistant to abrasion than those made of 150 yarns. This weave of 150 yarns is more open than the 225 and 450 yarn weaves.

6.45   Treated Fabric Tapes

Treated fabric tapes are slit to required sizes from varnished cotton fabrics and treated glass cloths previously described in Pars. 6.25 and 6.26 and their subparagraphs. The tapes generally reflect the inherent characteristics of the fabrics and cloths and the various resin or other applied treatments. Specifications and standards directly applicable to the treated tapes include the following:

NEMA Standards Publications

VF1-1957	Black straight-cut varnished cotton fabric
VF2-1957	Yellow straight-cut varnished cotton fabric
VF3-1957	Black bias-cut varnished cotton-fabric tape
VF4-1957	Yellow bias-cut varnished cotton-fabric tape
VF5-1956	Silicone-varnished glass fabric
VF6-1954	Black-varnished glass cloth
VF7-1958	Yellow-varnished glass cloth
VF8-1961	Silicone elastomer (rubber)-coated glass fabric
VF10-1964	Glass-polyester B-stage materials
MX2-1964	Varnished polyester nonwoven fabric
VF40-1963	Continuous-filament woven-glass fabric used in flexible electrical insulation
48-139	Yellow varnished nylon fabric
48-140	Yellow varnished silk fabric

Applicable ASTM specifications (titles abbreviated)

D 2148	Bondable silicone rubber tapes
D 2754	High-temperature glass-cloth PSA tapes
D 2443	Lacing twines and tapes
D D119	Rubber-insulated tapes
D 295	Varnished cotton fabrics and tapes
D 2361	Vinyl chloride PSA tapes
D 335	Woven-cotton tapes for electrical purposes
D 580	Woven-glass tapes
D 259	Tests and tolerances

Similar documents for fabrics listed in Pars. 6.25 and 6.26 are also useful sources of information.

The variety of fabrics and tapes available to the user is justified by the demands of many applications required in electrical insulation systems. The requirement for compatibility of all the members of an insulation system with each other and the environment is an important factor leading to the need for tapes of different substrates and treatments. Cost is also a constant consideration. Since operating temperatures higher than 105° C are so common, the use of glass fabric or glass-polyester fabrics has been particularly notable (see Par. 6.45.2). Special constructions and materials (such as composites, mats, and prepregs, discussed in Par. 6.46) also serve requirements not met by standard tape materials.

6.45.1   Cotton-fabric tapes

The cotton tapes are used in wrapping motor coils, bus bars, joints and terminals of wires and cables, and in general wherever high electrical strength is required. The substrate is either a seamless straight- or bias-woven fabric. The treatment is a yellow or black organic-base varnish. Low-loss, high-grade tapes may be made using an oleoresinous varnish. The bias-cut tapes have an inherent stretchability permitting close conformability to irregular shapes.

The black bias-cut varnished tapes as covered in VF 3 (NEMA) have nominal thicknesses of 7, 10, and 12 mils, with widths of ½, ¾, 1, 1¼, 1½, and 2½ in. Tensile breaking strength as shown in any test should not be less than 32 lb/in. of width. Elongation of 2 percent for 7-mil tape with a load of 5 lb/in. of width and 2.5 percent for 10- and 12-in. tape with a 6-lb load. Dielectric strength of tape "as made" is 1,100 to 1,150 volts/mil after 96/23/96 conditioning.[43] Elongation reduces the dielectric strength about 50 percent or less depending on quality of tape. Black bias-cut tape may be dry-packed or immersed in electrical insulating oil, per ASTM 1818.

Yellow bias-cut tapes (as covered by VF 4) have similar properties except that elongation and dielectric strength requirements are less, and packing in oil is not mentioned.

Mica dusting is available as is adhesive coating for all types of cotton tapes.

6.45.2   Glass-fiber cloth tapes

Similar in many respects to the cotton-fabric tapes, the glass-fiber cloth tapes differ by being inorganic in nature and by exhibiting vastly superior thermal endur-

---

[43] Conditioning for 96 hr at 23°C and 96 percent RH.

ance. Tapes with temperature classifications of 130° C, 155° C, 180° C, and 200° C are available.

The glass-fabric substrates are woven in various yarn specifications, reflected in tape thickness, with the finer yarns used to weave the thinner fabrics. The applied treatments may be black or yellow varnish for Class 130° C tape, a silicone varnish for Class 180° C, a polyester treatment for Class 155° C, an epoxy for 155° C, and a silicone resin for 200° C.

Some examples of glass-fabric tapes are as follows:

1. *Glass-fabric tape*   Epoxy-treated; Class 155° C; compatible with encapsulants and varnishes; good tensile strength; dielectric strength of 1,400 to 1,700 volts/mil; thicknesses of 3, 5, 7, 10, and 12 mils.

2. *Glass-fabric tape*   Polyester varnish; Class 155° C; good heat resistance, tensile strength, and dielectric strength; thicknesses of 3, 5, and 7 mils.

3. *Glass fabric tapes (B-stage)*   Semicured prepreg tapes with epoxy, polyester, and silicone treatments (prepreg materials were discussed in Par. 6.29). Class 130° C rating can be given to the epoxy-treated tape, 155° C to the polyester, and 180° C to the silicone. Tapes fit tightly when applied and cure to a very strong continuous covering. They include high-tensile-strength types used for banding purposes for armatures and the like.

4. *Glass-polyester fabric tapes*   A combination of glass yarn filler and polyester yarn woof available in a variety of widths, thicknesses, and treatments, including epoxy and polyester resins and black varnish. Stretch properties permit good conformability to parts requiring insulation, as well as a reduction in thickness. Temperature class depends on treatment, but for most types it appears to be 155° C. B-stage (prepreg) tapes are also made in this construction.

Despite the need for materials capable of operation at temperatures over 105° C, not all high-temperature treatments are suitable for *all* applications. The limitations of silicones, when used in rotating machines such as totally enclosed motors, have already been mentioned in Par. 6.25.1. Under rotational vibration, abrasive silicon dust may be released and cause damage to commutators.

6.46   Special Tape Constructions

Some tapes are available with friction coating or an adhesive coating. These constructions help in holding wires in place during winding operations. For example, self-adhering tapes with polyester resin coating are useful in d-c motor assembly. (In some instances, however, caution is indicated. Interlayer separators used for coated tapes packed in rolls may hinder machine taping and wrapping operations.)

Tapes made of composites may be considered as special constructions. Examples of various composite materials have been given in previous sections of this chapter. It will be recalled that a composite is made from two or more discrete materials so bonded as to form a unified structure that possesses characteristics not shown by any of the individual constituent materials. With the use of suitable adhesives or other bonding techniques just about any sheet or film material can be combined to form composite electrical insulation.

One of the most widely used composites consists of a 100-percent rag paper combined with polyester film. NEMA Standard VF 20 covers six duplex and five triplex constructions, varying in the thickness of the constituents. The duplex construction combines a layer each of paper and film; the triplex "sandwiches," a layer of film between two layers of paper. A nonstandard type reverses the sandwich construction—a layer of paper between two layers of film. The rag-paper/polyester film composites are much used for slot insulation because of their excellent formability. They can be bent and flexed without any impairment of their insulating properties. These composites accept varnish and impregnating compounds.

Among other composites materials that can be included in the category of special tape constructions are the following: A triplex combination of fully-saturated polyester mat, polyester film, and partially-saturated polyester mat; a fully-saturated polyester mat on both sides of a polyester film; acetate film with kraft or rag paper; rope paper with various films; kraft paper and films; and fish paper and films. Composites that incorporate metal foils may be used in wire and cable constructions as RF shielding.

Mats or nonwoven fabrics have been described in Par. 6.31. The randomly oriented fibers (usually polyester or glass) are impregnated with epoxy, polyamide, polyimide, or silicone resins to form a wide selection of tapes and fabrics. Because of the random orientation of the fibers, these materials are of equal strength in all directions—a characteristic that facilitates formability and application by hand or by machine wrapping.

A typical mat tape (prior to treatment with an impregnant) may be 5 mils thick, exhibit a tensile strength of 9 lb/in., and show an elongation of 12 percent minimum, at the break. The applied varnishes or other impregnants penetrate quickly and deeply into the wrapped coils to provide excellent protection for the entire unit. For certain applications high unidirectional tensile strength is desirable. In such tapes, the mat is reinforced with parallel continuous fibers. A reinforced tape of this type may be 5.5 mils thick, have a tensile strength in the direction of the reinforcement of 30 lb/in., and an elongation of 20 percent minimum at the break.

A polyester mat tape may also be treated with a bondable epoxy in the B-stage. The tape is applied "dry" and when cured provides a sealed coating for the unit being insulated. The tape remains flexible after full cure and allows non-damaging assembly of coils into slot cells. A tape of this type 8 mils thick has tensile strength of 30 lb/in. and elongates at break by 15 percent. Dielectric strength is 1000 volts/mil at 105° C and 130° C and 950 volts/mil at 150° C.

In some applications, B-stage tape can be used to advantage to produce a rigid insulation. For example, a protective armor on coils may not be subjected to flexing during installation. Here, the partly cured resin or varnish treatment on the tape is heated for 15 minutes or longer at 150° C (or at some other suitable temperature) until an infusible rigid insulation is obtained that meets the requirements.

REFERENCES

(1) Clark, Frank M., *Insulating Materials for Design and Engineering Practice*, Wiley, 1962, p. 252.
(2) *Symposium on Fibrous Materials*, J. H. Ross, ed.
(3) *Handbook of Asbestos Textiles*, Asbestos Textile Institute.
(4) Koves, G., *Materials for Structural and Mechanical Functions*, Hayden, 1970.
(5) Hsu, Raymond T., "Future of High-Temperature Electrical Pressure-Sensitive Adhesive Tape," 8th IEEE/NEMA Electrical Insulation Conference, 1969.

CHAPTER 7

TUBES, SLEEVING, RODS, FIBERS, YARNS, AND CORDS

7.1 Introduction

Included in this chapter is a broad variety of insulations characterized by a common concept of length. The materials discussed are sleeving, tubing, rods, fibers, and the like. To a considerable extent, flexibility is also a common concept. The combination of length and flexibility provides an application factor that obviously meets a great number of functional design requirements. For example, flexible sleeving insulation is slipped over conductors and the combination may be bent or shaped to fit into the desired space or to a desired configuration. As another example, cords, twine, lacing tapes, and the like are used for banding of armatures, tying down windings, holding lead wires in place, cable harnessing, and for a miscellany of similar purposes.

Heat-shrinkable flexible sleeving or tubing provides an important variant in flexible materials. The heat-shrinkable type is used to encase, cover, or encapsulate a device, such as a capacitor. Sufficient flexibility is needed to permit slipping the sleeving over the device, but close adherence to device-configuration is accomplished by heat-shrinkage of the sleeving. No bending or shaping to fit is necessary.

7.1.1 Definitions

The term "sleeving" is used to denote a flexible tubular form made by braiding of cotton, glass, or other yarns. The braided product is coated with a varnish or other material such as a silicone rubber. Tubing is used to denote a product made by extrusion of a resin or an elastomer, also by special constructions of polymeric films, such as spiral winding or welding, followed by suitable treatments.

Since the early 1920's there has been a proliferation of materials capable of conversion into sleevings or tubings. So there are many kinds of both. Perhaps there are more kinds than are necessary to meet the demands, but at least this situation offers the user a wide choice in selecting competitive materials.

Rods made of laminated materials are covered in Chapter 6 (Paragraphs 6.11 and 6.13). They are described in the NEMA standards applying to tubing. Here the rods are made by extrusion or molding to produce fluorocarbon rods, for one example. Also included are molded shapes having cross sections that fit them for some specific application.

Fibers are the basic building blocks of woven or knit fabrics. Fibers may be "staple"; these are relatively long, such as the continuous glass fibers. Yarns are made of fibers by spinning or twisting the latter into long lengths. Cords, strings, and twines are made by twisting or spinning yarns together. Fabrics are made by weaving or braiding fibers or yarns.

7.2   Coated Sleevings—Classification

NEMA Standards Publications VS1-1962 and VS1a-1967 cover the physical and electrical requirements and the testing of flexible coated sleeving used for electrical insulation. ASTM D 372 and D 350 cover much the same subjects. Applicable military specifications are MIL-I-21557, MIL-I-31908, and MIL-I-18057A. NEMA VS-1 divides coated sleevings into six types as described below:

Type 1—Organic base materials, impregnated or coated with an organic substance. Suitable for Class 105 C (Class "A")

Type 2—Inorganic base material, impregnated or coated with a thermosetting organic substance. Class 130 C (Class "B")

Type 3—Inorganic base material, impregnated or coated with a thermoplastic material. Class 130 C (Class "B")

Type 4—Inorganic base materials, impregnated or coated with silicone resin or other equal-performance resin. Class 200 C (exceeds Class 180 C, or Class "H")

Type 5—Inorganic base material, impregnated or coated with silicone elastomeric or other equal-performance elastomerics. Class 200 C (exceeds class 180 C)

Type 6—Inorganic base material, impregnated, coated, or impregnated and coated with epoxies, polyesters, or polyurethanes. Class 155 C (Class "F")

In addition, three "grades" are specified, A, B, and C. All three are applicable to types 1 through 5, whereas only two grades, A and C, apply to Type 6. Each of these grades is based on dielectric breakdown values. Grade A calls for the highest value—7,000 volts. Grade B calls for 4,000 volts, and Grade C for 2,500 volts. Full details are available in the NEMA standards cited above (V-S1 and V-S1a).

7.2.1   Coated sleevings—applications

The uses for coated sleeving are many. One of the most common is as insulation over the lead wires of motors, transformers, and many other pieces of electrical apparatus. The problems presented by these uses fall into three categories. First are the hazards of placing the sleeving into position on the wire, which often involves some rough handling. The sleeving needs to be flexible enough so that it is easy to slide over the wire. And it must not be harmed by the installation operation which must be done as quickly as possible if labor costs are to be controlled. After the device has undergone all the "dry" operations, it is usually varnish-impregnated or otherwise treated. A baking operation then cures or dries the impregnating varnish or other applied material. During this operation and afterward, the sleeving must maintain its essential characteristics and prove that it has the ability to withstand the hazards of operating conditions for a satisfactory period of use.

The various standards and specifications of NEMA, ASTM, Underwriters' Laboratories and other controlling organizations contain a wealth of data setting forth performance requirements and testing methods. The technical literature of the various manufacturers also give much valuable data, including sizes, colors, and "put-ups" available for the various standard types and grades. ("Put-ups" describe the available lengths, method of packaging, lists of stock items, and indications of items considered as "special.")

Characteristics of outstanding importance, in addition to those previously mentioned, are dielectric strength, compatibility with magnet wire insulation, and thermal endurance. NEMA Standard VS-1a relates specifically to this subject.

Many textiles can be braided or woven into the base sleeving, the most used now being glass fiber. Cotton and rayon, formerly the most used, are still specified, if in reduced quantity. Asbestos, fused quartz, silica, and ceramic fibers (alumina-silica) are the choice for higher-temperature products. A polyamide fiber similar in chemical composition to Nomex (Du Pont) is also available.

The inside diameter and the wall thickness of the sleeving are important. A table in NEMA Standard VS1-1962 shows the maximum and minimum inside diameters for the size of sleeving. This "size" is a number in a series from 24 to 0. Above size 0 the designation is in fractions of an inch from $\frac{3}{8}$ in. to 1 in. in seven steps. Some types and grades are available with an inside diameter as large as 3 in. — for example, vinyl-coated fiberglass sleeving.

Size is determined by inserting gage rods of specified diameters into the sleeving for 5 in. If the minimum-diameter rod fits snugly, the sleeving diameter is that of the rod. If the minimum size rod fits loosely, the next size rod is then inserted for 5 in., and so on.

Wall thickness is determined by inserting a gage rod that fits snugly and then using a micrometer to measure the outside diameter of the sleeving containing the rod.

Table 7.1 gives selected data that may help in specification of coated sleeving. Experimental trials are essential to solve most selection problems.

7.3   Uncoated Sleevings

In addition to the coated sleevings previously described, basic (uncoated or lightly treated) sleevings are also available. Owing to their flexibility and braided construction, they readily absorb most impregnating and treating compounds and meet many special application requirements. To the characteristics imparted by the applied coating or treatment, the basic sleeving provides an insulating capability equal to that of an equivalent air space.

Basic glass-fiber sleevings are in the greatest demand because of their inherent properties. Next in use are the cotton sleevings, formerly the most widely used. Asbestos braided sleevings are required where the cognizant Underwriters' Laboratories' Asbestos Grade is specified. As would be expected, glass-fiber and asbestos sleevings offer heat resistance superior to that of cotton.

Light treatments are applied to prevent fraying, help maintain the round tubular form, and facilitate further application of impregnants and other treatments.

Table 7.1   Selected Data for Various Coated Sleevings

Nomenclature: E — Excellent; G — Good; F — Fair; P — Poor; U — Unusable
W/T — Electrical values dependent on wall-thickness space factor;
UL — Underwriters' Laboratories.

Coating	Temperature classification, °C	Brittleness temperature,* °C	Average short-time dielectric breakdown†	Resistance to burning‡	Push back	Flexibility	Applicable Standards
Organic on cotton or rayon	105	−10	A, B, C	—	—	G	ASTM D372, NEMA VS-1 MIL-1-3190
Polyvinyl chloride	130	−50	8.0	G	G	E	Same plus MIL-1-21557
Polyvinyl chloride	130	−50	4.0	G	G	E	Same as item 1
Lecton**	130	−20	4.0	G	U	P	Same as item 1
Lecton UL**	130	−20	4.0	G	U	G	Same as item 1 plus UL for hermetic motors
Epoxy	155	−60	5.0	G	G	E	—
Polyurethane	155	−25	A, B, C	G	G	E	NEMA VS-1, MIL-1-3190
Polyacrylic copolymer	155	−50	8.0	F	U	G	Same as item 1
Same	155	−50	4.0	F	U	B	Same as item 1
Polyacrylic copolymer	155	−50	2.5	F	F	G	Same as item 1
Same	155	−50	1.5	F	F	G	Same as item 1
Acrylic polymer	155	−50	W/T	E	G	E	ASTM D-372, NEMA VS-1
Silicone varnish	200	−68	7.0 4.0 2.5	G	G	E	Same as item 1
Silicone rubber	220	−85	8.0	G	E	E	ASTM D-372, NEMA VS-1 MIL-1-18057A
Silicone rubber	220	−85	8.0	G	E	E	Same plus UL for internal wiring

Table 7.1   Selected Data for Various Coated Sleevings (Cont'd)

Coating	Temperature classification °C	Brittleness temperature,* °C	Average short-time dielectric breakdown†	Resistance to burning‡	Push back	Flexibility	Applicable Standards
Silicone rubber, double wall	220	−85	18.5	G	U	G	—
Silicone rubber, super wall	220	−85	12.5	G	E	E	Exceeds all commercial and military standards
Silicone rubber	220	−85	4.0 and 2.5	G	E	E	ASTM D-372, NEMA VS-1
Viton B**	230	−45	2.5	E	G	E	Same as item 1
Pyre M.L.**	240	−30	4.0	E	U	P	Same as item 1
None	240	−100	W/T	E	E	E	—
None (heavy wall sleeving)	240	−100	1.25	E	E	G	Heavy wall permits UL minimum insulation
Saturated only (normal basic sleeving)	240	−68	W/T	E	E	E	ASTM D-372, NEMA VS-1
Saturated only (1/32-in. wall sleeving)	240	−68	1.5	E	E	G	Same plus heavy wall permits UL minimum insulation

* ASTM D-876.
† Applicable ASTM D-350 grade (see Par. 7.2) or in kilovolts.
‡ ASTM 350.
** Proprietary tradenames for E. I. du Pont de Nemours & Co., Inc., varnish enamels and coatings. Lecton is an acrylic resin, Viton B is a fluoroelastomer, and Pyre M. L. is a polyimide resin (now known as Kapton). Other data are compiled from various commercial sources.

7.3.1   Glass-fiber uncoated sleevings

An amazing number of different flat and round sizes and constructions using 450's, 225's, and 150's yarns are made.[1] Inside diameters (or widths) range from 0.018 to 5.5 in. The smaller sizes are usually round in shape, and the large sizes are flat. Above an inside dimension of 0.053 in. both round and flat shapes are available up to sizes practicable for packing, shipping, and handling. Wall thicknesses range from 0.006 to 0.047 in. Some six different types of treatments are used. From one manufacturer alone, some 2,000 styles, sizes and treatments of basic glass-fiber sleevings are available.

Fraying is prevented by application of a light coat of an insulating compound, or by a heat treatment that permanently heat-sets the fabric. Compatibility with silicone insulation systems is obtained by means of a light coat of silicone varnish. A special treatment provides nonfraying sleeving in six colors as well as clear. For use where polyester, epoxy, and phenolic compounds must adhere, a water-repellent chrome finish is applied to heat-cleaned sleeving.[1]

7.3.2   Cotton uncoated sleevings

Untreated cotton sleevings can be made in a large range of sizes and constructions but demand has centered attention on sizes to fit No. 24 to No. 00 AWG wire sizes. The tubing may bear any one of size numbers starting with 00 up to 12.

7.3.3   Asbestos uncoated sleeving

Asbestos sleeving is braided of asbestos yarns conforming to ASTM Specification D 299; it is uncalendered and without sizing. Underwriters' Laboratories Asbestos Grade is used for $\frac{1}{64}$ and $\frac{1}{32}$ in. wall thickness and $\frac{1}{16}$ in, up to $\frac{1}{8}$ in. inside diameter. Larger sizes of $\frac{1}{16}$ in. wall thickness are made of commercial grade unless the Underwriters' Grade is specified. Fourteen sizes of inside diameters are required in sleevings to cover AWG wire sizes No. 18 to No. 0000. An additional 10 sizes supply the inside diameters up to 2 in., inclusive.[2]

A woven asbestos sleeving is also available; it is used in large diameters (up to $22\frac{1}{2}$ in. for covering bus bars.[3]

7.4   Nonrigid (Flexible) Tubings

Nonshrinkable and shrinkable types of nonrigid tubings are used as electrical insulations and to cover and protect splices, connectors, cables, and components (such as capacitors, for one example). Although many materials can be used to extrude nonrigid tubing of either type, the space available here will be primarily denoted to description of four polymers of principal importance to the user of tubings, since these materials account for the largest part of the market and they can meet most requirements. Other tubings will be described in sufficient detail to reflect their principal field of application and their performance characteristics. The most common manufacturing process for tubing is extrusion.

The four most common polymers are the vinyls, the olefins, silicone rubbers,

---

[1]   The strand numbers, such as 450, 225, and the like, represent $\frac{1}{100}$ of the yardage of raw glass fiber per pound. Thus the smaller diameter fiber is 450 as there are 45,000 yd/lb. A large-diameter is 150 or 15,000 yd/lb.

and the fluorocarbons. Each may be obtained in a wide variety of sizes, colors, and performance characteristics. The extruded tubing may be packaged on spools, in coils, or as cut lengths. Printing can be applied to the outer surface if desired. Most of the product is used as continuous wall tubing but spiral cut tubing is available as left or right spirals.

7.4.1   Specifications and standards

Unlike its action in respect to sleevings, NEMA has not set standards for tubings. However, ASTM has two standards for vinyl tubings: D 922, "Standard Specifications for Nonrigid Vinyl Chloride Polymer Tubing," and D 876, "Standard Methods of Testing Nonrigid Vinyl Chloride Polymer Tubing." Both standards are rather voluminous and, although confined to vinyl tubings, contain much information that can be used when considering tubing made of other polymers.

ASTM designation D 1675 presents "Standard Methods of Testing Electrical Grade Polytetrafluoroethylene Tubings." There are also a number of military and commercial specifications that will be mentioned where applicable. The Underwriters' Laboratories, Inc. has a component approval plan applicable to tubing that will also be referred to where necessary.

7.5   Vinyl Tubing

ASTM D 922 lists six grades of vinyl chloride polymer or its copolymers with other materials for use as electrical insulation tubing. These are as follows:

Grade A	General-purpose
Grade B	Low-temperature
Grade C	High-temperature
Grade AFR	General-purpose, fungus-resistant
Grade BFR	Low-temperature, fungus-resistant
Grade CFR	High-temperature, fungus-resistant

The following characteristics are specified by D 922: Flammability, tensile strength, effect of elevated temperatures, oil resistance, brittleness temperature, resistance to penetration at elevated temperatures, insulation resistance, dielectric breakdown at high humidity, stress relief, corrosive effect (on copper wire), and fungus susceptibility. A comparison of ASTM, UL, and MIL requirements for 105°C vinyl tubing is given in Table 7.2. Many of these requirements are also applicable to tubing of other polymers. Hence, performance data for the latter will be given only where needed for proper evaluation and application.

7.5.1   Shrinkable vinyl tubing

Shrinkable vinyl tubing may be obtained in a variety of colors, dimensions, and performance characteristics. An important element in these characteristics is the ability of the tubing to meet military specification MIL-1-631 and to obtain Underwriters' Laboratories acceptance for temperatures up to 105° C and voltages up to 600 V.

There are hundreds of applications for this type of tubing. Examples include covers and encapsulations for components of wide variety; covers and insulation for joints, connections, and other "live" parts; tab disconnects; internal and external

Table 7.2 Comparative Specification Requirements for Vinyl Tubing, Grade C, 105°C

Property	ASTM D-922, Grade C		Underwriters' Laboratories		MIL-1-631, Grade C	
	Conditioning	Requirement	Conditioning	Requirement	Conditioning	Requirement
Flammability	Room	Average burning not over 15 sec.	5-15 sec. flame applications	Not burn longer than 1 min. after	None	Max. 15 sec.
Tensile strength	Room	2,000 psi	60 days @ 230°C 60 days @ 113°C	Min. 1500 psi Min. 70% of above	C96/23/50**	Min. 1,800 psi
Effect of elevated temperature	Loss of elongation Loss of weight	Not exceed 35% Not exceed 10%	See dielectric strength	—	—	—
Oil resistance	Oil at 105°C, 4 hr	Not exceed 10% ±no oil used	—	—	—	—
Brittleness temperature	D 746	Not above −10°C	−10°C	No cracks, six turns on mandrel	50% brittle temperature, impact method	Max. −10°C
Resistance to penetration	Special tester in oven	See Table 7.2.1	—	—	—	—
Insulation resistance	1% salt water, 24 hr	Not less 10,000 megohm-cm	—	Min. 2.11 to 0.59 megohms/M ft, #24 to 2½"	None	Volume resistance 1×10^{10} ohm-cm
Dielectric breakdown	Room	See Table 7.2.2	60 days @ 230°C 60 days @ 113°C	Min. 2,500 V increase to breakdown	C96/23/0 C96/23/96	Min. 8000 V/mil Min. 85% of above
Dielectric breakdown at high temp.	96 hr/23°C, 96.5% RH	Not less than 85%; see Table 7.2.2	See above	Min. 70% of above breakdown	—	—
Stress relief (change in length)	15 min. in 150°C glycerine	Not >18% for AWG 24 to 20 Not >14% for AWG 18 to 10 Not >9% for larger AWG sizes	—	—	—	—

Table 7.2 Comparative Specification Requirements for Vinyl Tubing. Grade C, 105°C (Cont'd)

Property	ASTM D-922, Grade C		Underwriters' Laboratories		MIL-1-631, Grade C	
	Conditioning	Requirement	Conditioning	Requirement	Conditioning	Requirement
Corrosive effect	700 hr at 70°C in sealed tube	Resistance in copper wire not > 2%	60 days at 113°C	No deleterious effects on Cu	High swelling oil, 720 hr, 70°C, high humidity	Max. 2% resistance change
Fungus susceptibility	Exposure to four fungi	Not greater than one trace growth	—	—	12 mo. at 23°C and 50% RH in fungus culture	No fungus growth
Elongation*	Method A Method B	Not more than 35% Not more than 15%	60 days @ 23°C 60 days @ 113°C	100% min. Min. 70% of above	C96/23/50	Min. 200% Max. 350%
Heat shock	—	—	1 hr @ 136°C	No cracking	400 hr @ 130°C	Elongation 35% max. change
Lengthwise shrinkage	—	—	—	—	2 hr @ 130°C	10% max.
Softening temperature	—	—	—	—	None	Min. 70°C
Deformation	—	—	2 hr @ 121°C	Max. wall decrease 50%	—	—
Flexibility	—	—	60 days @ 113°C	No cracks; recovers round after creasing	200 hr @ 100°C	No cracks @ 180°C around ⅛" rod
Dielectric constant	—	—	—	—	Dry, C96/23/0 Wet, C96/23/96	Max. 7.0 Max. 8.0
Dissipation factor	—	—	—	—	Dry, C96/23/0 Wet, C96/23/96	Max. 14 Max. 16
Bursting strength Increasing Pressure	—	—	—	—	C96/23/50	Min. 29 to 96 psi
Continuous Pressure	—	—	—	—	C96/23/50	Min. 15 to 48 psi

* The average loss of ultimate elongation after exposure to elevated temperatures shall be not greater than 35% for Grade C. *Loss of weight* after exposure to elevated temperature shall be not greater than 15%. Method A heating and measurement of dimensions, Method B heating and weighing and measurement of dimensions.

** Values are for duration of conditioning in hours; temperature in deg. C, and percent relative humidity; thus 96/23/50 is for 96 hr at 23°C and 50% RH.

Table 7.2.1   Resistance to Penetration at Elevated Temperature for Grade C Vinyl Tubing

Nominal wall thickness, in.	Average temperature of failure, °C
0.016 and 0.020	≥ 70
0.025, 0.030 and 0.035	≥ 75
0.040, 0.045, 0.055 and 0.060	≥ 85

Table 7.2.2   Dielectric Breakdown Requirements for Grade C Vinyl Tubing

Wall Thickness, in.	Dielectric breakdown, kV
0.012	9.0
0.016	12.5
0.020	14.0
0.025	15.7
0.030	17.1
0.035	18.0
0.040	19.2
0.045	20.4
0.050	21.5
0.055	22.8
0.060	24.0

motor leads; bus bars; cabling (bundling) of wires. In addition, the shrinkable form of tubing can be used to shape or form a wire or cable to a desired configuration.

If rapid production rates are desirable, tubing in the form of cut pieces offers an advantage. Pieces of tubing can be placed and heat-shrunk to fit within a matter of a few seconds. (See Par. 7.10 *ff* for further discussion.)

PVC (polyvinylchloride) tubing is widely used as a heat-shrinkable type because of its excellent inherent "elastic memory," a characteristic that enables a material to recover its *original* form and dimensions when subjected to a suitable degree of heat.[2] Two production methods are followed for commercial PVC tubing:

1. The original dimensions are enlarged (stretched) by means of special formulations; this technique enables the tubing to *shrink* after application and heating, thus conforming closely to the configuration of the part over which it has been placed.

2. The tubing is also subjected to a high-energy electron-beam radiation that not only maintains the elastic memory characteristics, but also improves other properties, such as resistance to cold flow, splitting, and extreme high temperatures (melting). (As explained elsewhere, the radiation treatment develops a cross-linked structure in the PVC polymer that results in the improved physical properties.)

Selection problems for shrinkable tubing is complicated by the lack of any

---

[2] The essential point in the phenomenon of elastic memory is that a given material retains its *volume* through the entire process. If the material is distorted from its original shape in certain directions (that is, by certain dimensions), it *shrinks* in opposite directions upon being heated, but the original volume is unchanged.

general over-all standard. Available manufacturers' data reflect the characteristics of the various proprietary products. Diameters, flexibility, color, and other pertinent data do not follow a uniform nomenclature for the products of all makers. So the user is forced to marshal all his known requirements and then conduct a search for tubing that will do what he needs.

Despite the lack of overall standards, the requirements of the Underwriters' Laboratories or of the military may be decisive in stipulating performance requirements of the tubing.

Selection of the proper size of tubing is important. If special sizes can be avoided, costs can be kept under better control. It is amazing that only a few standard sizes are offered by tubing manufacturers even if the needs of the user can usually be met by one or another of them.

Some trial and error methods are handy in selecting the needed tubing. Table 7.3 has been compiled from data published by selected makers. Although limited, it offers a very broad selection of tubing characteristics.

A special vinyl thin-wall, heat-shrinkable tubing with Underwriters' Laboratory approval and meeting MIL-1-631 requirements is available with 0.010-in. to 0.025-in. walls, ⅟₁₆-in. to 6-in. diameters, and in clear and black grades.[4]

7.6 Fluorocarbon and Related Tubings

TFE fluorocarbon (tetrafluoroethylene polymer) is extruded as tubing by a batch process.[5] It can be mechanically expanded so that it has elastic memory and can be heat-shrunk. A modified form of TFE-extruded tubing is irradiated to attain elastic memory and can also be heat-shrunk. Both types provide excellent service over a temperature range from $-67°$ to $+250°$C.[6] Tubings are also extruded from *FEP fluorocarbon* (fluorinated ethylene propylene) and can be made in heat-shrinkable form.

Both polymers can be combined with other polymers as well as with each other to form composite tubings. FEP tubing can also be used as a liner for synthetic rubber tubings (such as Du Pont Neoprene or Hypalon). Such constructions are principally meant for cabling wires. The outer synthetic sheath provides mechanical strength protection; the FEP liner helps to maintain the full usefulness of temperature service range of the tubing.[5]

A combination of TFE over FEP may be used so that the TFE heat shrinks and the FEP liner melts and an encapsulating effect is produced that is solid or nearly solid and able to resist hazards of pull and vibration. It may be used as high as $450°$ F.[5]

Both TFE and FEP tubings are recognized in cognizant military specifications and by the Underwriters' Laboratories.

Trifluorochloroethylene is made into tubing with a minmum upper temperature limit of $175°$ C. It is transparent in thin-wall construction, will not burn, has high tensile strength, very low temperature of embrittlement of $-170°$ C, has fair dielectric strength, and high volume resistivity of 10^{17} ohm-cm.

Polyvinylidene fluoride, PVF (Kynar by Pennwalt), has the best cut-through performance of all the fluorocarbons although the dielectric strength is of the order of 260 volts/mil. It is an "in between" material in terms of its service temperature range, which is $-65°$ to $+150°$C.[5]

Table 7.3   Selected Data for Four Categories of Shrinkable Tubing (PVC, Polyolefin, Silicone Rubber, and TFE)—Properties and Applicable Tests

Note: An attempt to tabulate in the space available here the many varieties of commercially produced tubing would be impractical. To help the reader in his selection processes, seven typical tubings have been listed, the data being taken from representative manufacturers' catalogs. (See list of "Sources" below.) These data should provide the reader with guidelines to general identification of various tubings both generically (type of material, processing, and the like) as well as by commercial designation.

Tubing*	Tensile strength	Ultimate elongation	Specific gravity	Brittle temperature	Dielectric strength	Volume resistivity
Flexible PVC Irradiated (1)	2,600 psi ASTM D876	250% ASTM D876	1.23 ASTM D792	−25°C ASTM D746(B)	750 V/mil ASTM D876	10^{12} ohm-cm ASTM D257
Semirigid PVC irradiated (2)	3,000 psi ASTM D876	150% ASTM D876	1.45 ASTM D792	−10°C no fail ASTM D-746(B)	800V/mil ASTM D876	10^{11} ohm-cm ASTM D257
Flexible clear vinyl (3)	2000 psi MIL-1-23053	200% MIL-1-23053	1.40 MIL-1-23053	−10°C no fail MIL-1-631	700 V/mil 0.025" wall MIL-1-23053	— —
Ultra-thin 0.006" wall PVC (4)	2,900 psi ASTM D-149	120% min. ASTM D876	1.35 ASTM D792	−20°C MIL-1-631	1900 V/mil ASTM D149	5×10^{15} ohm-cm ASTM D257
Semirigid polyolefin (5)	2,600 psi ASTM D638	400% ASTM D638	1.25 Type 1 ASTM D792	No cracks −55°C	1300 V/mil ASTM D876	10^{16} ohm-cm ASTM D257
Silicone rubber (6)	600 psi ASTM D412	200% min. ASTM D412	1.35 max. ASTM D792	No cracks −75°C	400 V/mil ASTM D876	10^{11} ohm-cm ASTM D257
TFE general purpose (7)	5000 psi ASTM D638	400% ASTM D638	2.2 ASTM D792	No cracks 4 hr at −67°C	1500 V/mil ASTM D876	5×10^{18} ohm-cm ASTM D257

Table 7.3 Selected Data for Four Categories of Shrinkable Tubing (Cont'd)

Tubing*	Flammability	Corrosion resistance	Fungus resistance	Water absorption	Oil resistance	Standard color
Flexible PVC irradiated (1)	Self-exting. 5 sec. ASTM D876	Within 1.2% increase MIL-1-631	Inert MIL-1-631	0.6% ASTM D570(A)	Passes 24 hr @ 25°C	Black
Semirigid PVC irradiated (2)	Self-exting. 15 sec. ASTM D876	Within 2% MIL-1-631	Inert MIL-1-631	1.0% ASTM D570(A)	Passes ASTM D876	Black
Flexible clear vinyl (3)	Self-exting. 45 sec. MIL-1-631	Passes MIL-1-20353	Inert MIL-1-23053	—	Passes MIL-1-23053	Clear
Ultra-thin 0.006" wall PVC (4)	Self-exting. ASTM D876	None MIL-1-631	Inert MIL-1-631	0.5% 24 hr ASTM D570	No penetration ASTM D570	Clear
Semirigid polyolefin (5)	Self-exting. JC 98	Excellent Bell Labs UL 224	Inert MIL-1-7444	<0.1% ASTM(A) D-570	—	Color coded for size
Silicone rubber (6)	Self-exting. AMS 3636	None AMS 3636	Inert MIL-1-7444	1.0% max. ASTM D570	Passes MIL-1 7808	Black
TFE general purpose (7)	Non-burn	Noncorrosive	Inert MIL-1-7444	<0.1% ASTM D570	No change ASTM D543	Milk white

*Sources: (1) Flexible irradiated polyvinylchloride tubing, "Thermofit," Raychem Corp., Redwood City, Calif.; (2) Semirigid irradiated polyvinylchloride tubing, Raychem Corp.; (3) Heat shrinkable flexible extruded vinyl tubing, Clear HT-105, L. Frank Markel & Sons, Norristown, Pa.; (4) Ultra-thin wall, heat-shrinkable PVC tubing, Gilbreth Co., Philadelphia, Pa.; (5) Semirigid polyolefin Tubing, Type I, Raychem Corp.; (6) Silicone rubber, flame-resistant, heat-shrinkable, Raychem Corp.; (7) TFE general purpose, flexible, high-temperature fluorocarbon tubing, modified poly-tetrafluoroethylene, Raychem Corp.

A heat-shrinkable semirigid tubing is made by subjecting a specially formulated PVF to massive doses of electron-beam radiation. This tubing is a clear, strong, tough product that will not cold flow, does not burn, and never melts nor splits.[6]

A composite dual-wall tubing is composed of an outer wall of polyvinylidene and an inner wall of polyolefin in equal thicknesses. This tubing (Kypol by Rachem) is made heat-shrinkable by irradiation. It combines the superior dielectric properties of the polyolefin with the excellent physical and thermal properties of the polyvinylidene.[6]

7.7   Polyolefin Tubing

Polyolefin tubing is made in several kinds, which differ in flexibility. A rigid dual-wall tubing is made of a cross-linked outer wall and a non-cross-linked inner wall. When heated, the outer cross-linked wall shrinks without melting while the inner wall melts. Tubing is color-coded. Operating temperatures range from $-55°$ to $+110°$C. The tubing meets the Bell Laboratories' critical corrosion-resistance standards. Diameters recovered after heating are 0.023 in. to 0.400 in. in eight available sizes. As supplied, the tubing is rigid.

Semirigid polyolefin tubing is made by subjecting high-density material to massive doses of electron beam radiation, thus forming a cross-linked, three-dimensional gel network. The irradiated polymer is formed into tough, semirigid tubing that will not melt. The tubing will shrink by 50 percent above 135°C. Its operating temperature range is from $-55°$ to $+135°$C. Type 1 is color-coded by size; type 2 is clear. Eight sizes are available in type 1 and ten in type 2. Range of diameters after shrinkage is 0.023 in. through 0.250 in. for type 1, and 0.023 in. through 0.500 in. for type 2.

Softer and more flexible types are also available that meet Underwriters' Laboratories, Inc., requirements for 1 type, UL file No. E 35586. After shrinking, available size range is 0.023 in. through 2.00 in.

7.8   Silicone Rubber Tubing

Silicone rubber is made shrinkable in the form of soft, flexible tubing. It retains flexibility and does not become stiff and hard after the shrinking that follows heating from 150° to 370°C. The silicone rubber is a non-thermoplastic, chemically cross-linked material and is supplied in a stabilized expanded condition with a shrinkage ratio of 1.6 to 1.0. It shrinks in the radial dimension while maintaining 95 percent of its longitudinal dimension. Service temperature range is from $-55°$ to $+260°$C. Dielectric strength ($\frac{1}{16}$-in. thick sample) is 400 volts/mil. Dielectric constant is 2.90 and volume resistivity, 10^{15} ohm-cm. Oil resistance is good and ozone resistance excellent. This tubing is also noncorrosive and is self-extinguishing. It is available in ten sizes (recovered dimensions) of $\frac{1}{16}$ in. through $1\frac{1}{2}$ in. diameter. The wall thickness ranges from 0.020 in. through 0.110 in. The tubing is specified by its recovered internal diameter.

7.9   Miscellaneous Tubings

In addition to the four categories of tubing described in the preceding chapters and the related tabular data, several other categories are available with special properties and constructions. Sufficient data and description are given here so the user may receive general help in the selection problem. Full details may be secured from the fabricators of the tubings. Although not directly applicable, the data given in preceding sections may also be utilized as a general guide to what may be expected of the tubings described below.

7.9.1   Neoprene synthetic rubber tubing

A highly flexible, abrasion-resistant electrical insulation tubing (Raychem's Thermofit NT) is made from a specially formulated neoprene. Rendered heat-shrinkable by irradiation, it is primarily used as a jacket for cables and wiring harnesses. It is furnished in 13 sizes from diameters of 0.143 in. to 4 in., in black only. The operating temperature range is $-55\,^\circ$C to $+90\,^\circ$C. Tubing does not split and has longitudinal shrinkage of only 10 percent.

7.9.2   Butyl tubing

Heat-shrinkable butyl rubber tubing is fabricated in black color and diameter sizes of ¼ to 4 in. It is used where resistance is needed to propellents, hydrazine zinc, nitrogen tetroxide, and nitric acid. (MIL specifications P-27402, 26539 and 7254 are applicable.)[6]

7.9.3   Polyester film tubing

Polyethylene terephthalate polyester film is formed into a heat-shrinkable tubing having a very thin wall. The fabrication process involves the thermal welding of the film without the use of an adhesive. The fabricated tubing has a temperature use range of $-60\,^\circ$C to $130\,^\circ$C. Wall thickness is only 0.002 in. before shrinking and 0.0023 in. to 0.0025 in. after shrinking. The tubing is used as protective covering for dry-reed relays, coils, and capacitors. It is available in 17 standard sizes from ⅛-in. to 6-in. diameters. After shrinking, reduction in diameter is 40 to 50 percent; in length, 10 to 15 percent. This tubing provides good resistance to heat, solvents, moisture, and physical stress under operational hazards. Its color is transparent.[4]

Polyester tubing is made of spirally-wound polyester film with a polyester adhesive both in nonshrinkable and shrinkable forms. There are three types: one with 50 percent shrinkage in diameter and length for service to 210$\,^\circ$F and two with 30 percent shrinkage in diameter and length, one of these being for short-term service up to 300$\,^\circ$F and the other for maximum service life at 300$\,^\circ$F. Colored tubing is available. Wall thickness is from 0.0015 to 0.006 in. Diameters are from 0.015 to 2.0 in.[7]

A polyester-fiber nonwoven mat is combined with a polyester film to form a nonshrinkable composite tubing. The mat is on the exterior and the film is inside. The mat surface readily absorbs compounds such as varnish, resins, and potting materials. This composite tubing has a wall thickness of 0.00055 in. and is available in AWG sizes No. 14 to No. 0 and ⅜-in., ⁷⁄₁₆-in., and ½-in. diameters. The tubing can be folded or bent sharply without change in dielectric strength. It is white in color. Coloring comes from a vinyl coating or a varnish base. Operating temperature range is $-55\,^\circ$C to $130\,^\circ$C.[4]

7.9.4   Polyimide tubing

Polyimide tubing (Du Pont's Kapton) is the basis of two types of thin-wall construction, one using a special adhesive and the other with the polyimide tubing coated with FEP. The tubing is not heat-shrinkable. It is capable of use for a long time at 200 ° C and a short time at 425 ° C and is available in 0.015- to 1-in. diameter with walls 0.003 to 0.008 in. It has high chemical and radiation resistance.[7]

7.9.5   Other "special" tubings

Sheets of the material desired are provided with a "zipper closure." This can be wrapped around a bundle of wires to form a cable. The zipper can be closed to complete the tube. Vinyl, polyethylene (plain or shrinkable), impregnated glass cloth with silicone are the polymers so used. Polysulfone, vinylidene fluoride, nylon, polycarbonate, polyurethane, crepe paper of plain or thermally upgraded papers are some of the other materials for special purposes of this sort.

7.10   Caps

Caps are usually short pieces of heat-shrinkable tubing, cut to the desired length, with one closed end and one open end. The caps can be applied quickly and are used to cover or encapsulate the ends of wires and crimped electrical connections and other similar applications. The heat-shrunk caps present a neat appearance, provide excellent insulation, and have the ability to withstand the effects of any handling incurred in additional fabricating operations. Many types of caps are made. Some will be described below.

7.10.1   Dual-wall polyolefin caps

In this cap, the inner wall is a non-cross-linked meltable polyolefin. The outer wall is a cross-linked polyolefin that may be flexible or semirigid as desired. When the cap is applied and heated, the outer wall shrinks while the inner wall melts. All the voids are filled by the melted polyolefin flowing under the pressure exerted by the shrinkage of the outer wall.

Five sizes are offered as standard by one maker. Diameters as supplied are $\frac{1}{8}$, $\frac{3}{16}$, $\frac{1}{4}$, $\frac{3}{8}$, and $\frac{1}{2}$ in. Lengths are 0.87, 1.00, 1.12, 1.25, and 1.50 in., respectively. The color is black.

Underwriters' Laboratories approve the use of these caps for continuous 600-volt service.[8]

7.10.2   Single-wall polyolefin caps

Made heat-shrinkable by irradiation, these polyolefin caps will not melt nor cold flow. They are available in five sizes that are color-coded to facilitate selection. The characteristics are as given for irradiated polyolefin tubing in previous paragraphs.[8]

7.10.3   Closed-end polyester caps

Spirally wound polyester film is used to make closed-end tubing and caps. A wide range of sizes from 0.1 in. to 2.0 in. is provided in a selection of colors. In addition to heat-shrinkable types, a nonshrinkable type is also made.

7.10.4   Dual-wall closed-end composite caps

A heat-shrinkable closed-end construction comprised of a polyester-film outer wall and a meltable polyethylene film inner wall is available in the same sizes and colors as the item in Par. 7.10.3.

7.11   Yarns, Twines, and Cords

Yarns are twisted assemblages of fibers. Certain twisted assemblages are also known as "twines." Two or more yarns may be twisted to form larger and stronger yarns, called cords. As the process of twisting yarns increases, the products become ropes or hawsers.

The electrical insulation of a device may contain parts that are held in place by being tied down. For example, lead wires may be tied to the coil; end turns in rotating equipment may be held in place by tying; armatures may be banded over the commutator leads to hold them in place; wire and cable harnesses may be made by lacing a laid assemblage of insulated wires so that the assemblage can hold its shape and the relative position of the wires comprising it. It is assumed that the ties will remain permanently in place. This means that an engineering selection is required to determine a suitable yarn, twine, or cord. If the ties are removed after serving as aids during assembly, the selection is of no electrical importance; it has such importance only if the ties will be in place during tests under voltage.

The characteristics of a yarn or twine depend upon the fiber used and the manner in which the twisting is done. Glass yarns or cords are coated so as to withstand the hazards of application and knotting.

The fibers used in cordage for electrical insulation are linen, cotton, glass, nylon, polyester, asbestos, and polyamide (high-temperature). The polymer used will impart its properties to the cordage. All of the fibers listed possess good electrical properties.

A few of the yarns often used and usually found in stock are briefly described in the following paragraphs.

Waxed linen twine   Made of 100-percent linen fiber treated with microcrystalline wax and a fungicide. Class 105°C. Diameters: 0.045 in., 0.056 in. 0.064 in.

Cotton twine   Five common sizes with diameters of $\frac{1}{32}$ in., $\frac{3}{64}$ in., $\frac{1}{16}$ in., $\frac{5}{64}$ in., and $\frac{3}{32}$ in. Class 105°C. Breaking strength for $\frac{1}{16}$ in. diameter about 50 lb, depending on moisture and ambient conditions. Twine intended for use in hermetic motors must be purified and its waxes removed.

Glass tying cord   For uses above 105° C. Cord should be suitably treated before use. Neoprene coating imparts excellent knotting properties. Table 7.4 gives dimensions and various strength data.

Table 7.4   Continuous-Filament Glass Cords (Neoprene-Treated)

Nominal diameter, in.	Average yd/lb	Minimum average breaking strength, lb		
		Straight	Knot	Loop
0.011	3,240	11	3	1.4
0.033	638	51	17	7.2
0.040	387	77	32	11
0.062	193	146	56	23
0.085	98	250	81	44
0.094	84	288	93	53
0.110	61	355	121	71
0.128	42	461	175	106
0.165	27	605	248	160

Table 7.5.   Silicone-Treated Glass Tying Cord

Approx. diameter, in.	Breaking strength, lb.
0.011	11
0.016	
0.020	
0.024	
$\frac{1}{32}$	51
$\frac{3}{64}$	
$\frac{1}{16}$	
$\frac{5}{64}$	146
$\frac{3}{32}$	
$\frac{1}{8}$	461

Silicone-treated glass tying cord can be used up to 208°C and is available in sizes approximating those in Table 7.5. A vinyl-coated glass cord is available in various colors and a broad range of sizes. It conforms to MIL-I-3158 as "electro-vinyl cord." A red-pigmented polyester-impregnated glass cord is used for tying stator coils. It is self-bonding during cure. As supplied, it is soft and flexible.

REFERENCES

(1) Technical data, Suflex Corp., Woodside, N.Y.
(2) Technical data, Atlas Asbestos Co., North Wales, Pa.
(3) *Handbook of Asbestos Textiles,* Asbestos Textile Institute.
(4) Technical data, 3M Company, Dielectric Materials and Systems, St. Paul, Minn.
(5) Technical data, Pennsylvania Fluorocarbon Co., Inc., Clifton Heights, Pa.
(6) Technical data, Raychem Corp., Menlo Park, Calif.
(7) Technical data, Niemand Bros., Inc., Elmhurst, N.Y.
(8) Technical data, L. Frank Markel & Sons, Norristown, Pa.

CHAPTER 8 | BULK INSULATING MATERIALS

8.1 Introduction

This chapter will deal with insulating materials that are not only produced but applied in bulk form. In broad terms, this definition will encompass molding and extrusion compounds, encapsulating compounds, rubbers, and elastomers.[1]

The number, kind, and variety of such materials are extensive. It is not the purpose here to present an encyclopedic survey of these insulations, but rather a more sharply focused approach that seeks to provide (1) an insight into their molecular structure and chemical composition as a prerequisite to a practical understanding of their working properties, (2) summaries of salient properties, and (3) indications of optimum application areas. The materials discussed are of organic type (with one or two exceptions that may be classified as organo-inorganic, such as the silicones). Inorganic materials, such as mica, are treated in Chap. 11.

The information supplied here may be used as a guideline to materials selection but essentially in a first-stage screening approach, either a "negative" or a "positive" one. In a negative approach those materials are eliminated that are not suitable for the purpose at hand because of one or more unwanted characteristics. In the positive approach, selection may be made from the materials that have escaped prior elimination and do possess the desired characteristics.

It should be noted that most of the materials described here are good insulators. Their electrical properties may, in fact, not differ sufficiently to make these serve as the basic factor of selection. Governing parameters are likely to be other than the electrical—for example, physical, thermal, or chemical. Other likely selection factors would include costs and "workability" (processing or fabrication characteristics).

---

[1] Many of the electrical properties given in this chapter have been obtained from the *Modern Plastics Encyclopedia*, vol. 47, No. 10A, 1970-1971, and are used here by permission of the publishers, McGraw-Hill, Inc., New York.

The discussion in this chapter follows an alphabetical arrangement of the materials involved (except that polyamide-imide and polyimide follow in sequence). Any attempt to arrange the discussion by various categories of chemical origin or application would be confusing since almost any given resin, elastomer, or compound finds use in more than one form. The amount of information for any given material has been governed by three criteria: (1) the extent of general-purpose electrical insulation use, (2) the importance of special-purpose electrical insulation properties, and (3) the presence of noninsulating properties useful in combination with electrical insulating functions.

Table 8.1 lists some typical functional requirements for bulk insulations that occur in application problems.

8.2   Molding, Extrusion, and Encapsulating Compounds

This section is, in fact, an "umbrella" grouping for all of the bulk insulations other than the rubbers, elastomers, and inorganics. Generally, the applications for molding compounds are for solid insulating parts. For extrusion compounds the applications are in the areas of wire and cable extruded insulation, and for sleeving, tubing, and the like. The term "encapsulating compounds" in itself is not precise and is used to cover functions other than encapsulation, such as embedding, potting, casting, and impregnation (for definitions of these terms see Par. 8.2.9).

Table 8.1   Check-list of Typical Requirements for Bulk Electrical Insulation*

Source: Compiled by Alex. E. Javitz

1. Combination of good electrical and physical properties for standard molded parts.
2. Outstanding mechanical and/or structural properties, in addition to acceptable electrical properties for special-function molded parts.
3. High-temperature requirements ($> 155°C$) in molded parts, wire and cable extrusions, encapsulating compounds, and other uses.
4. Extreme high-temperature requirements ($>200°C$).
5. Low-temperature capability, including ability to function in ambient cryogenic temperatures and in direct contact with cryogenic fluids.
6. Resistance to thermal shock over a wide range of temperatures, in addition to acceptable electrical properties.
7. Special dielectric properties, such as low loss, high dielectric constant, stability of properties over a wide range of frequencies.
8. Radiation resistance, combined with other requirements, both electrical and physical.
9. Resistance to chemical effects caused by solvents, acids, alkalis, and gaseous fumes, as well as to effects caused by exposure to water, steam, hot oils, refrigerants, and the like.
10. Fire-retardant properties; self-extinguishing properties.
11. Friction-, abrasion-, and cut-through resistance.
12. Self-lubricating and wear-resistance characteristics. (See requirements under item 11 above, which may be related.)
13. Resistance to weathering, ultraviolet rays, atmospheric pollutants, and miscellaneous corrosive effects. (See item 9 above, which may be related.)
14. Low-weight and low-volume requirements.
15. Suitable workability (satisfactory molding, fabricating, and like characteristics; adaptability to available manufacturing equipment or to standard industry equipment).
16. Flexibility and resiliency.
17. Special conditions of extraterrestrial service, aerospace or hydrospace (underwater).

* Where electrical properties are not mentioned, it is assumed that properties acceptable for the application of concern are also required.

8.2.1   ABS   (Acrylonitrile-butadiene-styrene)

The three constituents of this terpolymer can be combined to provide a diversity of properties. The terpolymers can be combined, in turn, with other resins (for example, PVC, polycarbonates, and polyurethanes) to form various composites, blends, and the like. The term ABS can be therefore considered as applicable to a family of versatile thermoplastics that offer a choice of outstanding performance abilities in particular areas.

Of the three monomers that form ABS compounds, acrylonitrile has the structure $CH_2 = CHCN$; butadiene, $CH_2 = CH - CH = CH_2$; and styrene, $C_6H_6 - CH = CH_2$ (phenylethylene). The acrylonitrile provides chemical resistance; butadiene, impact and toughness; and styrene, rigidity. The presence of styrene also helps in processing.

There are so many forms of ABS compounds that even extended data tables would only indicate the range of attainable properties. In summary, however, electrical properties may be considered as good. Temperature range is from $-40°F$ to $+225°F$, although most grades have a distortion temperature of $200°F$. Low water absorption and generally good chemical properties are noted.

Several special-purpose grades are available, such as the self-extinguishing, expandable, cold-forming, high-impact, high-heat-resistant, and plating (for metal coating). Molded parts may be cemented, painted, metallized, electroplated, and machined. Applications have been made for telephone housings, and the like. (Applicable specifications: ASTM D 1788-Rigid ABS Plastics.)

8.2.2   Acetals

There are two acetal resins: One is the homopolymer made by polymerization of formaldehyde; the other is the copolymer based on trioxane. The two are similar in appearance and exhibit similar properites. Both have acceptable electrical properties and both have some outstanding characteristics in mechanical, thermal, or chemical areas. The acetals are comparable to metals, such as aluminum or brass, in being strong, rigid, and able to withstand water immersion indefinitely. The copolymer can be used in hot water up to $180°F$. Although acids and alkalis can attack both the homopolymer and the copolymer forms, neither is attacked by organic solvents.

Acetals are classified as thermoplastics and can function in a temperature range from $-40°$ to $+220°F$. Glass-reinforced grades will function up to $240°F$. Mechanical strength is very high. Recovery from deflection is quick and complete; springs or snap-fitted parts are suitable applications. A low coefficient of friction is another useful characteristic. A composite of the homopolymer with TEF fluorocarbon fibers provides excellent self-lubricating characteristics, low sliding friction, and high load-bearing ability. In particular, where an electrical design requires a combination of good electrical properties and structural or mechanical properties typical of metals, the acetals may repay close examination.

The acetal linkage in these polymers is shown by the scheme,

$$>C<\genfrac{}{}{0pt}{}{OR}{OR}$$

in which R represents an organic radical, such as hydrogen. The acetals are dehydration products of formaldehyde, or

The presence of oxygen atoms helps to form inert polymers since these atoms are structurally positioned where they cannot react with other atoms. The acetal linkage is shown in diethyl ether, $CH_3CH(OC_2H_5)_2$.

ASTM D 2133 Specification for Acetal Resin Molding and Extrusion Materials is applicable.

8.2.3   Alkyds

The alkyds form a versatile family of thermosetting resins which are used, when formulated as molding compounds, in electrical, electronic, and automotive electrical devices. The alkyds offer distinctive advantages through their inherent properties and composition and their excellent molding characteristics.

Chemically, the alkyds are polyesters formed by the reaction of dibasic acids with polyfunctional alcohols. The resins so produced are the basis of the molding compounds when combined with other resins, a wide variety of fillers, fire-retardant additives, and stabilizers. A very wide range of properties is made possible through widely differing combinations of ingredients.

The basic resin reaction may be carried out with acid anhydrides such as phthalic anhydride. Maleic acid may also be used, as well as fumaric, sebacic, or adipic acids. The polyfunctional alcohols may be ethylene glycol, glycerine, or pentaerythritol, among others. Fillers include clay, asbestos, glass fibers, cellulose, nylon fibers, or various combinations of these materials.

To secure fast curing at a wide range of molding temperatures and pressures, various catalysts, lubricants, and the like are added. There is no liberation of volatiles during cure and the post-molding shrinkage is small. Since the alkyds exhibit soft flow characteristics, fast cures are obtained. The nature of the basic resin and the formulation of the molding compound would determine both the molding requirements and the ultimate performance properties of the finished piece. Molding compounds fall into four general forms:

Putty—An extruded mineral-filled ribbon is used to enclose capacitors, resistors, coils, and transformers. Molding temperatures are low: 270° to 330°F. Pressures are 400 to 800 psi. Compression- or plunger-type molding can be used.

Granular—A granular compound, mineral-filled, is used for connectors, sockets, automotive ignition parts, and the like. Mold temperatures (compression or plunger molding) are 270° to 330°F and pressures, 1,000 to 1,500 psi.

Rope—The combination, for example, of low-viscosity resin and medium-length glass fibers results in a molding compound extruded into rope from which pieces are cut for compression and hot- or cold-plunger molding. These are used for good arc resistance, molded-in inserts, and parts with reproducible change in dielectric strength and dissipation factor with temperature. Mold temperatures run from 290° to 320°F, pressures from 1,000 to 7,000 psi.

Bulk—Resin with long glass-fiber filler is used for highest strength and shock resistance. The compound is formed by hand into pellets, balls, or the like, for molding by compression, pot transfer, or hot-plunger methods. Typical applications include switchgear parts and portable tool housings. Mold temperatures run from 270° to 330° F and pressures from 1,500 to 10,000 psi.

Mechanical, thermal, and chemical properties are good and reflect the formulation of the molding compound. Arc resistance and maintenance of properties at high humidity and high temperatures are marked. Tensile strength runs from 3,000 to 7,000 psi; compressive strength, from 11,000 to 23,000 psi; and impact Izod, from 0.30 to 10.0 ft-lb/inch of notch. Heat-resistance distortion point is above 400° F.

Electrical properties are acceptable. Typical values for a putty and rope form compound are (putty data precedes):

Arc resistance ASTM, D495: 180 sec and 180 sec
Dielectric constant, at 10^6 Hz: 4.6 and 5.0
Dissipation factor, at 10^6 Hz: 0.019 and 0.020
Dielectric strength step by step, volts/mil: 325 and 300 to 350
Volume resistivity, dry, at room temperature (73°F), ohm-cm: 6×10^{12} to 1.5×10^{15} (approx.)

Applicable ASTM specifications for various properties include D1398, 1615, 2195, 2455, 2456, 2689, and 2690. Military specification MIL-14-5 also applies.

Alkyd resins modified with oils such as tung and with various other resins are used widely as coatings. Several types of insulating varnishes based on alkyds are used, primarily the glycerol and phthalate types.

8.2.4   Acrylics

Acrylic resins are based on the polymerizing ability of acrylic acid and its monomer derivatives. The resin most used is methyl methacrylate,

$$(CH_2 \!\!=\!\! C \!-\! \overset{\displaystyle \overset{O}{\|}}{C} \!-\! OCH_3)$$
$$|$$
$$CH_3$$

The acrylic acid has the formula, $CH_2 = CH - COOH$. It is an acid form recognized as one permitting easy polymerization.

The resins for molding powders are unmodified or modified. The modifiers are added to increase mechanical properties such as strength and toughness. Cast sheet is available in a variety of forms and dimensions. In fact, cast sheet accounts for the largest volume use, followed by moldings and coatings.

The unmodified resin, when made and fabricated with care, transmits 92 percent of incident light. The resin resists irradiation from mercury vapor, fluorescent, and incandescent lamps. It withstands outdoor exposure very well (which is a major reason for its use in airfield lighting equipment).

The maximum recommended continuous service temperature is dependent on formulation. A typical low-cost material exhibits a service range from 175° to 200° F and a premium material from 200° to 230° F. The burning rate (ASTM D

635) is 0.7 in./min. A grade having Underwriters' Laboratories SE-1 rating for fire-retardant or self-extinguishing plastics is available. The smoke density is very low. Outstanding is the arc resistance (there is no tracking as determined by ASTM D 495). An acrylic-PVC combination, in addition to being self-extinguishing, offers increased physical properties.

Electrical properties are good but not outstanding except for arc resistance. The resin is polar. The dielectric constant is 3.5 and the dissipation factor, 0.02 to 0.08. The dielectric strength and insulation resistance are acceptable (dielectric strength is 500 volts/mil and volume resistivity is 10^{18} ohm-cm).

It appears that when it is desired to take advantage of the outstanding optical or electrical-thermal properties of the acrylics, the other properties can be accepted for carefully designed applications.

8.2.5   Amino resins

Amino resins of principal electrical concern are urea-formaldehyde, melamine formaldehyde, and aniline formaldehyde. Of these, the ureas are most frequently used in electrical applications, followed by the melamine and then the anilines.

Amino indicates the presence of one or more NH_2 group in a molecule. When one of these groups is reacted with formaldehyde in the presence of an acid or alkali catalyst, the end result is a thermosetting resin usually in the form of a molded article such as a switch part. The intermediate or B stages of the reactions are thermoplastic and soluble and are used in making adhesives, coatings, and binders for laminates. The family of thermosetting amino resins are known as "aminoplasts." They are very much like the phenol-formaldehyde resins known as "phenoplasts." Since the two families are similar, choice between them depends upon which resin has the property vital to the performance of the part made of it. The choice may also depend on relative costs.

The chemical structure of urea,

$$CO \Big\langle {}^{NH_2}_{NH_2}$$

has two NH_2 groups per molecule. The mol proportions of reactants in the production of the basic resin are chosen to provide predetermined properties in the final urea-formaldehyde. Molding powders are made with all the ingredients necessary for any special performance requirements. The usual filler is alpha cellulose or wood flour. Urea resins are not difficult to mold by conventional machines and methods, but a large shrinkage is involved.

In addition to being strong, the urea moldings possess good mechanical, thermal, and chemical properties. They do not collect dust, since they are relatively free of static build-up. The moldings are mar-resistant and have excellent appearance, gloss, and color.

Melamine, $C_3N_2(NH_2)_3$, has three NH_2 groups per molecule. It is reacted with formaldehyde in much the same way as urea and the reaction products are much the same and exhibit similar properties. The melamine resins, however, have superior hardness and withstand boiling water very well. They can be combined with various fillers such as clays, asbestos, glass fibers, and silicas. The glass-filled resins have good impact strength and high heat resistance; they also maintain form and dimensions and retain good electricals.

Aniline, $C_6H_5NH_2$, has one NH_2 group per molecule on a heterocyclic ring. It is reacted with formaldehyde to form a group of resins similar to the ureas and melamines. The aniline resins are combined with phenolics for many of their electrical uses. They have low dielectric losses, resist water, and are adaptable for RF service in such applications as coil forms, terminal boards, and connectors.

Electrical properties of urea and melamine resins are much the same as is evident from the values tabulated below:

	Ureas	Melamines
Volume resistivity		
(50% RH, 23°C), ohm-cm	10^{12}–10^{13}	0.8–2.0×10^{12}
Dielectric strength		
(⅛ in., step by step), volts/mil	220–300	240–270
Dielectric constant:		
60 Hz	7.0–9.5	7.9–9.5
10^3 Hz	7.0–7.5	7.8–9.2
10^6 Hz	6.8	7.2–8.4
Dissipation factor:		
60 Hz	0.035–0.043	0.030–0.083
10^3 Hz	0.025–0.035	0.015–0.036
10^6 Hz	0.25–0.35	0.027–0.045
Arc resistance		
(ASTM D495), sec	80–150	110–140
Flammability	Self-extinguishing	Nonburning

Applicable specifications are provided in ASTM D 1597 (Test for Melamine Content of Nitrogen Resins) and ASTM D 1727 (Test for Urea Content of Nitrogen Resins).

8.2.6   Cellulosics

The cellulosics comprise a large family of thermoplastics. They are derived directly from cellulose, which is a natural polymer. In this respect, the cellulosics differ from other plastics, which are polymerized from a monomer base. The starting material is alpha cellulose from cotton linters or wood pulp. An outline of the chemistry of cellulose was given in Chap. 6. The cellulosics of interest here are the esters (the acetates, butyrates, propionates, and acetate-butyrate) and ethyl cellulose, which is an ether. The roster is completed with cellulose nitrate, the first commercial plastic but of little, if any, electrical use because of its high flammability.

With suitable modifications, the cellulosics can be produced in the form of molding compounds, fabricated sheets, rods, tubes, and films. Various coating compounds can also be made, such as lacquers, conformal coatings, and barrier and release coatings for tapes.

All of these forms are useful as electrical insulations, but probably the most important are the films and sheeting (as described in Chapter 6) and the protective and conformal coatings. Molded and fabricated shapes are used where it is primarily desired to take design advantage of some nonelectrical property of the cellulosic while meeting minimal acceptable electrical properties.

Many applications may present an extensive number of compounds to choose from, since most of the electrical characteristics they offer are so similar that any one compound might be selected without marked advantage. The same is true to a considerable extent of the mechanical properties. Choice may be dictated by the availability of some particular property, for instance, the ability of the compound to meet Underwriters' Laboratories "self-extinguishing" rating. Another determining factor could be the "flow" temperature, that is, the point at which sufficient fluidity is obtained for optimum fabrication processes.

As a group, the cellulose esters have acceptable electrical properties, although not outstanding. The volume resistivities are good with the propionate being highest at 10^{12} to 10^{16} ohm-cm for molded specimens. For both the acetates and the mixed ester (acetate-butyrate) the value is 10^{11} to 10^{15} ohm-cm. The figures given here are for tests at 50 percent RH and at $23°$ C.

The dielectric strength, as shown by step-by-step tests for ⅛-in. molded specimens, is 250 to 600 volts/mil for acetate, 300 to 450 for propionate, and 250 to 400 for acetate-butyrate. The dielectric constant is high. For moldings tested at 10^3 Hz, the value for acetates is 3.4 to 7.0; for propionates, 3.6 to 4.3; and for acetate-butyrate, 3.4 to 6.4. The dissipation factor for all three esters is also high—about 0.01 to 0.07 at 10^3 Hz. This high dissipation factor limits the use of cellulosics to application at high frequencies. The high dielectric constant supports use of cellulosic film (principally the acetates) as a capacitor dielectric.

One negative characteristic common to all the cellulosics is their poor resistance to irradiation. Other properties of the cellulosics can be summarized as follows:

1. The acetates are available in a wide range of flow characteristics. They exhibit good hardness qualities, stiffness, and dimensional stability. Their unit cost is the lowest, but they have the highest specific gravity. Some formulations are self-extinguishing.

2. The butyrates also are available in a wide range of "flows." They have excellent dimensional stability, are tough at low temperatures, and are available in formulations specially devised for outdoor use.

3. The propionates are available in very hard flows but not in soft ones. They have good dimensional stability and are harder and have higher tensile strength than the acetates and the butyrates. Outdoor formulations are available.

Ethyl cellulose is the reaction product of cellulose with caustic and ethyl chloride. The resulting ether has 45.5 to 49.5 percent ethoxy content. It is much like the esters but is relatively nonpolar. Dielectric strength is higher and dielectric constant and dissipation factor are lower. For molded specimens at 10^3 Hz, the dielectric constant is 3.0 to 4.1 and the dissipation factor, 0.002 to 0.020. Density of ethyl cellulose is the lowest of all cellulosics. The resin is compatible with many resins, waxes, and oils. It is therefore easy to modify for many specialized applications.

ASTM covers cellulose plastics with a number of specifications: D707—Cellulose Acetate Butyric Molding and Extrusion Compounds; D706—Cellulose Acetate Molding and Extrusion Compounds; D1562—Cellulose Propionate Molding and Extrusion Compounds; D787—Ethyl Cellulose Molding and Extrusion Compounds; D786—Specification for Cellulose Acetate Plastic Sheets.

8.2.7 Chlorinated polyether

Chlorinated polyether (a thermoplastic resin) is well known for its corrosion resistance, resistance to many solvents, and freedom from thermal degradation at molding and extrusion temperatures. It is available in powder form for molding and coatings. There is a well-documented history of its reaction to exposure to over 300 chemicals. Among these are refrigerants of the R types, tests having been made at temperatures of 150° F.

Its chemical structure is

$$-O-CH_2-\underset{\underset{CH_2Cl}{|}}{\overset{\overset{CH_2Cl}{|}}{C}}-CH_2-O-$$

The polymer is linear, crystallizes rapidly, and has a narrow melting range. It is continuously resistant at 290° F. Mechanical properties are as good as the electrical. Selected electricals are as follows:

Volume resistivity (50% RH, 23° C), ohm-cm:	10^{12}
Dielectric strength (short-time, $\frac{1}{8}$-in. thick sample), volts/mil:	400
Dielectric constant:	
60 Hz	3.1
10^3 Hz	3.0
10^6 Hz	2.9
Dissipation factor:	
60 Hz	0.01
10^3 Hz	0.01
10^6 Hz	0.01

For applications that require good electrical insulation capability and resistance to many chemicals, this polymer provides a good combination of these properties.

8.2.8 Diallyl phthalates

The allylic family of chemical compounds take their name from the presence of the allyl radical, which has the empirical formula, C_3H_5. This radical with the addition of an OH group forms allyl alcohol, $H_2C = CH - CH_2 - OH$, which can be reacted with dibasic acids or their anhydrides to form a number of valuable polymers, some of which make good electrical insulations or dielectrics. Two of these exhibit outstanding electrical properties: diallyl phthalate (DAP) and diallyl isophthalate (DAIP). These polymers are very much alike with the chief difference being the superior thermal endurance of DAIP.

DAP is formed from phthalic acid and DAIP from its isomer, isophthalic acid. The respective structures are shown below:

Phthalic acid   Isophthalic acid

Complete cure or polymerization is of the vinyl type, in which no water or volatiles are released, contrary to conditions found in polymerization of other thermosets. In the allyls it is possible to fully accomplish three-dimensional cross linking and so assure excellent properties in the finished molded part. There is very little change in specific gravity from the uncured to the cured stage; consequently, there is very low shrinkage, and both the dimensional stability and reproducibility of the parts are outstanding.

Obviously, if the full attainment of the highest performance characteristics is to be secured, the kind and amount of the catalyst plays a vital part. The molding cycle of time, temperature, and pressure must permit complete curing.

The outstanding advantage of the parts made of DAP and DAIP is their retention of their electrical, mechanical, chemical, and thermal characteristics under prolonged exposure to severe humidity and temperatures. For DAP filled with short glass fibers, the continuous temperature exposure is $350°$ to $500°$ F; for DAIP this figure rises to $450°$ to $540°$ F. Other advantages are excellent dimensional stability and the fact that inserts stay put, no vapors are liberated, and molding is speedy and easy.

Fillers combined with the DAP and DAIP basic resins are used to impart desired characteristics to the molded part. Such fillers include polyacrylon nitrile, polyethylene terephthalate, asbestos, short or long glass fibers, various minerals, and cellulose. To initiate and complete the full cross-linking polymerization previously mentioned, a catalyst of the organic peroxide type is added, for example, t-butyl perbenzoate.

DAP and DAIP resins may also be used in varnish, impregnating, and casting resins for protective functions on capacitors, resistors, transformers, coils, transistors, and other electrical/electronic devices.

Other allylics are diallyl maleate (DAM) used in electrical moldings, diallyl chlorendate (DAC) added to molding powders for flame resistance, bis-allyl carbonate-diethylene glycol for optically clear castings, and triallyl cyanurate, which increases thermal endurance when added to polyesters.

Allylic monomers are used in cross-linking polyesters, in making preforms, in thermoforming, and in processing glass cloth and laminates.

Combinations of DAP or DAIP with other resins to secure special characteristics are possible. As an example, TFE tetrafluoroethylene added to DAP and glass fiber gives good electricals with improved resistance to wear. Carbon added to DAP produces a conductive resin used in potentiometers.

Selected mechanical properties of glass-filled DAP include a mold shrinkage (inch/inch) of 0.001 to 0.005, specific gravity of 1.75, tensile strength of 6,000 to 11,000 psi, compressive strength of 25,000 to 35,000 psi, impact strength (ft-lb/inch notch, Izod test) of 0.4 to 15.0, water absorption (24-hour test, ⅛-in. sample) of 0.12 to 0.35 percent. The glass-filled DAP is also self-extinguishing, insoluble in organic solvents, and is only slightly affected by strong acids and alkalies. Selected electrical properties are as follows:

Volume resistivity (50% RH, $23°$C), ohm-cm:	10^{13} to 10^{16}
Dielectric strength (⅛-in. sample, step-by-step), volts/mil:	380
Dielectric constant:	
60 Hz	4.3–4.6
10^3 Hz	4.1–4.5
10^6 Hz	3.4–4.5
Dissipation factor:	
60 Hz	0.01–0.05
10^3 Hz	0.004–0.009
10^6 Hz	0.009–0.014
Arc resistance (ASTM D 496), sec	130 to 180

8.2.9   Epoxy resins

Epoxy resins as used in day-to-day commercial work are "cured" or "hardened" materials. They are made by polymerizing the basic epoxy resins with the addition of "hardeners." In this section, the material described is the basic epoxy resin as it results from the reaction of the stated materials without the addition of curing or nardening agents.

The basic epoxy resins consist of four families. All of them are thermosetting and possess characteristics that make them most useful as electrical insulation.

The word epoxy is formed from the Greek "ep(i)," meaning "from" or "close to," and "oxy," meaning oxygen. It may be taken to mean "based on oxygen." The chemical formula is characterized by the presence of the group

$$
\begin{array}{c}
\text{O} \\
/ \quad \backslash \\
CH_2 \!-\!\!-\! CH
\end{array}
$$

This is a cyclic ether containing one atom of oxygen and two of carbon.

The first family of basic epoxy resins is that obtained by reacting epichlorohydrin with bisphenol A in the presence of a base. Sometimes called a conventional epoxy or an epi-bis epoxy, it has the following structure:

The molecule as shown has an epoxy group at each end of the chain. The subscript n denotes the number of repeating parts or units of the structure contained within the brackets.

The characteristics of the basic epoxy resin varies with the magnitude of n. If it is small, the resin is liquid (room temperature). An increase in value results in epoxy resins that are hard and tough solids.

Since the character of the resin depends upon the number of repeating units, it is desirable to have an index that will indicate this and serve to identify the resin. One such index is known as the epoxy or epoxide equivalent. It is the number of grams of the material shown by the structure that contains one gram of the epoxy or epoxide group,

$$
\begin{array}{c}
\text{O} \\
/ \quad \backslash \\
CH_2 \!-\!\!-\! CH_2
\end{array}
$$

Another way of expressing this index is by the weight of the epoxy groups contained in 100 grams of the structure material.

The members of this first family of basic epoxy resins have innumerable uses. Commercial epoxy resins may be made that are suitable for use as coatings, adhesives, potting compounds, embedding compounds, encapsulating compounds, molding compounds, laminating compounds, impregnants, foams, paints, and others. They may be colored, filled, and mixed with other resins for use in composite compounds to obtain a vast number of excellent resin materials.

The cycloaliphatic family of basic epoxy resins has the epoxy groups located on a ring structure. Produced by the peroxidation of cyclic olefins, their outstanding advantage is the ability to form hardened epoxy resins having outstanding weatherability.

The cresol novolac family of basic epoxy resins is produced from a novolac resin formed by the reaction of O-cresol and formaldehyde under acid conditions. This novolac resin is then reacted with epichlorohydrin to obtain the epoxy cresol novolac resins. The latter, with the addition of hardeners and accelerators, form epoxy resins of improved thermal properties, high resistance to solvents, and fast reacting systems.

The phenol novolac family of basic epoxy resins is produced from a novolac resin formed by reacting phenol and formaldehyde under acid conditions. This novolac resin is then reacted with epichlorohydrin to form the epoxy phenol novolac basic resin. Properly hardened, it produces resins of outstanding high temperature endurance, excellent electrical and mechanical properties, and good resistance to solvents. Applicable specifications are ASTM-D1763, Standard Specification for Epoxy Resins, which covers the basic resins, and ASTM-D1652, Standard Method of Test for Epoxy Content of Epoxy Resins.

Epoxy compounds for encapsulating, embedment, and the like, play a particularly important part in the electrical/electronic field. Compounds vary with the type of basic resin, hardeners, catalysts, special-function additives, and processing techniques. A great diversity of compounds is available from commercial sources, and many other are available on a custom-made basis to meet specific requirements. As mentioned in Par. 8.2, there is a lack of precise terminology in respect to these compounds, both in commercial practice and in the technical literature. It would be helpful to define the various functional compounds so that optimum use can be made of each. Definitions are given in Table 8.2.

Some design guidelines for potted and encapsulated applications are included in Appendix A at the end of the volume.

8.2.10   Fluorocarbons

There are two principal members of the fluorocarbon family that are of prime importance to the electrical/electronic field: Polytetrafluoroethylene, generally known as TFE-fluorocarbon, and fluorinated ethylene propylene, known as FEP-fluorocarbon. Both are characterized as electrical insulations having the lowest dielectric constant and dissipation factor of any known solids. Both have outstanding thermal stability and chemical inertness. Although TFE will not melt and FEP will, both can now be fabricated into a wide variety of forms, components, parts, and insulating materials.

Table 8.2 Definitions for Embedment and Related Compounds

Embedment compounds: The military specification MIL-Q-16923, "Insulation Compound, Electrical, Embedding," may be used in relation to nonmilitary as well as military applications. The specification states that the compound "shall preserve the electrical properties of the equipment to which it is applied by sealing against such environmental conditions as moisture, dirt, fumes, fungus, or other deleterious substances which may be encountered [in service.]" The compound is also required to be capable of mecanically supporting the part or assembly embedded without requiring a case or other external support after cure has been completed. Finally, the compound is required not to cause any deterioration of materials used in the part or assembly being embedded. Temporary molds are discarded.

Encapsulating compounds: These materials are those applied as coatings to components, devices, or various assemblies, in such manner that the coatings conform to the configuration of the unit being coated. For this reason these coatings are frequently called "conformal coatings." The encapsulant may be in the form of a thixotropic liquid into which the unit being coated is dipped until encapsulation is completed under specified process conditions. Encapsulants are also applied in the form of powders by means of the fluidized-bed method. In fact, any resin system capable of forming a suitable conformal coating may be applied by several means.

Potting compounds: These compounds are broadly the same as the embedment compounds, except that they are used to seal a given component or the like *within* a permanent case.

Impregnating compounds: Functionally, these compounds perform essentially the same duties as the embedment and encapsulating compounds: they seal and protect a given unit against external deteriorating effects; they provide structural support after cure has been completed; and they are required not to react adversely with any of the associated materials in the impregnated unit. However, the *manner* of application differs. Instead of the compound forming an embedment *around* a given unit, it is required to flow or enter into all parts or elements of the unit to prevent reentry of all air and moisture as well as to hold all elements (such as turns of a winding) securely in place. The impregnation would be expected to increase the insulation or dielectric level and increase the heat transfer and heat radiative ability of the unit. Highly desirable would be the ability of impregnating compounds to cure to an infusible void-free state without loss of volatiles. However, most impregnating compounds presently available do not meet such requirements.

Casting compounds: This term is used interchangeably with embedment compounds, since in each case it refers to an embedment process within a temporary mold that is discarded after cure has been completed. More precisely, the term "casting compounds" may be used as a generic description of all embedment, encapsulating, potting, and impregnating compounds.

In structure, TFE is a homopolar polymer containing only monomer units of TFE, as follows:

$$
\underset{\text{Monomer}}{\overset{\displaystyle F \quad F}{\underset{\displaystyle F \quad F}{C = C}}} \rightarrow \underset{\text{Polymer}}{\overset{\displaystyle F \quad F \quad F \quad F \quad F \quad F}{\underset{\displaystyle F \quad F \quad F \quad F \quad F \quad F}{C-C-C-C-C-C}}}
$$

On the other hand, FEP is a copolymer containing tetrafluoroethylene monomer units in combination with hexafluoropropylene monomers, in accordance with the following structure:

Monomer
TFE

Monomer
HFF

Copolymer FEP

Fluorine is an intensely active element, but in combination with carbon, as shown above, the end polymers possess chemical inertness, thermal stability, and relatively high density (as compared with polyethylene, for example).

The carbon-fluorine ratio is about the same in the two polymers. It is postulated that the molecule is twisted into a helix, with the fluorine atoms around the carbon backbone. This structure is responsible for the properties of the polymers.

Electrically, the polymers have nearly the same characteristics, but as already pointed out, they differ in melt viscosity and melting point. Since FEP will melt, it can be fabricated under well-known melt processes of extrusion and injection molding. Since TFE does not melt, or at least has enormous melt viscosity, it cannot be fabricated by the conventional molding methods. Methods common in the processing of ceramics or powdered metals are therefore applied to TFE parts fabrication. A very wide assortment of components and devices are produced. The availability of TFE resins in several types of small particles (600, 300, and 20 micron average size) has made it feasible to apply these processes. Glass, graphite, or bronze fillers may be used to alter the properties of the fluorocarbon finished parts. These polymers are inert to solvents and most chemicals. Sodium attacks them. Water absorption is nearly zero.

The electrical properties are outstanding in exhibiting low dielectric constant and low dissipation factor over a very wide range of frequency and temperature. The outstanding mechanical property is the low coefficient of friction; other characteristics are adequate but not outstanding. The polymers do not resist irradiation and are quickly destroyed by even moderate levels.

The electrical characteristics are as follows:

	TFE	FEP
Volume resistivity,		
ohm-cm, 50% RH, 23°C, ⅛ in.	$> 10^{18}$	$> 2.0 \times 10^{18}$
Dielectric strength,		
short time, volts/mil	480	500–600
Dielectric constant		
60 HZ	2.1	2.1
10^3 HZ	2.1	2.1
10^6 Hz and above	2.1	2.1
Dissipation factor		
60 Hz	< 0.0002	< 0.0003
10^3 Hz	< 0.0002	< 0.0003
10^6 Hz and above	< 0.0002	< 0.0003
Heat distortion temperature,		
66 psi, °F	250	162

ASTM Specification D1675 covers Electrical Grade Polytetrafluoroethylene (TFE) Tubing.

8.2.11 *Fluorocarbons containing chlorine*

A variant of TFE which contains four fluorine atoms is polychlorotrifluoro-ethylene (known as CTFE) in which a chlorine atom replaces fluorine on every other carbon according to the following scheme:

The CTFE polymer retains many of the properties of TFE, and also of FEP, but with some lower electrical properties and some superior molding and physical properties. The dielectric constant becomes somewhat lower with increasing frequency and the dissipation factor peaks at about 10^4 Hz. It is lower at 60 Hz and markedly lower at frequencies higher than 10^4 Hz. In contrast to the special fabricating methods required for TFE, the CTFE molding resins can be used with conventional compression, transfer, injection, and extrusion molding techniques.

Selected electrical properties for CTFE are as follows:

Volume resistivity, 50% RH, 23°C, ohm-cm	1.2×10^{18}
Dielectric strength, ⅛-in., step-by-step, volts/mil	500–600
Dielectric constant	
60 Hz	2.6–2.8
10^3 Hz	2.6–2.7
10^6 Hz	2.3–2.4
10^9 Hz	2.3
Dissipation factor	
60 Hz	0.0012
10^3 Hz	0.023–0.027
10^6 Hz	0.009–0.017
10^9 Hz	0.004

Water absorption is practially nil. CTFE is also resistant to most solvents, but hydrazines attack it as does sodium. It will not burn.

8.2.12   *Fluoropolymer ETFE*

ETFE fluoropolymer is a copolymer of ethylene and tetrafluoroethylene. The TFE part of the molecule is more than 75 percent of the total weight. The molecular structure is represented by the formula below:

$$
\begin{array}{cccc}
H & H & F & F \\
| & | & | & | \\
-C- & C- & C- & C- \\
| & | & | & | \\
H & H & F & F
\end{array}
$$

The actual structure may exhibit a different arrangement.

The specific gravity is 1.70 as compared with 2.15 for TFE. The mechanical properties resemble those of TFE except that the tensile strength is higher, 5,000 to 7,000 psi as compared with 3,000 to 5,000 for TFE. It is harder and the coefficient of friction is higher, 0.4 compared with 0.05 for TFE with a 100-psi load at 1 ft/min.

Development of ETFE was initiated with wire insulation as the application goal. The resin is melt-extrudable and can be applied readily over aluminum or copper wire plated with tin, nickel, or silver. Other uses include molded parts, flat cable, tubing, and heat-shrinkable tubing. ETFE can be cross-linked by radiation, which implies that it is more resistant to radiation than TFE; the latter, as noted previously, is quite poor in this respect.

The usable temperature range is from $-100°C$ to $+150°C$. After cross linking, the range may be extended to $+180°C$. The melting point is $270°C$. The heat distortion point is $105°C$ at 66-psi load and $70°C$ at 264-psi. ETFE is non-flammable or self-extinguishing. Its performance under ASTM test D 635 warrants a "nonburning" rating. The resin needs 30-percent oxygen to sustain burning. Heat of combustion is only 5,900 BTU/lb. It is chemically inert. No solvents are known below $200°C$. The electrical properties, as given by the producers, are as follows:

Resistivity	
volume, ohm-cm	10^{16}
surface, ohm-sq	10^{14}
Dielectric strength (10-mil film)	
volts/mil	2,000
Dielectric constant (unchanged with	
frequency and temperature)	2.6
Dissipation factor	
10^2 Hz	0.0006
10^3 Hz	0.0008
10^6 Hz	0.005

For lowest losses at high frequencies, TFE is much better.

A similar copolymer, in this case combining ethylene and CTFE, has appeared more recently, with much of the same application field. Its properties remain typical of fluoropolymers, but it features better mechanical properties and greater ease of processing. At this writing, evaluation work is progressing in several directions.

Two other fluorine-containing polymers are available—polyvinyl fluoride and polyvinylidene fluoride. These will be described in Par. 8.2.33.

8.2.13   Ionomers

Ethylene is the major component of ionomer polymers. A metal ion (believed to be magnesium or zinc) is present. The ionomers contain carboxylic groups that create ionized cross-linked long-chain molecules. The ionization mechanism develops properties usually associated with normal cross-linked structures. The properties of the ionomers are dependent on the amount of the carboxylic acid and on the effects introduced by the metal ion. In essence, the unique character of these polymers lies in the fact that the organic and inorganic components are linked both by covalent and ionic bonds.

Ionomers are being investigated for a variety of uses. Of interest to this discussion is their use as wire insulation. They are tough, flexible, and resistant to oil, grease, and organic solvents. They have high melt strength, can be extruded, and, in particular, are indicated for closely controlled insulating and jacketing operations. Abrasion resistance is good, and so is resistance to stress-cracking, except in environments where combinations of detergents and alcohol may be present. Their outstanding basic mechanical properties stem solely from the inherent cross-linked structure. No catalysts for curing are necessary and, for most uses, plasticizers are not required. However, stabilizers and antioxidants are added when certain special requirements are desired, such as good weatherability, flame resistance, or improved processing characteristics.

Ionomers have a density range of 0.935 to 0.960. Thermally, the upper temperature limit is 160°F. Low-temperature performance is good. For some of the ionomers brittleness is not evident until a low point of -200°F is reached. A practical use range for insulation could be set for -40°F to $+100$°F. Average electrical properties are good, depending on the composition, shape and thickness of the insulation. Volume resisitivity is of the order of 10^{16} ohm-cm; dielectric strength (for thin sheets) is 1000 volts/mil; dielectric constant remains essentially unchanged at 2.4 to 2.5 at frequencies from 10^3 to 10^9 Hz. For the same range of frequencies, the dissipation factor is 0.007.

Ionomers may be processed on conventional thermoplastic molding equipment. As already noted, they may be extruded and, in addition, they may be injection molded and blow-molded into various forms. They may be co-extruded with other polymers, including low-density polyethylene, with which they share several similar characteristics, such as chemical inertness and toughness.

The ionomers are at a relatively early stage of application research, but the unusual crosslinking mechanism on which they are based seems to offer possibilities for the emergence of a family of polymers useful in the solution of special application problems in insulation as well as other fields.

8.2.14   Methyl-pentene polymer

Methyl-pentene polymer is a thermoplastic olefin that has a crystalline melting point of 464°F and retains form stability near this point. Density is only 0.83, apparently the lowest known for thermoplastics. Transparent as well as opaque grades are available.

Electrical properties are in the broad range of the polyolefins, but with several exceptions, methyl-pentene being the superior member of the group. This is true both of the dielectric constant, which is 2.12 for frequencies up to 10^6 Hz, and for the dissipation factor, which is 0.00007 at 60 Hz and 0.000025 at 10^6 Hz.

In certain other properties methyl-pentenes do not show up as well. They are not as tough as the engineering thermoplastics (such as polycarbonates, polysulfones, and acetals). Nor are they able to withstand high-energy radiation or exposure to sunlight for any extended time. However, on balance, the excellent electrical properties and such physical characteristics as very low density offer interesting application prospects.

Indicated uses include those of cable insulation and in coaxial connectors for microwave devices. The low density provides a favorable cost factor where volume is one of the major design considerations.

8.2.15   *Phenolics*

Phenolics are formed by the reaction of phenol with formaldehyde in the presence of a catalyst. The reaction is of the condensation type, with water as well as the resin being produced. Two types of catalyst are required: alkalines and acids. Reactions may be in three stages, two stages, or one stage, as explained below. The three-stage reaction requires the presence of an alkaline catalyst, such as ammonia or sodium carbonate, and proceeds as follows:

1. The "A" stage produces a so-called *resol*, which is the basic condensation product or resin and water. It is fusible and soluble in polar solvents.
2. The "B" stage results in a so-called *resitol*, which is the resin obtained by heating the resol. The resitol is still fusible, but is soluble only with difficulty.
3. The "C" stage culminates the process with the phenolic resin in its fully hardened, infusible, and insoluble form. It is known as a *resite*.

The above description is rather a simplified explanation of a complex example of polymer chemistry. The techniques of such chemistry make it possible to vary ultimate properties by various modifications and combinations of starting materials and reaction processes. The introduction of fillers, of diverse nature, can be used to further refine or enlarge the properties of the final resin in such directions as mechanical strength, moisture resistance, shock resistance, electrical properties, and the like, as well as in fabricating characteristics.

The use of an acid catalyst and excess phenol causes a more rapid condensation reaction and the production of thermoplastic one-stage resins known as novalaks. These resins can be cured to a cross-linked thermosetting form by the addition of hexamethylene tetramine, which supplies both the basic alkaline catalyst (ammonia) and the necessary formaldehyde. The thermosets so produced can be described as two-stage resins. As with the three-stage resins, suitable modifications in the process can be introduced to produce resins with desired properties. Here, too, fillers can be used to obtain specially desired properties.

The family of phenolic resins is extensive and can be only briefly described within the space available here. In an age of new polymeric materials it is important to realize that the phenolics (first known as Bakelite, after their inventor L.H. Bakeland) have held a dominant position for many years and are still the largest-volume thermosets in use. For electrical applications the range of end-uses is both diverse and wide, from simple electrical parts to special-purpose complex devices. Most of the phenolics are used as molding compounds, of which there are many grades and types, depending on the basic resin, fillers, modifiers, and the like. In 1970, some 135 million lb were used in electrical controls, switchgear, communications equipment, appliances, wiring devices, and many other applications.

Phenolics are also used widely in forms other than molding compounds, such as insulating varnishes, various coatings, casting and potting resins, and they are widely used in various plastics laminates, such as the paper-phenolic, glass fiber-phenolic, cloth-phenolic, and others. Copper-clad phenolic laminates are used extensively as printed-circuit boards. (See Chap. 6 for detailed discussion of laminates.)

The patent and technical literature on phenolics has been prolific. Commercial catalogs, handbooks, and the like provide data on an extensive list of compounds, laminates, and other forms, identified by standard or special function and described by type of filler. With all this variety it would be difficult to summarize any particular spectrum of properties. The phenolics can be best described as the "work-horse" of the electrical industry with a broadly useful combination of electrical and physical properties and good mechanical and fabricating characteristics.

For the molding compounds, either compression- or transfer-molding methods are used. Special processes are used in casting and potting. Methods for applying laminates (also tube and rod stock) are described in Chap. 6. Coatings will be discussed in Chap. 10.

The most frequently used fillers are wood flour, asbestos, glass fibers or cloth, mica, chopped fabric, paper pulp, and various combinations. For some purposes, not strictly electrical, such as space-vehicle ablation shields, phenolics filled with inorganic fibers have been able to withstand brief exposure to temperatures in the 10,000° F range.

For a typical general-purpose molding compound (wood-flour filled phenolic) the electrical properties are as follows:

Volume resistivity (50% RH, 23°C), ohm-cm	10^9 to 10^{13}
Dielectric strength (⅛-in. thick, step-by-step) volts/mil	200 to 400
Dielectric constant	
60 Hz	5.0 to 13.0
10^3 Hz	4.4 to 9.0
10^6 Hz	4.0 to 6.0
Dielectric factor	
60 Hz	0.05 to 0.30
10^3 Hz	0.04 to 0.20
10^6 Hz	0.03 to 0.07

ASTM D 494, D 834, and D 796 cover testing procedures for phenolic moldings.

8.2.16   Phenoxy resins

The phenoxies comprise a family of high-molecular-weight resins based on bisphenol A and epichlorohydrin with terminal groups of bisphenol A. They are thermoplastic and have a continuous temperature rating of 170° F and a low heat-distortion temperature of 188° F. Mechanical and electrical properties are acceptable but not outstanding, except for a high degree of ductility. These resins are reported as remaining ductile under relatively high loads and over a temperture span from about $-75\,°F$ to $+212\,°F$.

The phenoxies are injection- and extrusion-molded. They can be used to produce various thermosets through cross-linked interactions with other materials, including the isocyanates and melamines.

8.2.17 Polyallomers

Polyallomers are produced through copolymerization of polyolefin monomers with the aid of anionic catalysts. The propylene-ethylene copolymer is a typical polyallomer. It exhibits excellent characteristics as wire and cable insulation. The mechanical, electrical, chemical, and thermal characteristics are all favorable for this application.

General electrical properties are similar to those of polyethylene and polypropylene. Volume resistivity at 50 percent RH and $23\,°C$ is 10^{15} ohm-cm; dielectric strength, $\frac{1}{8}$-in. thickness, is 800 to 950 volts/mil; and dissipation factor at 60 Hz is 0.0005 and at 10^{6} Hz, 0.0005.

The polyallomers are highly crystalline and exhibit low density. Modifications in starting formulations are used to produce changes in final properties as may be desired. A tendency to oxidation in the presence of copper can be inhibited by special formulations.

8.2.18 Polyamides (nylons)

The major polyamide molding compounds are commonly known as nylons 6/6, 6/10, and 6. The numbers used in this nomenclature refer to the number of carbon atoms in the starting acids as explained below. Other nylons identified as 7, 8, 9, 11, 12, and 13/13 vary in their commercial importance. Some are still in the development stage. The chemical structure of the polyamides is formed by the condensation of dibasic organic acids and diamines. Essentially, the structure is one of linear chains in which amides alternate with hydrocarbon groups. The 6/6 and 6/10 nylons are diamine-dibasic acid types. Nylon 6 is an amino acid type. Specifically they may be described as follows:

Nylon 6/6 has a long-chain molecule. It is produced by the reaction of adipic acid and hexamethylenediamine. The repeating unit is

$$\left[---HN(CH_2)_6\ NH\overset{O}{\overset{\|}{C}}\ (CH_2)_4\ \overset{O}{\overset{\|}{C}}--- \right]_n$$

Nylon 6/10 is made by using sebacic acid in place of adipic acid. The repeating unit is

$$\left[---NH(CH_2)_6NH\overset{O}{\overset{\|}{C}}(CH_2)_8\overset{O}{\overset{\|}{C}}--- \right]_n$$

Nylon 6 is made by polymerization of E-caprolactam (a cyclic amide produced from amino acids). The repeating unit is

$$\left[-NH(CH_2)_5\overset{\overset{\displaystyle O}{\|}}{C}-\right]_n$$

In general properties, the nylons are marked by high impact strength, tensile and flexural strengths, low coefficient of friction (self-lubricating properties), and good fabricating characteristics. Although the electrical properties are acceptable, they are offset by nylon's hydrophilic tendencies—obviously a serious problem in many electrical insulation applications. Nylon's use in electrical products, therefore, falls primarily in the area of *mechanical* parts such as coil forms, cable clamps, fasteners, and the like, as well as protective devices such as cable jacketing.

Nylon 6/6 is the most used. Introduced about 30 years ago, one of its first electrical uses was as a magnet wire enamel. It quickly became used as a coating material over other insulations of many kinds of wires. It is tough, strong, has a slippery surface, withstands severe impact, resists most solvents, and has acceptable electrical and thermal characteristics. Typically of the nylons, it is adversely affected by absorbing water. In this it is better than nylon 6, but nylon 6/10 is better than 6/6.

Fillers are employed to provide nylon compounds for special functions. For combined high-temperature and high-strength service, glass fillers are added; for self-lubricating parts, such as bearings, MgS_2 is added. Other composites are available.

The more recent development by Du Pont of its Nomex high-temperature polyamide has had a very strong impact on the electrical insulation field. Its primary use has been as a paper made from the basic resin fibers, both as a discrete material and in a growing number of composite forms with other materials. Nomex is discussed in Chap. 6 with other materials of the same category.

8.2.19   Polyamide-imide

The formation of polyamide-imide resins is the result of a reaction between trimellitic dianhydride and diamino diphenyl methane. The resulting compound contains amide groups plus some imide groups and free carboxyl (acid) groups. When heated to drive off all solvents, the amide groups and the carboxyl units form many more imide groups. The ratio of amide to imide groups is nearly one, with most of the carboxyls used up in the reaction. Although the initial reaction product is soluble in polar organic solvents, the *final* amide-imide polymer is highly resistant to solvents and exhibits high thermal stability. The structural scheme is similar to that of the polyimides (see Par. 8.2.20), except for the presence of the trimellitic anhydride and the diamino diphenyl methane instead of the acid anhydride and diamine shown for the polyimide, as follows:

Trimellitic anhydride   Diamino diphenyl methane

Magnet wire enamel made of the amide-imide polymer shows a thermal endurance of 220° C (428° F). For molded parts, the figure is 260° C (500° F). Electrical properties of the molded material are as follows: Volume resistivity is 0.8×10^{15} ohm-cm; dielectric strength (50 percent RH, 23 °C) for a ⅛-in. thick sample is 400 volts/mil; dielectric constant at 10^5 Hz is 3.8 to 4.1; and dissipation factor is 0.0005 to 0.0007.

Most of the wire mills use N-methyl-pyrrolidone as the solvent in the application of the polyamide-imide enamels.

8.2.20   Polyimides

Polyimides are linear aromatic polymers with most unusual properties. They are obtained by reacting an acid anhydride and an aromatic diamine. The anhydride is tetra carboxylic pyromellitic dianhydride:

The aromatic diamine most used is diaminodiphenyl ether:

The infusible and solvent-resistant polymer has a linear aromatic structure:

The polymers are available as molding compounds, fabricated parts, films, and wire coatings. Variation in properties of the polymers can be done by use of fillers. Some polyimides have no melting point and must be fabricated by machining, punching, or other means.

The thermal endurance is amazing. Continuous use at 260°C (500°F) is possible and limited time excursions to 485°C (905°F) can be withstood. Solvent resistance is outstanding. Wire enamels can be made by the use of N-methylpyrrolidone as the solvent. Bases, strong alkalis, and aqueous ammonia, however, attack the polymer. Mechanical properties are good and well-maintained at high temperatures such as 260°C. Specific gravity is also high—1.43.

Electric properties are good and remain high at temperatures that usually destroy most polymers. Volume resistivity at 50 percent RH and 23°C is $> 10^{16}$ ohm-cm. Dielectric strength for $\frac{1}{8}$-in. thickness and short-time test is 560 volts/mil. Dielectric constant is 3.4 at frequencies from 60 Hz to 10^6 Hz. Dissipation factor at 10^3 Hz is 0.002; at 10^6 Hz, 0.005.

Because of their unusual properties and relatively recent commercial development, the literature on polyimides is particularly useful for current information, and the reader is referred to such sources in the chapter references.

8.2.21 Polyaryl ether

The presence of the radical *aryl* in its nomenclature indicates that a given resin has a structure consisting mostly of aromatic monovalent hydrocarbons. One example is the phenyl radical in polyphenylene oxide (see Par. 8.2.26); others are polyaryl sulfone (Par. 8.2.22) and polysulfone (Par. 8.2.30). All these resins are thermoplastic but have high strength over a broad range of temperatures and at a comparatively high temperature. Combined with good electrical properties, they all seem to be useful members of the electrical insulation family.

The overall mechanical, thermal, chemical, and electrical properties of polyaryl ether itself are acceptable. It can be plated and coated. It has the high impact strength of 8 ft-lb per inch of notch, and it molds very well. Specific electrical properties are as follows:

Volume resistivity (35% RH, 23°C), ohm-cm	1.5×10^{16}
Dielectric strength ($\frac{1}{8}$-in. thick), volts/mil	430
Dielectric constant	
60 Hz	3.14
10^3 Hz	2.85
10^6 Hz	3.10
Dissipation factor	
60 Hz	0.0057
10^3 Hz	0.0041
10^6 Hz	0.007

Arc resistance, by ASTM D 495, is 180 seconds plus.

8.2.22   Polyaryl sulfone

Polyaryl sulfone thermoplastic resin consists of aryl units linked by oxygen and sulfone groups. The structure is as follows, with two aryl groups and one sulfone:

$$
\bigcirc - \underset{\underset{O}{\overset{O}{\|}}}{S} - \bigcirc
$$

The polymer chain of the polyaryl sulfone is completely aromatic. Its molecular configuration gives it strength at high temperatures and resistance to oxidative attack.

The advantages of high-temperature performance, solvent resistance, strength, weight saving, and good electrical insulating properties enable the material to be used where ceramics or metals have been used.

The tensile strength at $73°$ F is 13,000 psi and at $500°$ F it is 4,100 psi, as per ASTM D 638 test. The heat-deflection temperature at 264 psi is $525°$ F. The resin is self-extinguishing. It is resistant to hydrolysis as shown by samples exhibiting no change after 1,000 hr of $300°$ F saturated steam.

Molding powders and pellets are available for injection molding and extrusion.

Electrical properties are as follows:

Volume resistivity (50% RH, $23°$ C), ohm-cm	3.2×10^{16}
Dielectric strength (1/16-in. thick; short-time), volts/mil	356
Dielectric constant	
60 Hz	3.94
8.5×10^{9} Hz	3.24
Dissipation factor	
60 Hz	0.0030
8.5×10^{9} Hz	0.010
Arc resistance (D 495), sec	67.0

The present price of this thermoplastic (*circa* mid 1972) is high compared with those of the thermosetting resins. It is hoped by the producers that increasing volume use may bring prices down. This is not an uncommon event in the history of some of our advanced materials.

8.2.23   Polycarbonates

Polycarbonates may be thought of as a group of dihydric phenols linked through carbonate linkages. The general formula may be written as:

$$
\left[C_6H_4 - C - (CH_3)_2 C_6H_4 - O - \underset{\underset{O}{\overset{}{\|}}}{C} - O \right]_n
$$

This may be represented as:

$$\left[-\!\!\left\langle\bigcirc\right\rangle\!\!-\!\overset{\overset{\displaystyle CH_3}{|}}{\underset{\underset{\displaystyle CH_3}{|}}{C}}\!-\!\!\left\langle\bigcirc\right\rangle\!\!-\!O-\overset{}{\underset{\underset{\displaystyle O}{\|}}{C}}-O- \right]_n$$

One way to produce the polycarbonate polymer, which may be considered as an ester, is to react carbonic acid and bisphenol. The latter is the reactant used in production of epoxies. Since carbonic acid, H_2CO_3 or

$$-O-\overset{}{\underset{\underset{\displaystyle O}{\|}}{C}}-OH-$$

is a weak acid, phosgene, the symmetrical chloride of carbonic acid,

$$\overset{\overset{\displaystyle Cl}{|}}{\underset{\underset{\displaystyle Cl}{|}}{C}}\!=\!O$$

is used to provide the carbonate group.

This is the first polymer to incorporate the carbonate linkage. Developed in Germany, it was introduced commercially in 1960. Other polycarbonates might be made by using other phenols and other ketones or by replacing the

$$-\overset{\overset{\displaystyle CH_3}{|}}{\underset{\underset{\displaystyle CH_3}{|}}{C}}\!-$$

bridge with other radicals.

Although polycarbonate is a thermoplast, it is capable of continuous use at $130°C$ ($266°F$) and has good resistance to temperatures as high as $150°C$ ($302°F$). Support insulators for bus-bars are approved by Underwriters Laboratories at $125°C$.

The combination of excellent mechanical properties such as very high impact strength and tensile strength with good thermal, chemical and electrical properties make this polymer outstanding for use in many places where good electrical insulation must be combined with exceptional performance ability in other characteristics.

Polycarbonates may be reinforced by the addition of fillers such as glass fiber. They may be colored. The material is amorphous and responds well to machining. Most of the polymer is used as molding compounds although film and other forms are available.

Polycarbonates are insoluble in aliphatic solvents, soluble in aromatic hydrocarbons and chlorinated hydrocarbons. They resist dilute acids, but are attacked by alkalis and amines. They are not subject to "silver migration" degradation in the presence of silver contacts or other silver parts. Molded parts resist electrolytic action.

The electrical properties are nearly immune to change by humid conditions. The dielectric constant changes slightly with temperature and frequency. The dissipation factor has a minimum value at about $90°C$ $(194°F)$ rising for higher and lower temperatures. Volume resistivity falls about four decades from $25°C$ $(77°F)$ to $175°C$ $(347°F)$. Dielectric strength is about half that of the olefins but is adequate since moisture does not produce a marked change.

Volume resistivity at 50 percent RH and $23°C$ is 3×10^{16} ohm-cm. Dielectric strength of ⅛-in. material is 375 to 400 volts/mil. Dissipation factor at 60 Hz is 0.0009; at 10^3 Hz, 0.0021; and at 10^6 Hz, 0.010.

ASTM D 2473 covers polycarbonate plastic molding, extrusion, and casting materials.

8.2.24   Polyesters

The term polyesters is used as a broad designation for a great many resins and compounds formed by the reaction of an organic acid and an alcohol. Polyesters occupy an important position in electrical insulation and dielectrics as films, casting and potting resins, molding compounds, and laminates. When the reaction is between a carboxylic compound and a polyfunctional alcohol, each containing *more* than two functional groups per molecule, the reaction products are *unsaturated* or branched thermosetting polyesters. When each of the reactants contains only two functional units per molecule, the reaction products are *saturated* or linear thermoplastics.

The unsaturated thermosets are made with maleic or fumaric acids and are used primarily as molding or casting compounds and in laminates. Glass-fiber reinforcements are added. The saturated polyesters are used in fibers and films. They are made with terephtalic acid. In either type of polyester the properties of the various derived resins and compounds depend on the starting materials and the nature of the reaction conditions; the use of stabilizers, inhibitors, and the like.

The widely used "polyester film" with many applications in electrical insulation either alone or in composites with other films, papers, and the like, is a condensation product of ethylene glycol and terephthalic acid (see Chap. 6).

8.2.25   Polyethylene (PE)

Leading all other plastics materials in volume output, polyethylene plays a major role in electrical insulation, among other uses. Wire and cable insulation, the chief application, takes up about 11 percent of PE's current output.

Polyethylene consists only of ethylene molecules joined end-to-end to form a giant hydrocarbon molecule. Ethylene as the repeating unit has the chemical structure,

$$
\begin{array}{cc}
\text{H} & \text{H} \\
| & | \\
-\text{C}- & \text{C}- \\
| & | \\
\text{H} & \text{H}
\end{array}
$$

The degree of polymerization, DP (or the chain length or the molecular weight) reflects the fact that the growth of the chains is a random process. Here the arithmetical average of all the chain lengths obeys a known relation, but the distribution of these chain lengths depends upon the statistical laws governing the process.

Three parameters describe the polyethylene resins. These are density, melt index, and molecular weight distribution (MWD):

Density is an indication of the crystallinity of the polyethylene. Low density means a low degree of crystallinity and high density a high degree of crystallinity. *Melt index* is an arbitrary index of the fluidity of the molten polyethylene as covered by ASTM D1238-65T or later issue. The *MWD* is the ratio of the *weight* average molecular weight to *number* average molecular weight. It may be expressed as M_W/M_N. If the ratio is less than 5/1, the polymer is designated as "narrow"; if greater than 5/1, it is "broad."

Polyethylene is made by one of two basic processes (or modifications thereof), described briefly as follows: (1) the original *high-pressure process,* using pressures of 15,000 to 45,000 psi, and (2) the *low-pressure process* (a later development) in which catalyst systems permit the process to be carried out with pressures of 1,500 psi or lower.

The *high-pressure process* produces a *low*-density polyethylene having a high degree of side chain branching. The *low-pressure process* produces a *high* density polyethylene having few side chain branches. It is more linear and of increased crystallinity.

The degree of crystallinity controls most of the mechanical, thermal, and chemical parameters. The electrical properties are very nearly the same for both low-density and high-density PE. The other characteristics vary and the selection decision for any given application becomes a trade-off process. Both processes can be controlled to produce some "tailor-made" compounds for special requirements. Mixtures of high- and low-pressure PE may be used for such compounds. Additives and fillers of various types can be used to control properties. Chlorinated PE, for example, contains chlorine compounds that act as fire retardants. Carbon black is generally used to impart weather-resistant properties to wire or cable used in open environments. It is also effective as protection against the effects of ultra-violet radiation.

Copolymerization with other monomers such as vinyl acetate or ethyl acrylate is used to increase flexibility and impact strength. Ethylene vinyl acetate, EVA, is important in electrical uses.

Cross linking by irradiation with high-energy electron beams or by chemical means increases thermal endurance and other parameters. Complete cross linking makes the PE a thermoset. Introduction of a gas (nitrogen) produces a foam of reduced dielectric constant.

ASTM specification D1248 covering the requirements for polyethylene molding and extruding compounds divides the polymer into three types according to density. Each of these types is subdivided into three classes according to composition and use:

Type 1—Density range, 0.910–0.925 gm/cc
Type 2—Density range, 0.926–0.940 gm/cc
Type 3—Density range, 0.941–0.965 gm/cc
Class A—Natural color; dielectric use

Class B—Colors, white and black; dielectric use

Class C—Black (not less than 2-percent carbon black); weather-resistant

As to mechanical properties, low-density PE (LDPE) has a density of about 0.920 gm/cc, approximately 55 percent crystallinity, a yield stress of 1,700 psi, and a crystalline melting point of 109 °C. High-density PE (HDPE) has a density of about 0.960, approximately 85 percent crystallinity, a yield stress of 5,100 psi, and a crystalline melting point of 133 °C.

As already noted, the electrical properties of the low-density and high-density polyethylenes do not differ markedly. Volume resistivity at 50 percent RH and 23°C is $> 10^{16}$ ohm-cm. Dielectric strength, ⅛-in. thickness, is from 450 to 1,000 volts/mil, depending on the test method. Dissipation factor is very low over an extremely wide range of frequencies and is less than 0.0005. Dielectric constant ranges from 2.25 to 2.35, with the high-density material showing the higher value. Of great importance is the stability of these parameters over a wide variation of environmental factors.

Polyethylene with its long hydrocarbon chain and high degree of electrical symmetry is nonpolar. The inherent moisture resistance of PE maintains the electrical properties unchanged under wet and humid conditions.

Cross-linked polyethylene is discussed under rubbers and elastomers (Par. 8.3.4) since it is vulcanizable and has a number of properties similar to those of the elastomeric materials.

8.2.26   Polyphenylene oxide

Polyphenylene oxide (PPO) is another of the more recently developed engineering thermoplastics (such as the acetals, polycarbonates, and polysulfones) that provide some outstanding mechanical and physical properties combined with good electricals. Chemically, it is derived from 2,6 xylenol as poly-2,6-dimethylphenylene oxide of the following structure:

In its *unmodified* form, PPO has a variety of special uses, many of them in direct competiton with metals, particularly because of the good machining characteristics. Heat deflection temperature is 375°F at 264 psi. Dielectric constant over a range of 60 to 10^6 Hz is 2.58. Dissipation factor at 60 Hz is 0.00035, and at 10^6 Hz, 0.0009. A special grade is available for service at 10^9 Hz.

The *modified* grades are the most widely used in at least two UL-approved grades. Molded PPO has been applied in connectors, sockets, TV yoke coil frames, coil bobbins, housings, and structural parts. Glass fibers have been used to add to the mechanical properties.

The advantages of the modified resin are its low specific gravity of 1.06, excellent dimensional stability under load, low creep, and very low moisture absorption. The temperature coefficient of expansion is low at 3.3. in./in./°F $\times 10^{-5}$.

The low creep is maintained under high-temperature fluctuations. Glass-fiber fillers help in this temperature-stability characteristic. Tensile strength (yield point) ranges from 9,600 to 17,000 psi; Izod impact strength is 1.4 to 1.8 ft-lb/ per inch of notch. The physical properties are somewhat better for the *unmodified* resins.

The electrical properties for the modified resins (no fillers added) are: Volume resistivity at 50-percent RH, 23°C, is 10^{17} ohm-cm. Dielectric strength (⅛-in., step-by-step) is 500 to 550 volts/mil for one UL grade and 400 to 500 volts/ mil for the other. Dielectric constant over the frequency range of 60 to 10^6 Hz is 2.58 to 2.64. Dissipation factor at 60 to 10^6 Hz is 0.0006 for one grade, whereas for the other the values are 0.008 at 60 Hz and 0.001 at 10^6 Hz.

The UL-approved modified grades are rated as Group 1, self-extinguishing. The heat deflection temperature at 264 psi is 265°F. The brittle point is −275°F. Impact and tensile strength are maintained after aging at 240°F. The UL service rating is 212°F. Both grades comply with MIL-P-46129n(MR).

8.2.27   Polyphenylene sulfide

Polyphenylene sulfide is an aromatic polymer consisting of para-substituted benzene rings and sulfur. Its melting point is 550°F. There are no solvents known below 375°F. It has excellent thermal stability and good mechanical and electrical properties. It is used in molding compounds (plain, asbestos filled, and glass filled). Its overall properties show that it is adapted for use in coatings by the slurry process or the fluid-bed technique. Dielectric strength (⅛-in. thick, short-time) is 590 volts/ mil; dielectric constant from 60 Hz to 10^6 Hz is 3.4. Dissipation factor from 60 to 10^6 Hz is < 0.0005.

8.2.28   Polypropylene

Propylene is an unsaturated hydrocarbon (an olefin), normally a gas with three carbon and six hydrogen atoms per molecule. The formula may be written as $H_2C{=}CHCH_3$. As such it is a vinyl monomer containing the vinyl group $H_2C{=}$ CH—. The monomer may be polymerized by use of suitable catalysts at the low pressures used in the polymerization of ethylene. The result is the polymer polypropylene. This is very much like polyethylene with nearly the same electrical characteristics but higher thermal endurance. Polypropylene has good mechanical characteristics that can be improved by copolymerization and by the addition of inorganic fillers, such as talc and asbestos, and of flame retarders. The crystalline melting point is 355°F for the unmodified polymer.

The polymerization of the propylene monomer may result in three structural configurations of the molecule. These are described with respect to the tacticity or placement of the asymmetric atoms in the chain. Tacticity has an important bearing on the physical properties of the polymer.

Atactic configuration has a random placement of side chains with respect to the vinyl backbone. The result is lowered crystallinity. *Isotactic* configuration has a stereo-regular structure with all the side chains on the same side of the plane of the polymer background. It exhibits maximum crystallinity. *Syndiotactic* configuration has the side chains alternating regularly above and below the plane of the backbone.

Isotactic polypropylene with its high degree of crystallinity resembles ASTM Type 3 high-density polyethylene in its low moisture absorption, excellent electrical characteristics, good rigidity, and resistance to environmental stress cracking.

Unmodified polypropylene with a density of 0.902 to 0.906 has a volume resistivity at 50 percent RH and $23\,^{\circ}$C of $> 10^{16}$ ohm-cm. Dielectric strength ($\frac{1}{8}$-in. thick) is 450 to 660 volts/mil, depending on test method. Dielectric constant is 2.2 to 2.6 from 60 to 10^6 Hz. Dissipation factor at 60 Hz is < 0.0005, with an increase at higher frequencies.

A stabilizer is usually added to polypropylene polymers to prevent degradation. ASTM D 2146 covers propylene plastic molding and extrusion materials.

8.2.29 Polystyrene

Styrene and polystrene are the basis for a very large family of resins and plastics. Volume output is quite large—one of the largest in the plastics industry. The resins are used in many kinds of modified polymers. copolymers, and blends. For example, high-impact polystyrene is usually a copolymer of styrene and butadiene. Note also the terpolymer of acrylonitrile-butadiene-styrene (ABS) described in Par. 8.2.1.

Styrene, the monomer, is formed by combining benzene with ethylene to form

$$-H_2-C-CH-\hexagon$$

and loss of hydrogen. Polymerization proceeds by an exothermic reaction. The process is controlled to produce the polystyrene polymer of desired characteristics. For example, higher molecular weights are desirable for molding compounds, lower ones for coatings. The polystyrene polymer has the following structure:

$$CH_2-CH-CH_2-CH-CH_2-CH$$

Polystyrene has excellent appearance characteristics. It is available in clear, transparent form, and also can be pigmented. Added to the material's excellent electrical characteristics, these advantages promote use in applications like appliance housings.

8.2.30 Polysulfone

Polysulfones are a family of tough, rigid, high-strength thermoplastics that maintain their properties over a temperature range of $-150°$ to $340°$F. The chemical structure has as its most distinctive feature the diphenylene sulfone unit which is responsible for the outstanding thermal endurance of the polymer and its resistance to oxidation. The structure takes the following form:

The excellent thermal properties are inherent in the polymer. No additives are needed to produce them.

Heat-deflection temperature is 345°F at 264 psi. Physical properties are good down to $-150°$F. Creep and cold-flow characteristics are quite satisfactory. The polymer is classified as self-extinguishing Group II by Underwriter's Laboratories, which rate it as a 302°F (150°C) material.

Polysylfone is available as molding compounds, rods, sheets, tubes, film, and slabs. An extrusion compound is available as insulation for wires and cables. It is resistant to mineral acids, alkalis, salt solutions, detergents, and hydrocarbon oils. However, polar organic solvents attack it. Water does not, however, and this characteristic makes it feasible to steam-clean polysulfone parts.

Physical properties are impressive. It appears that polysulfone is the strongest thermoplastic over a wide temperature range. It has a high tensile strength and high modulus of elasticity in tension and flexure. Tensile strength is 10,200 psi. Flexural strength is 15,400 psi. Percent elongation is in the range of 5 to 6 percent. Izod impact strength is 1.3 ft-lb per inch of notch. Initial physical properties are maintained under exposure to temperatures of up to 300°F for extended periods. Reinforcement with glass fibers adds to these properties. Availability in clear and opaque forms is still another advantage.

Polysulfone has excellent electrical properties that are stable over a wide temperature range up to 350° and after immersion in water or exposure to high humidities. Specific values are as follows: Volume resisitivity, at 50% RH and 23°C, is 10^{14} ohm-cm. Dielectric strength (⅛-in. thick) is 425 volts/mil. Dissipation factor at 23°C and 60 Hz is about 0.001 and remains at this level up to 250°F where it begins to rise, reaching 0.002 at 320°F, and continuing to rise thereafter. Over the frequency range from 60 to 10^{10} Hz the dissipation factors rises from 0.001 to a maximum of about 0.0035 at 10^6 Hz, dips, and then rises to a level of 0.005 at 10^{10} Hz. The dielectric constant does not change significantly with temperature. At room temperature and 60 Hz the dielectric constant is 3.1 and reaches a minimum of 2.7 at 350°F. Arc resistance, ASTM D 495, is 122 sec.

8.2.31   Polyurethane

A reaction product of isocyanates and hydroxyl units, the polyurethanes can be produced in a variety of forms, from rigid foams to liquid coatings. Of primary electrical interest are the casting compounds, which feature light weight and resiliency, and the wire and cable insulation, discussed in Chap. 9.

The urethane elastomers are discussed in Par. 8.3.18.

8.2.32   Polyvinylchloride (PVC)

Polyvinylchloride (usually known as PVC) is another of the large-volume plastics in the field. It is polymerized by several different methods from the basic vinyl chloride monomer shown below:

Monomer   Polymer

The different polymerization processes determine the particular properties of the respective polymer, but, more importantly, PVC is an extremely versatile thermoplastic resin in its adaptability to a wide range of modifications. Such modifications employ plasticizers to impart flexibility, stabilizers to prevent heat degradation, lubricants to facilitate processing, and fillers for a diversity of reasons, including cost reduction of the final compound, better electrical properties, and many general-purpose needs as, for example, pigments if a color is desired.

The wide use of PVC in wire and cable insulation, both as primary dielectric and as jacketing, is of major interest in this discussion. Rigid PVC also finds some use in insulating parts. Molding compounds are available for various electrical components. In paste forms PVC can be applied as an insulation coating known as "plastisols." PVC exhibits excellent chemical and moisture resistance. Physical properties are good. Temperature range is moderate, however, from $65°$ C to $105°$ C, depending on the compound. For electrical insulation the upper limit would be selected. Excellent electrical properties are typical, as indicated in the following data for a flexible filled compound: Volume resistivity at 50 percent RH, $23°$ C, is 10^{11} to 10^{14} ohm-cm. Dielectric strength (short-time, $\frac{1}{8}$-in. thick) is 200 to 900 volts/mil. Dielectric constant at 10^3 Hz is 5. Dissipation factor at 10^3 Hz is 0.09 to 0.16.

Various copolymers are made with PVC. The most widely used combines PVC and polyvinylacetate. This copolymer combines the very hard chloride with its relatively high softening temperature and the soft, tacky acetate with a softening temperature near room-temperature level. The copolymer exhibits good mechanical and electrical properties. One application is a wire-insulating coating at low temperatures on the order of $-20°$ C.

There is a considerable variety of other vinyl compounds, some of which are useful in electrical insulation, especially polyvinyl acetal, polyvinyl butyral, and polyvinyl formal. These are primarily used as magnet-wire coatings. Polyvinyl carbazole is employed as a paper capacitor dielectric. PVC has also been modified with polypropylene and ABS resins. Combined with polyethylene, PVC contributes fire-retardant properties to the final product (see Par. 8.2.25).

The importance of PVC in electrical insulation is indicated by the fact that some 400 million pounds are used annually for wire and cable insulation alone.

8.2.33 Polyvinylidene fluoride

Another member of the fluorine-containing family important in the electrical insulating field is polyvinylidene fluoride (usually known as PVF_2). This polymer has an unusual combination of properties—mechanical, environmental, thermal, and electrical. It is used effectively in wire and cable insulation, coil forms, tubing, sleeving, and other parts. Heat-distortion temperature is $300°$ F at 66 psi. Service range extends down to $-80°$ C.

Among the particular advantages of PVF_2 is its resistance to abrasion and radiation. It is rated self-extinguishing. Fabrication processes can use injection- as well as extrusion-molding methods. PVF_2 could be described as a special-purpose insulating material of excellent properties for aerospace systems, computers, instrumentation, and the like.

In addition to its availability as a molding and extrusion compound, PVF_2 is available in film form. Also available in film form is polyvinyl fluoride (PVF). Both film materials are discussed in Chap. 6.

8.2.34 *Silicones*

The silicone resins are based on polymers having a molecular backbone of alternate atoms of silicon and oxygen. Attached to the silicon atoms of the chain are organic groups. These may be methyl, phenyl, or vinyl. The basic structure is polydimethyl siloxane:

$$
CH_3\!-\!\underset{\underset{\textstyle CH_3}{|}}{\overset{\overset{\textstyle CH_3}{|}}{Si}}\!-\!O\!-\!\left[\underset{\underset{\textstyle CH_3}{|}}{\overset{\overset{\textstyle CH_3}{|}}{Si}}\!-\!O\right]_n\!-\!\underset{\underset{\textstyle CH_3}{|}}{\overset{\overset{\textstyle CH_3}{|}}{Si}}\!-\!CH_3\!-\!-
$$

The phenyl or vinyl groups are substituted for one of the methyl groups in the brackets. The substituted groups form repeating units of the chain.

The characteristics of the silicones depend upon the amount of the cross-linking and the nature of the organic groups. Liquids, resins, and elastomers can be made as the end-product by suitable control of these variables. Silicones so made tend to be inert and do not combine with other polymers such as the epoxies or polyesters. Intermediate products containing Si-OH groups (silanols) will react with other polymers.

The silicones have emerged into use at a time when developments in the field of insulation and dielectrics demanded materials with just their combination of a wide range of thermal stability, chemical inertness, and moisture resistance, all added to excellent electrical properties. Continuing developments in many fields indicate the need for materials of even greater performance abilities, and the silicones appear to be capable of extended development to meet some of these needs.

Adaptable to a diversity of fillers (such as glass fibers, mica, asbestos and fused silica), the inherent properties of the silicone resins are further improved by such additions to permit use in many electrical parts. Fire retardance is a specially valuable characteristic. Silicone resins used as insulating varnishes, impregnants, and binders find a particularly wide area of electrical applications. (See Chap. 6 for their use in plastics laminates, flexible composite insulation, insulating fabrics, and the like. See also Chap. 10, which deals with varnishes and similar materials *per se*. Silicone rubber is discussed in Par. 8.3.17.

8.3 Rubber and Elastomers

The distinction between the terms "rubber" and "elastomer" is tenuous. Both in industrial practice and in the technical literature the tendency is to use the two terms interchangeably. Another tendency is to use "elastomer" as a generic term to cover *all* types of rubber-like materials. However, it is probably accurate enough to say that (1) all materials having their origin in the sap (latex) of rubber trees, such as the *hevea brasiliensis*, can be considered as *natural rubbers*, that (2) all synthetic compounds that have been formulated to approximate the properties of natural rubber (plus or minus certain values) can be described as *synthetic rubbers*, and that (3) all polymers that possess or that have imparted to them rubber-like properties can be described as *elastomers*.

Whatever the problems of nomenclature may be, it is essential to bear in mind that all of these materials are polymeric in structure. They are characterized by long, coiled, high-molecular weight chains, cross-linked by certain chemical mechanisms to form molecular networks. A common characteristic of these materials is their resilience, that is, their ability to accept and recover from extreme deformation. Other properties they offer are flexibility, impermeability, electrical insulating capability, and damping characteristics.

Of particular significance to the design engineer is the fact that advances in high polymer chemistry have made possible a wide choice of synthetic materials in addition to the natural rubbers. In selecting and specifying optimum materials for given applications, the importance of compounding is paramount. With the wealth of materials available, and with an extensive source of knowledge at hand in the literature, the skilled compounder can provide the most suitable materials for a specified use.

The list of applications for rubbers and elastomers in electrical and electronic design includes the following:

1. Wire and cable insulation
2. Embedment, encapsulating, and potting compounds
3. Coatings for cloths, tapes, and similar electrical insulation
4. Adhesives
5. Molded or otherwise fabricated insulating parts
6. Mechanical parts and components used as associated elements in the materials systems of equipment and devices (for example, gaskets in transformers)
7. Accessory protective equipments, such as mats and hoods.

The electrical properties of rubbers and elastomers that are usually of concern are volume resistivity (ASTM D 257 test); dielectric strength, short time (ASTM D 149); dielectric constant (ASTM D 150); and dissipation factor (ASTM D 150). It is usually desirable to have a material with high resistivity and dielectric strength and low dielectric constant and dissipation factor (or power factor). Some materials, such as butyl rubber, are nonpolar; others, such as the nitrile rubbers, are polar.

The index to ASTM Standards lists some 100 of them that are applicable to rubbers and elastomers. ASTM Designation D 1418-69 provides recommended nomenclature for these materials. Letter symbols are used to codify and identify the primary chemical composition of the various materials, according to the following scheme:

M — Rubbers having a saturated chain of the polymethylene type
N — Rubbers having nitrogen in the polymer chain
O — Rubbers having oxygen in the polymer chain
R — Rubbers having an unsaturated carbon chain (for example, natural rubber and synthetic rubbers derived at least partly from diolefins)
Q — Rubbers having silicone in the polymer chain
T — Rubbers having sulfur in the polymer chain
U — Rubbers having carbon, oxygen, and nitrogen in the polymer chain.

Data on the chemical background and the major properties of the more important materials are given in the following paragraphs. The information is ar-

Table 8.3.   Cross-Reference to Rubbers and Elastomers by Common Abbreviations*

Abbreviation	Composition of rubber or elastomer	Discussed in paragraph indicated
ABR	Acrylonitrile-butadiene rubber (acrylic rubber)	8.3.1
Buna N	See under NBR	—
Buna S	See under SBR	—
DPR	Depolymerized (flowable) rubber	8.3.5
CO	Epichlorohydrin elastomers	8.3.6
CR	Polychloroprene rubber (neoprene)	8.3.13
ECO	Epichlorohydrin copolymer elastomers	8.3.6
EPDM	Terpolymers of ethylene, propylene, and a diene	8.3.7
EPT	Ethylene and propylene copolymer	8.3.7
IR	Isoprene synthetic rubber	8.3.10
FKM	Fluoroelastomers	8.3.8
NBR	Butadiene-acrilonitrile copolymer rubber	8.3.11
NBR/PVC	Butadiene-acrilonitrile/polyvinyl chloride blend	8.3.12
NR	Natural rubber *(hevea)*	8.3.10
RTV	Room-temperature-curing silicone rubber	8.3.17
SBR	Styrene-butadiene copolymer rubber	8.3.16

* See Par. 8.3 for ASTM list of letter designations for various rubbers and elastomers

ranged in alphabetical order of the chemical names or abbreviations, whichever are the more commonly used. Table 8.3. provides a cross-reference for the abbreviations and related chemical compositions. Cross-linked polyethylene, depolymerized rubber, and hard rubber are included in the discussion, although these materials are not covered by ASTM D 1418. However, their applications are similar to those for the other materials discussed.

8.3.1   ABR (Acrylonitrile-butadiene rubbers)

The ABR elastomers represent a series of copolymers based on acrylic acids and esters of such acids. Their electrical properties are only fair, and so their value in electrical insulation application would depend on how important their other properties are in a given design. Such properites feature markedly high resistance to ozone, heat, and oils. Special grades can be used in rapid-cycle high-temperature injection molding because of their resistance to hot sulfur-modified oils and elevated-temperature atmospheres encountered in such molding operations.

8.3.2   Butyl rubber

Butyl rubber, a copolymer of isobutylene (97.5 to 99 percent) and the balance of isoprene or butadiene, is made in several grades. The difference in grades is determined by the number of moles of isoprene per hundred moles of isobutylene. The defining term is "mole percent unsaturation." If the unsaturation value, say, is two, then there are two moles of isoprene for every 98 moles of isobutylene.

The largest use of butyl rubber has been in power cable usually compounded of a 1-percent mole unsaturated material. Although the use of butyl rubber has been lessened because of the development of other materials, there are probably still many installations of butyl cable. Other applications for butyl rubber in the electrical field include embedments of transformers such as current transformers required to resist exposure to weathering.

As an electrical insulation material, butyl rubber has excellent resistance to ozone, heat, and moisture. It has good electrical properties and is strong enough so that abnormal usage is necessary to damage or break it mechanically. Other advantages are its great impermeability to gasses, high-energy absorption, and chemical resistance (except to petroleum oils and solvents which attack all hydrocarbon rubbers).

Butyl rubbers can be compounded to exhibit high damping properties combined with low dynamic modulus. Such materials can be employed in equipment or devices where shock-resistant mounts or pads are needed.

For the 1-percent mole unsaturated compound the electrical properties are as follows: Volume resistivity is 2×10^{15} ohm-cm. Dielectric strength is 700 volts/mil. Dielectric constant at 60 Hz is 3.4, and the percent power factor at 60 Hz is 0.8. The dielectric constant is essentially constant over a wide frequency range, whereas the power factor shows a slight increase. The continuous-high temperature operating limit is $300°$ F. Brittle temperature is $- 80°$ F.

Isobutylene-isoprene copolymers are chlorinated to produce a modified butyl rubber which is more reactive and so will cure faster, exhibit higher heat resistance, afford lower compression set, blend better with other polymers, and cure in thick or nonuniform cross sections. Compounds that will cure at room temperatures are possible. Electrical properties are good. This copolymer is commercially known as chlorobutyl.

8.3.3   Chlorosulfonated polyethylene

Chlorosulfonated polyethylene is the reaction product of polyethylene, sulfur dioxide, and chlorine. It is similar to neoprene (see Par. 8.3.13) in many properties, except for superior resistance to oils, strong oxidizing chemicals, and weathering. Electrical properties are also generally superior to those of neoprene. It is self-extinguishing. Resistance to ozone and corona effects are outstanding. Abrasion resistance is good. Temperature range is $-10°$F to $+300°$F. Special compounds will give intermittent service at up to $325°$F.

When properly compounded, this elastomer does not support growth of mold, fungus, and the like. Since it withstands the effects of direct burial and long-time immersion in water, it is quite suitable for underground power wire and cable installations. Insulation applications are varied (for example, in building wire, mining cable, and industrial power cable). It conforms to the U.S. Bureau of Mines Burning Test and has Underwriters' Laboratories approval for power cable up to 600 volts. A considerable variety of vulcanizable compounds can be formulated depending on the starting reaction materials and additives.

8.3.4   Cross-linked polyethylene

Thermoplastic polyethylene can be transformed into a vulcanizable thermoset by so-called "chemical cross-linking" techniques (in contrast to the cross-linking that results from irradiation). In the chemical approach, the basic polymer is compounded with a suitable catalyst, such as dicumyl peroxide, then subjected to heat, approximately 260° F. Under these conditions the free ethylene radical reacts with the catalyst to form the cross-linked structure of the thermoset.

The cross-linked polyethylene has a number of properties that place it at advantage in comparison with various rubber and elastomer materials. Particularly noteworthy are the good aging properties, resistance to ozone, abrasion, cut-through, and stress-cracking. Low-temperature properties are generally good. Resistance to aging can be further improved by the addition of anti-oxidants. The addition of carbon black enhances physical strength.

Extensive use of cross-linked polyethylene is found in wire and insulation. Electrical properties of the basic thermoplastic polymer are largely retained by the cross-linked material, which is available in low- and high-density grades depending on the starting polymer.

8.3.5   Depolymerized (flowable) rubber

Depolymerized rubber (DPR) is basically an isoprene rubber that has been reduced to a flowable state without requiring any solvents. It offers the economy of being applied without the need for any elaborate equipment. Various DPR compounds are available for either room-temperature cure or heat-curing. The room-temperature cure method produces little, if any, exotherm. The mass of the casting being cured has no effect on the exotherm.

The room-temperature curing compounds are available as two-part systems that are mixed in application. The heat-curing compounds are already mixed and are available as one-part systems. Accelerators can be added and cure temperatures increased as required to shorten the cure process. Potting and encapsulation are major applications.

Temperature service range for cast units (continuous use) is from $-150°$ to $+221°F$ ($-103°$ to $+105°C$). Electrical properties for a typical casting are as follows:

Volume resistivity (ASTM D 257), ohm-cm	1.3×10^{13}
Dielectric strength (ASTM D 149), volts/mil	380
Dielectric constant (ASTM D 150),	
at 60 Hz	4.1
at 10^6 Hz	4.0
Dissipation factor (ASTM D 156),	
at 60 Hz	0.007
at 10^6 Hz	0.0073

For the same casting, the tensile strength after 24 hr at 158°F is 130 psi. Elongation is 100 percent (zero set at break). Hardness is 42 on the Shore A durometer.

8.3.6   Epichlorohydrin elastomers

Two materials comprise the epichlorohydrin elastomer family. The first is the polyepichlorohydrin elastomer; the second is the copolymer of epichlorohydrin and ethylene oxide. The respective structures are as follows:

Homopolymer   (CO)   Copolymer   (ECO)

The specific gravity of the homopolymer is 1.36; of the copolymer, 1.27. When cured, these elastomers possess properties that are not commonly found in other elastomers or rubbers. In addition, they possess the properties normally associated with elastomeric materials.

The homopolymer has outstanding low gas permeability and ozone resistance. However, it has very little "snap" at room temperature. The copolymer has high resilience and flexibility over a wide temperature range of 350° down to −75°F. Both elastomers show resistance to oils, aliphatic solvents, and aromatic fuels. Both have low hysteresis. Hardness is about the same for both—a Shore A durometer reading of 73 over a temperature range of 70° to 350°F.

Possible applications for these elastomers include jacketing for wire and cable where service conditions involve their special characteristics. Other uses in electrical equipment are indicated for mechanical parts such as gaskets and the like.

8.3.7   EPDM hydrocarbon rubber

The abbreviation EPDM refers to a hydrocarbon rubber that is a terpolymer of ethylene, propylene, and a diene. It is a general-purpose rubber with properties that favor its use as wire insulation. It is vulcanized and processed into finished form by methods standard in the rubber industry.

Performance properties can be summarized as follows: EPDM is virtually immune to attack by ozone. It has outstanding resistance to heat. Compounds for service at 300° to 350°F can be made, with higher temperature levels for short-time use. The rubber is flexible at low temperatures, with impact brittle points at −90°F or below. Dynamic properties are constant over a wide temperature range. Mechanical properties are good. Chemical resistance is also good, but, typically of the hydrocarbons, EPDM is attacked by petroleum-based oils. Resistance to steam is excellent. Weathering resistance is good and so is resistance to sunlight. Flame resistance is a weak spot; it is poor.

Typical electrical values are indicated in the following data for $\frac{3}{64}$-in. insulation on a H12 AWG wire after 24 hr in 60°F water:

Insulation resistance: 25,500 megohms/1,000 ft
Insulation resistance (constant): 76,400 megohms/1,000 ft
Dielectric strength: 800 volts/mil
Dielectric constant at 60 Hz: 3.07
Power factor at 60 Hz, 80 volts/mil: 0.40 percent.

Changes in dielectric constant and power factor after tests in water at 167°F are slight. EPDM shows good corona resistance. A $\frac{3}{64}$-in. cable with cover grounded against a steel plate with 10,000 volts applied to the conductor showed no damage after 3,500-hr exposure.

A copolymer of ethylene and propylene (known as EPT) has properties similar to those of EPDM. It is vulcanized with a peroxide system, whereas sulfur is used with EPDM.

8.3.8   Fluoroelastomers

The elastomer described here is a copolymer of vinylidene fluoride and hexafluoropropylene. It is marked by its outstanding chemical resistance. It withstands the effects of oils, fuels, lubricants, most mineral acids, many aliphatic and aromatic hydrocarbons (such as carbon tetrachloride, toluene, benzine, and xylene). However, it is severely affected by low-molecular weight esters, ethers, ketones, certain amines, and hot anhydrous hydrofluoric or chlorosulfonic acids. Solvents that are commonly used in lacquers attack it.

Mechanically the rubber is adequate for many uses such as gaskets and O-rings, which must withstand extreme conditions of temperature and liquids that attack other rubbers.

High-temperature limit for continuous service is 450°F which drops to 48 hr at 600°F. Low-temperature limits depend greatly upon application. If dynamic properties are important, -10°F is the low limit. For static applications, much lower temperature are handled. The brittle point is about -50°F.

The electrical properties are fair. When compounded for use as a wire insulation for low-voltage, power-frequency applications and where resistance to fluids and to high temperatures are involved, the fluoroelastomers show insulation properties in the following approximate values:

Volume resistivity, ohm-cm	2×10^{13}
Dielectric strength, volts/mil	500
Dielectric constant (approx.)	15
Power factor, percent (approx.)	5

A wire insulated with the fluoroelastomer aged seven days at 300°F and tested at 75°F had the following characteristics:

Insulation resistance, megohms/1,000 ft	93
Dielectric strength, volts/mil	869
Dielectric constant at 10^3 Hz	7.16
Power factor at 10^3 Hz, percent	1.65

8.3.9   Hard rubber

The description "hard rubber" is applied to several different types of material. Here, it is used to describe a material compounded from sulfur, SBR hydrocarbon rubber, butyl rubber, and nitrile rubber. Various other ingredients are added to obtain special properties; for example, carbon black is a common additive. The ratio of sulfur content to that of the finished hard rubber is usually held to 47-percent sulfur

The hard rubber, when processed, can be given a most attractive jet black, shiny surface. It is often used as a panel material for electrical instruments, in which it combines an insulating and a structural function.

The uncured hard rubber compound can be sheeted and formed during cure.

Variations in compound properties depend on the extent of sulfur content and other ingredients.

8.3.10   Natural rubber and isoprene synthetic rubber

Natural rubber is of two types: *Hevea*, which is soft and resilient and characterized by a low heat buildup, and *gutta-percha*, a hard and tough material. The *hevea* is a stereospecific cis-1,4 polyisoprene whereas the *gutta-percha* is a trans-1,4 polyisoprene. Although the gutta-percha material finds some use in electrical insulation, this discussion will be concerned primarily with the more widely used *hevea.*

Synthetic isoprene rubber is essentially of the same chemical structure as *hevea.* Consequently, both may be considered as identical for most engineering applications. As with most hydrocarbon rubbers, both are attacked and swelled by petroleum oils, both have poor flame resistance, and both are subject to stress-cracking under exposure to oxygen, ozone, heat, light, and irradiation. Despite their chemical similarity, there are certain characteristics that *do* differ. For example, the synthetic is nonstaining whereas the natural stains. For use in pressure-sensitive adhesive electrical tapes, for example, the synthetic would be preferred. (For convenience, the natural *hevea* will be referred to here by its common abbreviation, NR, and the synthetic isoprene by IR.)

The fact that IR is a synthetic is the most important distinction between the two materials. It is possible to control the molecular weight of IR during synthesis and so produce a material with predetermined properties. This is not possible with NR, although many different grades are commercially produced through the many possibilities provided by the available compounding techniques. Resiliency is probably the most outstanding characteristic of both NR and IR materials. A ball of good NR compound can rebound 70 percent or more of the height from which it is dropped.

The room-temperature electrical characteristics of NR and IR are given here as typical for the unvulcanized material: Volume resistivity, 10^{15} to 10^{17} ohm-cm; dielectric constant at 10^3 Hz, 2.3 to 3.0; and dissipation factor at 10^3 Hz, 0.0023 to 0.0030.

The effects of moisture or water are particularly important in the electrical applications of rubbers since these materials are used for insulation in wire and cable conductors (also as encapsulants for transformers and other devices) as protection from the effects of direct immersion in water or the effects of environmental moisture.

In general the effects of immersion in water or the presence of moisture in the environment depend upon the composition of the compound. The nonrubber ingredients in a rubber compound tend to increase the absorption of water. The vulcanization process tends to decrease the absorption of water. Increase in temperature increases the absorption, and aging decreases it.

The specific effects of water or moisture in electrical parameters are as follows:

Insulation resistance of rubber in the unvulcanized or vulcanized state is not substantially affected by absorption of water.

Dielectric constant of unvulcanized and vulcanized rubber is increased by absorption of water since water itself has a high dielectric constant. The *amount* of water or moisture absorbed would correspondingly increase the dielectric constant.

The *dissipation factor or power factor* is also increased by absorption of water. The effect here, however, depends on the compound and its ingredients, rather than on the amount of water absorbed.

The *dielectric strength* in volts/mil as used in engineering is a logarithmic function of the thickness tested. At 60 Hz the electric discharge of the test voltage demands that the time and voltage used in the test must be controlled and considered in use of the data. The logarithmic relation of thickness tested is kept where temperature is increased. Water has a small effect on the dielectric strength. ASTM D 149 describes test methods for dielectric strength of rubber and rubber products.

8.3.11 NBR nitrile rubber

Butadiene-acrilonitrile copolymers comprise a family of rubbers known as NBR, also as Buna N. These materials are noteworthy for their resistance to oils, solvents, water, and hot air, but the electrical resistivities are too low to allow for their use as primary electrical insulation. With the addition of conductive fillers, however, the NBR rubbers find electrical applications as low-voltage conductive elements in such applications as static dissipation devices and certain types of heating devices. Volume resistivities of such conductive compounds are not in excess of 100 to 500 ohm-cm.

Blends with several resins are possible; for example, with phenolics or polyvinyl chloride.

8.3.12 NBR-PVC blends

A blend of NBR with polyvinyl chloride exhibits all the rubber-like properties of the nitrile rubbers and has good oil and ozone resistance. The presence of nitrile groups limits the electrical properties, but these properties are acceptable for certain applications. Among those suggested are retractable connector cords and heater cords.

When heated, the NBR/PVC blends harden rather than melt. Even under very heavy overloads NBR/PVC wire or cable insulation will remain in place.

Various fillers and plasticizers are used in compounding these blends. Pigments may also be added to obtain desired colors. Standard molding techniques may be employed.

8.3.13 Polychloroprene rubber (neoprene)

Polychloroprene rubber (known more commonly as neoprene) contains chlorine polar groups that limit electrical insulating use to low voltages (600 V) and to low frequencies (60 Hz). As a protective outer jacket for insulated wire and cable, however, neoprene can be used at all voltages. The most important properties of neoprene are not the electrical but the mechanical, thermal, and chemical.

Neoprene offers excellent resistance to abrasion, impact, and to damage from flexing and twisting. Heat build-up during flexing is low. Neoprene is tough and does not crack easily. Recovery from deformation is good; there is a relatively low residual effect.

Temperature range for continuous service is from $-10°$F to $+200°$F. Brittle point is about $-40°$F. Some formulations are capable of being used beyond this temperature range. Heat-aging, however, hardens neoprene and it loses resilience. Neoprene will burn when in contact with flames but will not support combustion.

Neoprene is resistant to petroleum oils but is attacked by hydrocarbons and chlorinated solvents. It is resistant to sunlight, ozone, and weathering, and can withstand long immersion in fresh or salt water and direct burial in earth. It can be specially compounded to provide resistance to corona and can be combined with other elastomers or rubbers (SBR, for example) to attain improved selected properties.

Typical electrical properties are: Volume resistivity, 10^{11} ohm-cm; dielectric strength, $150+$ volts/mil; dielectric constant at 10^3 Hz, 9.0; and dissipation factor at 10^3 Hz, 0.030.

8.3.14   Polyisoprene rubbers

Synthetic rubbers produced through polymerization of isoprene and use of associated techniques are very similar to natural rubber in their properties because their chemical structures are essentially of the same order. The isoprene synthetics have been already discussed in Par. 8.3.10 together with the natural rubber.

8.3.15   Polysulfide rubbers

Elastomeric polymers characterized by sulfur-containing backbones in their structure are available in a variety of types and compounds. Sulfur content is about 80 percent. These polymers are useful where exposure to oils, fuels, solvents, and ozone would cause deterioration of other rubbers. Cable jackets provide a typical application. A liquid polysulfide rubber that will cure at room temperature is useful for casting, embedment, potting, and coating of electrical devices. A wide range of physical and electrical properties may be obtained by selective compounding. The rubber-like castings remain flexible at low temperatures. Useful temperature range is from $-65°$F to $+230°$F.

For a cast polysulfide rubber, typical electrical properties are as follows:

Volume resistivity at 25°C, 50% RH, ohm-cm	4×10^{10}
Dielectric strength, volts/mil	340
Dielectric constant at 10^3 Hz	25
Dissipation factor at 10^3 Hz	0.07

The flexibility of polysulfide rubber and the outstanding properties of epoxy resins for casting and like applications can be combined in poylsulfide-modified epoxy casting systems. Improved physical and electrical properties can be varied by selective compounding. A liquid polysulfide/epoxy system with a 1:1 ratio of polysulfide to epoxy shows the following electrical properties:

Volume resistivity, ohm-cm	3×10^{13}
Dielectric strength (short-time), volts/mil	340
Dielectric constant at 10^3 Hz	5.0
Dissipation factor at 10^3 Hz	0.2

The presence of a large sulfur content gives the polysulfide rubbers a distinctly pungent odor.

8.3.16 Styrene-butadiene rubber (SBR)

A copolymer of styrene and butadiene (abbreviated as SBR) provides a general-purpose rubber. Although its use as insulation for wire and cable has declined, it is still used in low-voltage applications. SBR can be vulcanized and processed by standard rubber-industry methods. Historically, SBR was the first large-scale replacement of natural rubber. It was first known as Buna-S and then also as GR-S (government rubber, styrene type).

There are several types and subtypes of SBR rubbers, based on different compounding ingredients and polymerization techniques. The original SBR compound (the so-called hot type) is polymerized at $47°C$ to $49°C$ ($117°$ to $120°F$). The cold types are polymerized at $5°C$ ($41°F$). With the wide choice of compounding recipes, it is apparent that SBR properties can vary widely. It is important to consult with the compounder to assure procurement of the optimum compound for a given application.

The mechanical properties of SBR are generally good, with some better and some worse than those of natural rubber, but, on the whole, with lower tensile strength and lower tear strength. Also, the cold-type SBR compounds tend to have better mechanical properties than the hot-type compounds. Compared to natural rubber, SBR has higher heat conductivity, greater resistance to aging and oxidation, greater resistance to plastic flow at higher temperatures, and less tackiness at high temperature.

Chemical resistance of SBR is good except in petroleum oils. Chlorinated solvents also attack it. Resistance to water is good, but resistance to flame is poor. SBR can be combined with other rubbers to secure improvement in selected properties.

With so many compounds available, there is variation in electrical performance. The data given here has been selected from numerous sources to provide a rough guide to properties that can be expected:

Volume resistivity, ohm-cm	10^{15}
Dielectric strength	Depends on effects of electrical discharge during test
Dielectric constant at 10^3 Hz	2.9
Dissipation factor at 10^3 Hz	0.0032

When exposed to heat-aging, SBR hardens, unlike natural rubber, which softens. Use temperature of SBR in network cables can be as high as $260°C$ ($500°F$) for four-hour duration. Conductor temperature with SBR insulation is on the order of $60°$ to $90°C$ ($140°$ to $194°F$).

8.3.17 Silicone rubbers

Silicone rubber is probably the most important polymeric elastomer in electrical applications. The range and extent of silicone rubber is perhaps best indicated by reference to ASTM 1418-69, "Recommended Practice for Nomenclature for Rubbers and Rubber Latices," which provides a code designation for the silicone materials with the letter Q. In the previous ASTM document (1967), the identifying symbol was "Si"; both symbols are given in Table 8.4.

Table 8.4   Silicone Rubbers

Letter code 1418-69	Letter code 1418-67	Composition	Indicated use
MQ	Si	Methyl group only	General-purpose
VMQ	VSi	Methyl, vinyl group	General-purpose; high-temperature; low-compression set
PVMQ	PVSi	Methyl, vinyl, phenyl	High-strength; low-temperature; irradiation resistance
PMQ	PSi	Methyl, phenyl	Low-temperature
FMQ	FSi	Methyl, fluorine, fluorosilicones	Oil-, fuel-, and solvent-resistance

The above-listed silicone rubber polymers have the following common structural configuration with its typical SiO backbone:

$$
\begin{array}{ccccccc}
\mathrm{CH_3} & & \mathrm{CH_3} & & \mathrm{CH_3} & \\
| & & | & & | & \\
-\mathrm{Si}-\mathrm{O}-\mathrm{Si}-\mathrm{O}-\mathrm{Si}-\mathrm{O}- \\
| & & | & & | & \\
\mathrm{CH_3} & & \mathrm{CH_3} & & \mathrm{CH_3} &
\end{array}
$$

The methyl (CH_3) side chains may be replaced by other groups such as phenyl (C_6H_5) or vinyl ($CH_2{:}CH-$).

Silicone rubber compounds may be fabricated by various molding techniques, including compression-, transfer-, and injection-molding. When applied as wire and cable insulation, the autoclave or continuous vulcanization (CV) processes may be employed. The CV process is lower in cost when large volumes are involved. Some silicone rubber-insulated wire and cable may require a post-oven cure to develop optimum properties. Silicone rubber-coated glass cloth makes excellent high-temperature electrical insulating material. The coating process uses a dispersion of silicone rubber in a suitable solvent. Application may be by the "knife-coating" method or by dip-coating the cloth.

The characteristics of silicone rubber allow very many applications where it can be used to great advantage. For example, a series of general-purpose room-temperature-vulcanizing (RTV) compounds have been developed for use as potting, encapsulating, and embedment materials for electrical devices. Cure is effected at room temperatures, in contrast to the high temperatures required to vulcanize silicone rubber for use in molding operations.

When flexibility has to be maintained under conditions of extremely low temperatures and high-temperature insulation has to be provided as well, RTV silicone rubbers of methyl-phenyl composition are available as pourable liquids and thixotropic pastes.

The electrical properties of silicone rubbers are much the same for all types and forms. These materials retain a very large part of their properties over a wide range of temperatures and frequencies. Typical values are as follows: Volume resistivity, 10^{11} to 10^{17} ohm-cm; dielectric strength, 450 to 600 volts/mil; dielectric constant at 10^3 Hz, 3.0 to 3.5; and dissipation factor at 10^3 Hz, 0.001 to 0.002. Indicative of their relatively constant properties over a range of frequencies, a specific silicone rubber compound showed a dielectric constant of 2.93 at 10^2 Hz and 2.90 at 10^8 Hz. For the same compound, the power factor was 0.000056 at 10^2 Hz and 0.001 at 10^7 Hz.

The electrical properties change very little up to $250°$ C ($480°$ F). Long aging at this temperature likewise produce few changes. When failure occurs, it is generally caused by physical embrittlement, not by electrical stress. As a rule, compounds made for optimum retention of physical properties after heat aging will also show optimum retention of electrical properties.

The effects of ozone and corona are very slight when compared with the behavior of other rubbers. Properly cured silicone rubbers will not support the growth of fungus. Because there is no sulfur content, silicone rubber does not tend to cause staining or corrosion. Silicone rubbers are chemically inert materials. Radiation resistance is related to the phenyl content of the rubber; the higher the content, the greater the resistance.

The properties of any specific silicone rubber will depend on many variables far too numerous to enumerate here. When specifications are being prepared for a given application, the requirements should be thoroughly reviewed with prospective suppliers of the rubber.

8.3.18   Urethane rubbers

Polyurethane or isocyanate rubbers constitute a family of elastomeric materials that are the reaction product of diisocyanates and polyalkylene ether glycols. The reaction processes involve cross-linking and chain extensions that produce a wide variety of materials, in form and properties. These rubbers are available as "solids" or gums that can be compounded and processed by conventional rubber-making equipment. Curing produces tough but resilient solids. The urethanes that are available as liquids find application in casting, embedment, encapsulation, and the like, of electronic devices, particularly where both flexibility and light weight are major application factors. Flexibility adds an extra measure of protection to delicate devices. Unpigmented castings are translucent. Opaque, black, and colored castings can be made.

The typical physical properties of urethane rubbers indicate them to be tough and durable. They possess outstanding abrasion resistance, impact resistance, and greater load-bearing capacity than other elastomers of comparable hardness. A low, unlubricated coefficient of friction enables them to be used in bearings.

Temperature range of "solid" rubbers is from $-80°$ F to not above $+250°$ F. Resistance to water is good, even when water is present as an emulsion in oil. Flame resistance merits the "self-extinguishing" rating under ASTM D 635. Oil resistance is also good, but aromatic hydrocarbons and polar solvents attack the urethanes.

Electrical properties when used in encapsulation and the like at frequencies up to 10^5 Hz and temperatures up to $212°$ F are given in Table 8.5

Table 8.5   Electrical Properties of Urethane Rubbers

	Temperature; °F	Frequencies	
		100 Hz	10⁵ Hz
Power factor (ASTM D150), percent	75	4.7	5.9
	158	4.7	—
	212	12.6	3.9
Dielectric constant	75	9.4	7.8
	158	11.0	—
	212	11.5	9.9
Volume resistivity, ohm-cm	75	4.8×10^{11}	
	158	3.8×10^{10}	
	212	2.3×10^{10}	
Dielectric strength (ASTM D 194), volts/mil	—	450–500	
Arc resistance (ASTM D 495) sec	—	73 (no carbon tracking)	

BIBLIOGRAPHY

Note 1: *The information in this chapter has been largely drawn from technical data published by manufacturers of the various polymers and elastomers discussed. Since in many instances several manufacturers produce given materials, it would not be feasible to credit individual sources. For guidance to manufacturing sources, the reader is referred to the following:* NEMA, Electrical Insulation Div., New York, N.Y.; Society of the Plastics Industry, New York, N.Y. Society of Plastics Engineers, Greenwich, Conn., Rubber Manufacturer Association, New York, N.Y.; Modern Plastics Encyclopedia *(published annually)*, New York, N.Y.; Insulation/Circuits Directory/Encyclopedia *(published annually)*, Libertyville, Ill.

Note 2: *The following sources are suggested for basic information on polymer and elastomer chemistry:* Billmeyer, F. W., Jr., Textbook of Polymer Science, *Interscience, N.Y., 1962;* Flory, P. J., Principles of Polymer Chemistry, *Cornell University Press, 1953;* Morton M., Introduction to Rubber Technology, *Reinhold, New York, 1959;* Winding, C.C., and Hiatt, G.D., Polymeric Materials, *McGraw-Hill, New York, 1961.*

Note 3: *For information on current technical and scientific work in the fields of insulation and dielectrics, see the* Annual Reports *of the Conference on Electrical Insulation and Dielectric Phenomena, National Academy of Sciences/National Research Council. These reports are available from the NAS Printing and Publishing Office, Washington, D.C.*

<table>
<tr><td>CHAPTER
9</td><td># MAGNET-WIRE
AND WIRE-AND-CABLE
INSULATIONS</td></tr>
</table>

9.1 Magnet Wire

The term "magnet wire" describes a wide variety of nonstranded, thinly insulated wire used for windings and coils. This type of wire holds a strategic functional position in any device or piece of equipment in which it is used. More than any other component it must be capable of meeting a great many operating and environmental requirements. Magnet wire carries current at elevated temperature with voltage stress, and it also may be called upon to serve under such adverse conditions as corona, moisture, vibration, and radiation. Special tests have been developed to prove the multifaceted performance requirements for magnet wire. These tests will be described in this chapter, as well as examples of successfully used wires.

NEMA has been very active in recent years improving existing tests and developing new test methods and test equipment to meet user demands for higher quality and greater uniformity of magnet wire. Standardization of test methods is a major goal in these activities.[1,2] These methods are discussed in the following subparagraphs.

9.1.1 Thermal stability

Based on ASTM D-2307, which was established by reference to IEEE No. 57, a 20,000-hr temperature index has been set up by NEMA as the rating level for untreated film-insulated magnet wire in air.[3,4] NEMA specifies a heat-aging point of 5,000 hr as well as several other accelerated aging points, all plotted as log life compared to reciprocal absolute temperature for extrapolation purposes.

9.1.2 Enamel film build

NEMA has established the increased minimum enamel film build (addition to bare-wire diameter) on an exponential basis as follows:

$$log \text{ f.b.} = A - \text{AWG}/44.8 \qquad (9.1)$$

in which:

    f.b. = film build (min.), in mils (0.001 in.)
    A = a constant (0.518 for single, 0.818 for heavy, and 0.995 for triple grade)
    AWG = American Wire Gage size

9.1.3   Dielectric strength

NEMA's increased minimum dielectric strength for enameled magnet wire is also based on an exponential basis, as follows:

$$log\ V = A - B(\text{AWG}) \tag{9.2}$$

in which:

    V = minimum breakdown voltage, in volts (rms)
    AWG = American Wire Gage size
    A, B = constants

The constants A and B for Eq. 9.2 are listed in Table 9.1.

Table 9.1   Constants for Minimum Voltage Relationships, Eq. 9.2

AWG range	Grade	Test type	Constant A	Constant B
4–9	Heavy	Foil	3.618	0.0126
10–13	Heavy	Twist	3.939	0.0147
14–30	Single	Twist	3.706	0.0114
14–30	Heavy	Twist	3.962	0.0114
14–30	Triple	Twist	4.087	0.0114
31–44	Single	Twist	4.782	0.0473
31–44	Heavy	Twist	4.708	0.0363

9.1.4   Solubility

NEMA has shortened the solubility test to a 30-min exposure at 60° C in three solvents followed by a scrape test. The solvents are 50/50 ethyl cellosolve/xylol, perchlorethylene, and xylol. The scrape test calls for the same needle size as that used in the unilateral scrape test but with approximately one-half the loads required for unexposed wire.

9.1.5   Refrigerant extraction test

The present NEMA R-22 extraction test is based on a residue method. The wire is exposed to the refrigerant at 600 psi for 6 hr. The refrigerant is then evaporated and the container weighed for residue pick-up. Acceptable limits call for residues of less than 0.25 percent of film weight.

9.1.6   NEMA blister-voltage check

Blisters have been found even in successful hermetic wire types after release of pressure following the conclusion of tests at simulated use conditions of 600-psi pressure and working temperatures. NEMA has devised a voltage-check method for determining the severity of blistering. The twisted-pair specimen is subjected to 600-psi R-22 exposure for 72 hr followed by a pressure release and exposure to 150° C in air for 10 min. The breakdown voltage of the room temperature specimen (after these exposures) must not be less than the minimum voltage set in Par. 9.1.3.

9.1.7   Adhesion and flexibility

NEMA has combined these properties in one test. The sample of straight wire is rapidly stretched to a required elongation followed by a mandrel wrap as shown in Table 9.2 (for copper wire).

Table 9.2   Adhesion and Flexibility

AWG range	Elongation, percent	Mandrel wrap
4–9	30	None
10–13	25	5X
14–20	20	3X
21–30	15	1X
31–44	20	3X
45–56	To breaking point	—

Similar parameters are being studied for aluminum wire. Certain exceptions to Table 9.2 appear in NEMA MW-1000 for individual specifications.

9.1.8   Heat shock

NEMA heat-shock tests are run on the specimens made for the adhesion and flexibility tests described in the preceding paragraph. These specimens are exposed in air ovens to various temperatures for half-hour durations. The temperatures are determined by the thermal classification of the enamel, as shown in Table 9.3.

Table 9.3   Heat Shock Temperatures

Temperature index, ° C	Heat shock temperature, ° C
105	175
130	175
155	175
180	200
200	220
220	240

Exceptions for individual specifications appear in NEMA MW-1000. It should also be noted that although some enamels have excellent thermal stability when tested according to Par. 9.1.1, derating them is necessary because of heat-shock deficiencies.

9.1.9   Thermoplastic flow

NEMA has devised a thermoplastic flow (or cut-through) device that tests 20 samples under load at a temperature rise of 5° C per min. The wires are crossed under the load and a contact voltage is applied between wires to indicate faults. End-point minimum cut-through temperatures have been set for the various wires in Table 9.4.

Table 9.4   Thermoplastic Flow Limits

Temperature index, ° C	Minimum cut-through, ° C
105	170
130	170
155	200
180	300
200	300
220	400

9.1.10   Scrape resistance

NEMA has standardized on the unilateral scrape tester which makes one scrape per operation but does it under increasing load and stops when the bare wire is contacted. A 0.009-in. steel needle is used for the scraping head. Values (minimum) range from 300 grams for small wires and thin coatings to 1,800 grams for large wires and heavy coatings, and the test applies only to the AWG range of 10 to 30. For example, a single grade No. 18 wire must have a scrape of 550 grams whereas a heavy grade No. 18 wire must have a scrape value equal to or greater than 1,150 grams.

9.1.11   Springback

To assure proper winding conditions, the wire must be relatively soft annealed. NEMA has included the springback test with tighter controls to insure the meeting of these properties for desired winding properties. The test calls for specific weights and mandrel sizes, and the angle measured as springback is the sum of three turns of wire as it unwinds from the mandrel. Table 9.5 shows the parameters for the test and the constants for Eq. 9.3.

$$log \, SB = A + (B \times AWG) \tag{9.3}$$

in which:

SB = maximum springback, in degrees
AWG = American Wire Gage size
A, B = constants

Table 9.5   Springback Test

AWG range	Load, oz	Mandrel, in.	A	B
14–20	16	3¼	1.164	.0328
21–26	4	1⅞	1.086	.0306
27–30	2	¾	0.609	.0403

9.1.12   Elongation

The minimum limits have been raised by NEMA to improve winding conditions. The minimum percent elongation (for copper wire in the size range of 4 to 44 AWG) is given by:

$$E = 57.5 \, log \, log \, d + 7.55 \tag{9.4}$$

in which

E = elongation, in percent
d = diameter of wire, in points (0.0001 in.).

For extra fine wire (45 to 50 AWG), the equation must be altered as follows:

$$E = 61.4 \log \log d + 5.11 \qquad (9.5)$$

An approximate relationship for aluminum minimum elongation (sizes 15 to 38 AWG) is given by the following:

$$E = 54.9 \log \log d - 3.2 \qquad (9.6)$$

Aluminum sizes larger than 15 AWG do not fit Eq. 9.6.

9.1.13   Continuity

NEMA has developed a high-voltage d-c continuity bench-check test for wire sampling. The wire is run over a grooved sheave with high voltage d-c applied between wire and sheave to detect flaws in the insulation. The voltages and maximum allowable faults per hundred feet are as required in the MW-1000 specification, as are the test parameters. The criteria for 14 to 24 AWG are given in Table 9.6.

Table 9.6   Continuity Test

Wire grade	Voltage, d-c	Faults/100 ft
Single	1,000	10
Heavy	1,500	5
Triple	2,000	3

The test is currently being extended to finer wire sizes. Continuous continuity testing within magnet wire plants can be a helpful tool in the control of quality.

9.1.14   Overload test

The overload or burnout test developed by NEMA in the last few years has been a significant advance in rapid determination of thermal stability. NEMA standardized on a step-current test in which a twisted-pair specimen is subjected to a series of currents in increasing steps representing seven to nine times normal full-load levels. Temperatures range from 350° to 570° C, and end points are reached in a matter of minutes. NEMA is currently setting minimum overload limits expressed in "Figure of Merit" values. This is a method of giving more "weight" to time at higher current settings and helps to separate the various magnet wires into classes. Table 9.7 shows the kind of data gathered for different classes:

Table 9.7   Overload Tests

Temperature index, ° C	Typical figure of merit values
105	1–4
130	2.5–4
155	5–6
180	6–8
200	8–9
220	10–20

A definite advantage of the NEMA overload test is that only one test procedure is needed to test enamels from polyamide to polyimide (the lowest to the highest in overload characteristics). Polyimide has more than 28 times the burnout resistance of polyamide (some values are over 60 times those of polyamide).

9.1.15   Hot dielectric

The dielectric strength test at rated temperature has been adopted by NEMA. The minimum limits are set at 75 percent of the minimum requirements at room temperature.

9.1.16   Mandrel pull test

A test for evaluating windability has been developed by NEMA. This test automatically runs a specimen of magnet wire around a specified mandrel under tension and checks its continuity by high voltage d-c after each run. The test evaluates performance under severe winding conditions. To perform properly, an enameled magnet wire must have good adhesion and flexibility characteristics and must be resistant to abrasion and repeated sharp bending at a rapid rate.

9.2   Magnet Wire Types

Table 9.8 reviews magnet wires listed by the proposed revised military specification MIL W-583D and by NEMA MW-1000 specification and gives additional wires on the market as of this printing. The table shows construction, conductor, sizes and shapes, the appropriate MIL-W-583D type, with section number, the NEMA MW-1000 specification, the classification assigned to the wire type (often by NEMA-approved test but sometimes by practice), and projected uses for the wire.

A wire type exists in film-coated variety for every Class set up by IEEE Nos. 1, 57, 117, and other documents; the 200 and 220 classes are added to the IEEE list. In fiber types, less variety in respect to classes is available but all classes can be filled in both round and shaped wires.

The hermetic types shown by FH and the MW-C series 71–73 are especially resistant to the refrigerants: R-11, R-12, R-13, R-22, R-103, R-113 and the R-500 series of fluorocarbon liquids and gases (see Chap. 12 on liquid and gaseous dielectrics).

The newer polyimide and aromatic polyamide films and papers have not reached NEMA and military specification levels at this printing, although the aromatic polyamide paper is being considered by NEMA, and both materials are being used.

Several extremely high-temperature types with ceramic and glass insulation and protected conductors appear in the list. These range up to 650°C capabilities and belong in the IEEE Class C category. All are useful in high temperature and high radiation environments.

Table 9.9 gives various combinations of films and fibers available in round (O) and rectangular shapes (R) for added properties such as dielectric strength after bending and for mechanical protection afforded film by tough fibers.

The ultra-fine enameled wire (UFEW) is shown as a single entry although it is available in six enamel combinations: oleoresinous, polyester, polyurethane, polyvinyl formal, polyimide, and polyvinyl butyral bond coat over polyurethane base

Table 9.8 Magnet Wire Types

NOTES:

O.C. = Overcoat	Syn = Solderable acrylic plus nylon	C = Cotton
Bond = Bond coat or cement	C.A. = Cellulose acetate	P = Paper
O.R. = Oleoresinous or low-cost synth.	PES = Solderable polyester	S = Silk
P.U. = Polyurethane	FR = Friction coat	N.F. = Nylon fiber
P.E. = Polyester-terephthalate	T = Class 105 film (MIL)	G = Glass fiber
N.E. = Nylon-polyamide	B = Class 130 film or bond	Dg = Glass fiber plus polyester fiber
C.E. = Ceramic	L = Class 155 film (MIL)	G2 = Double glass
S.E. = Silicone	H = Class 180 film (MIL)	Av = Asbestos plus varnish
EP = Epoxy	K = Class 200 film (MIL)	CAF = Cellulose acetate fiber
TFE = Polytetrafluoroethylene (also PTF)	M = Class 220 film (MIL)	AWG = American Wire Gage
F = Formvar	V = Varnish (MIL), 155°C	NEMA = National Electric Mfgr. Assoc.
C.C. = Cement coat (bondable by heat)	O = Heat removable resin	MIL = MIL-W-583D Spec.
P.E.I. = Polyester imide	PVB = Polyvinyl butyral	Cu = Copper
PEAI = Polyester amide imide	FH = Hermetic formvar	Al = Aluminum
P.E.T. = Polyester-THEIC modified	SLW = Self-lubricating wire	Ni = Nickel
PAI = Polyamide imide (also AI)	UFEW = Ultra-fine enameled wire	Ag = Silver
SAC = Solderable acrylic (also SA)	P.A.P. = Nomex: polyamide paper (H.T.)	Fe = Iron
ML = Polyimide	P.I.F. = Kapton: polyimide film (H.T.)	

Description (see notes below)

Base	O.C.	Bond	Cond.	AWG sizes and shapes	MIL-W-583D Type	NEMA MW-1000	Class °C	Uses
O.R.	—	—	Cu	Rd 25-44	1-E	MW 1-C	105	Ignition coils
P.U.	—	—	Cu	Rd 25-44	2-U, U2	MW 2-C	105	Solder application
P.E.	—	—	Cu	Rd 4-44	10-L, L2, L3	MW 5-C	155	Coils
N.E.	—	—	Cu	Rd 14-44	3-N, N2	MW 6-C	105	Automotive
C.E.	S.E.	—	Cu	Rd 17-44	—	MW 7-C	180	Nuclear radiation
E.P.	—	—	Cu	Rd 4-44	13-TF, TF2, TF3	MW 9-C	130	Oil-filled transf.
TFE	—	—	Cu	Rd 8-44	4-T, T2, T3	MW 10-C	180	Low gassing
F	—	—	Cu	Rd 4-44	15-M, M2, M3	MW 15C	105	General-purpose, Class A
ML	N.E.	—	Cu	Rd 4-44	5-TN, TN2	MW 16-C	220	High temperature
F	—	—	Cu	Rd 14-44	—	MW 17-C	105	Rotating equipment
F	—	C.C.	Cu	Rd 14-44	6-TB, T2B, T3B	MW 19-C	105	Self-supporting coils
P.E.	N.E.	—	Cu	Rd 10-30	11-LN, LN2, LN-3	MW 24C	155	Rotating equipment

Table 9.8   Magnet Wire Types (Cont'd)

| Description (see notes below) | | | | AWG sizes and shapes | MIL-W-583D Type | NEMA MW-1000 | Class °C | Uses |
Base	O.C.	Bond	Cond.					
P.U.	N.E.	—	Cu	Rd 9-44	9-UN, UN2	MW 28-C	130	Rotating equipment
PEI or PEAI	—	—	Cu	Rd 4-44	12-H, H2, H3	MW 30-C	180	Hermetic motors
P.E.T. or PEI or PEAI	AI	—	Cu	Rd 4-44	14-K, K2, K3	MW 35-C	200	Wide variety of uses up to Class 200 and also hermetic motors
SAC	—	—	Cu	Rd 10-30	7-SA, SA2	MW 37C	105	Solder applications
SAC	NE	—	Cu	Rd 10-30	8-SYN, SYN2	MW 39-C	105	Motor windings
P.E.S.	—	—	Cu	Rd 10-30	—	—	155	Solder applications
P.E.S.	NE	P.E.	Cu	Rd 10-30	—	—	155	Solder applications
P.E.	AI	—	Cu	Rd 18-30	—	—	155	TV yoke coils
P.E.	AI	EP	Cu	Rd 18-30	—	—	130	Self-supporting coils
C.A.	—	—	Cu	Rd 36-51	—	—	105	Extra small coils
P.E.T.	NE	—	Cu	Rd	—	—	155	Motor windings
EP	—	EP	Cu	Rd, Sq. & Rect.	—	—	130	Self bonding
P.U.	FR	—	Cu	Rd 27-44	—	—	105	Universal coils
P.U.	—	PVB	Cu	Rd 15-44	—	MW 3-C	105	Solder-bond
P.U.	NE	PVB	Cu	Rd 8-31	—	MW 29-C	130	Self-supporting solder
F.H.	—	—	Cu	Rd & Rect. 8-30	—	—	105	Hermetic
SLW	—	—	Cu	Rd 10-44	—	—	Varied	Self-lubricating wire
ML	—	—	Cu	Rd 14-30	—	MW 71-C	220	Hermetic
PEI or PEAI	—	—	Cu	Rd 14-30	—	MW 72-C	180	Hermetic
PET or PEI or PEAI	AI	—	Cu	Rd 14-30	—	MW 73-C	200	Hermetic
C.E.	—	—	Cu-Ni or Ag-Ni	Rd 17-45	—	—	650	Very high temperature and radiation

Table 9.8   Magnet Wire Types (Cont'd)

Base	Description (see notes below)			AWG sizes and shapes	MIL-W-583D Type	NEMA MW-1000	Class °C	Uses
	O.C.	Bond	Cond.					
C.E.	O	—	Cu-Ni, Ag-Ni, Cu-Fe-Ni	Rd 18-40 Rect.	—	—	650	Very high temperature and radiation
C.E.	(Conv.) P.T.F.	—	Cu, Cu-Ni	Rd 17-50	—	—	220	High temperature and radiation
UFEW	—	—	Cu & Al	Rd 45-60	—	—	Varied	V.S. coils
F	—	—	Al	Rd 4-25	—	MW 15-A	105	General-purpose, Class A
P.E.	N.E.	—	Al	Rd 10-25	—	MW 24-A	155	Rotary equipment
P.E.T.	Al	—	Al	Rd 4-25	—	MW 35-A	200	Wide variety of uses up to Class 200
or PEI	Al	—	Al	Rd 4-25	—			
or PEAI	Al	—	Al	Rd 4-25	—			
PET	Al	—	Al	Rd 4-25	—	MW 36-A	200	Large equipment to Class 200
or PEI	Al	—	Al	Rd 4-25	—			
or PEAI	Al	—	Al	Sq. & Rect.	—			
PET	Al	—	Cu	Sq. & Rect.	—	MW 36-C	200	Large equipment to Class 200
or PEI	Al	—	Cu					
or PEAI	Al	—	Cu					
P.E.	—	—	Cu	Sq. & Rect.	17-L2, L4	MW 13-C	155	Large coils
EP	—	—	Cu	Sq. & Rect.	—	MW 14-C	130	Large oil-filled
F	—	—	Cu	Sq. & Rect.	16-T2, T4	MW 18-C	105	Large general-purpose
ML	—	—	Cu	Sq. & Rect.	18-M2, M4	MW 20-C	220	Large Class 220
Al	—	—	Cu	Sq. & Rect.	—	—	220	General-purpose, Class 220
Glass and P.E. fiber (varnished)			Cu	Sq. & Rect.	—	MW 46-C	155	
Glass and P.E. fiber (silicone)			Cu	Sq. & Rect.	—	MW 48-C	180	
Glass and P.E. fiber (silicone)			Cu	Rd 4-30	—	MW 47-C	180	
Glass and P.E. fiber (no bond) Fiber/bare Fiber/S. film Fiber/H. film			Cu	Rd 4-30	20- Dg, Dg2 BDg, BDg2 B2Dg, B2Dg2	MW 45-C	155	
Glass and P.E. fiber (varnished) Fiber/bare			Cu	Rd 4-30	20- DgV, Dg2V	MW 45-C	155	

Table 9.8   Magnet Wire Types (Cont'd)

Description (see notes below)				AWG sizes and shapes	MIL-W-583D Type	NEMA MW-1000	Class °C	Uses
Base	O.C.	Bond	Cond.					
Fiber/S. film					BDgV, BDg2V B2DgV, B2Dg2V			
Fiber/H. film								
Glass (varnished)			Cu	Rd 4/0-30	19- GV, G2V BGV, BG2V B2GV, B2G2V	MW 41-C	155	
Glass/bare								
Glass/S. film								
Glass/H. film								
Glass (silicone)			Cu	Rd	21- GH, G2H LGH, LG2H L2GH, L2G2H	MW 44-C	180	
Glass/bare								
Glass/S. film								
Glass/H. film								
Glass (varnished)			Cu	Sq. & Rect. 0-14	22- G2V B2GV, B2G2V	MW 42-C	155	
Glass/bare								
Glass/S. film								
Glass/H. film								
Glass (silicone)			Cu	Sq. & Rect. 0-14	23- G2H L2GH, L2G2H	MW 43-C	180	
Glass/bare								
Glass/H. film								
Cotton			Cu	Rd	—	MW 11-C C, C2, TC, TC2 T2C, T2C2	90	Oil-filled equipment, Class 105
Cotton/bare								
Cotton/S. film								
Cotton/H. film								
Silk			Cu	Rd	—	MW 21-C S, S2 TS, TS2 T2S, T2S2	90	Instrument coils
Silk/bare								
Silk/S. film								
Silk/H. film								
Nylon fiber			Cu	Rd	—	MW 22-C F, F2 TF, TF2 T2F, T2F2	90	Instrument coils
Nylon/bare								
Nylon/S. film								
Nylon/H. film								
Paper			Cu	Rd	—	MW 31C P TP	90	Oil-filled equipment, Class 105
Paper/bare								
Paper/S. film								

Table 9.8 Magnet Wire Types (Cont'd)

Description (see notes below)				AWG sizes and shapes	MIL-W-583D Type	NEMA MW-1000	Class °C	Uses
Base	O.C.	Bond	Cond.					
Paper/H. film						T2P		
Paper-Cotton P2C			Cu	Sq. & Rect. 0-14	—	MW 32-C	90	Oil-filled equipment, Class 105
Paper			Cu	Sq. & Rect. 0-14	—	MW 33-C	90	Oil-filled equipment, Class 105
Asbestos			Cu	Rd. Sq. & Rect.	AV*	—	130–180	
C.A. fiber			Cu	Rd	F, F2*	—	90	
Polyamide paper			Cu	Rd. & sq. 2/0-12	—	In process	220	
Polyamide paper			Al	Rd & sq. 2/0-12	—	In process	220	
Polyamide film			Cu	Rd. Sq. & Rect.	—	—	220	
Glass fiber plus glass powder and resin			Cu-Ni, Ag-Ni, Cu-Fe-Ni	Rd, Sq. Rect.	—	—	650	Extreme heat and radiation

* Refer to MIL-W-583C

Table 9.9   Combinations of Films and Fibers for Magnet Wires

Fibers	Enamels*						
	Formvar	Nylon-polyamide	Polyimide	Oleo-resinous	Epoxy	Polyester terepthalate	Polyurethane
Glass	O-R	—	O-R	O	O-R	O-R	—
Glass plus P.E. fiber	O-R	—	O-R	O	O-R	O-R	—
Cotton	O-R	O	—	O	—	—	—
Paper	O-R	—	—	O	O-R	—	—
Nylon	O	O	—	O	—	—	O
Silk	O	—	—	O	—	—	—
Nomex	—	—	**	—	—	—	—
Kapton	—	—	**	—	—	—	—
Asbestos	—	—	**	—	—	—	—
Cellulose acetate	**	—	—	**	—	—	—
Paper-cotton	**	—	—	—	—	—	—

* O-R = oleoresinous or low-cost synth; O = heat-removable resin.
** Suggested but not available

coat. These wire types require use of dust-free "white rooms" for making, handling, storage, winding, and treating.

Self-lubricated wires made by polymer modification are dry and oil-free and possess coefficients of friction below that of oiled nylon. They can be made by using any one of a variety of enamel types without sacrifice of properties. A second self-lubricated wire development, equally successful, is the use of dry polymeric lubricants formulated into enamels in a variety of film types.

Self-bonding of magnet wire by use of a heat-softened or solvent-softened outer coat on the base resin can gain desirable solidification of a coil without use of an additional varnish treatment.

Bond coats vary in thermal stability and in thermosetting characteristics. The search is on for a material with high thermal stability and sufficient cross linking to give high hot-bond strength similar to the harder varnishes. Some polyester bond coats approach this desired combination.

Solderable wires, using cellulose acetate, polyurethanes, solderable acrylics, polyamides, and solderable polyesters are useful where a multiple of taps or connections are made on small wire coils and where insulation removal is undesirable economically or technically. The connections can be made rapidly with ordinary lead-tin solders at soldering temperatures without first scraping the leads.

A variety of polyester resins exists with a range of thermal properties from Class 130 to Class 200 depending on the type of polyester (iso- or tere-) and on its modification. The latter is accomplished either by cyanuration or by imidization (esterimide or esteramide-imide).

Overcoating with amide-imide resin improves mechanical and chemical properties and makes a combination capable of hermetic use, in R-22 refrigerant, for example.

Use of aluminum conductors in place of copper often improves the thermal index for an enamel by as much as $15°$ to $20°C$, enabling an increase in the thermal classification in some cases.

Aluminum alloys with high aluminum content are available with improved mechanical properties over EC aluminum and with little sacrifice of conductivity. For example, at a certain level of anneal, one such alloy has 25 percent higher yield and tensile strengths with only 1 percent lower conductivity than EC aluminum.[1] This advantage allows higher winding tensions, a more compact winding body, and eliminates hot spots due to stretching in manufacture.

Use of protected conductors above 220 °C (see Cu-Ni and other constructions) is necessary to prevent oxidation of the base metal. Usually a thin layer (10 percent) of nickel over copper suffices to 400°C, but this must be increased to 20 percent up to 500° C. For 650° C use, a barrier layer, such as an iron lamination, can be inserted to stop migration of metal ions at the Cu-Ni interface.[2]

Twinned, triplet, and Litz wire (many strands of magnet wire interwoven or transposed to minimize skin effect) are available. The bifilar and trifilar types often are cemented together in parallel form and may or may not be further covered with textile threads such as silk, nylon, and rayon. These wires are often intended for high-frequency use, but some are useful in other applications.

Magnet-strip copper or aluminum has found largest use in the transformer field. Bare aluminum strip plus interlayer insulation such as paper, plastic film, Nomex (polyamide paper), mica paper, or other sheet insulation is the construction most used.

Coated strip has been developed, but at this printing has not been able to compete with the interleaved construction. Epoxy and polyamide-imide films on contoured edge strip have been offered in the past and may be useful in the future when greater demands are made on space and thermal stability.

Hollow-conductor magnet wires are available for forced cooling designs in large equipment. These conductors are in the square and rectangular shapes, and insulations run from epoxy enamel to glass, paper and cotton.

Recent highly technical interest in superconductors has produced a number of conductors with embedded Nb_3Sn filaments in copper and insulated with Formvar. The wires show zero resistance at temperatures close to absolute zero at which superconductivity results. The losses of the system are largely in the cryostat that produces the very low temperature, and no I^2R losses are evident. This state has prompted investigations into transmission, generation, and motor applications, not to mention the huge electromagnets being used in nuclear accelerators and atom smashers.

9.3   Wire and Cable Insulations

There are many types of wire and cable in use. It would appear that a commensurate variety of insulations would be required. Actually, grouping the wires and cables in use in logical categories would show that relatively few insulations dominate the field. Six major categories of wire and cable can be established: (1) Weatherproof line wire, (2) building wire and cable, (3) rubber and plastic power cables, (4) hook-up wire, (5) communications wire and cable, and (6) high-voltage power cable.

---

[1] Known commercially as Hytek 20 (Anaconda Wire & Cable trademarked material).
[2] Known commercially as Cufenic (Anaconda trademarked material).

The two insulations used on most of the wires in the first four categories are polyethylene (PE) and polyvinyl chloride (PVC). These insulations have captured most of the market. They are often used on the same wire, the polyethylene functioning as the primary insulation and polyvinyl chloride as the jacket.

In 1971 the consumption of low-density PE on wires and cables was estimated at 450 million pounds. High-density PE was used for a total of 35 million pounds. PVC consumption for the same year also reached 35 million pounds.

PE is used in several grades, in addition to the low-density and high-density materials, chlorosulfonated polyethylene, cross-linked polyethylene, and black (carbon black-filled) cross-linked polyethylene are employed.

PE has supplanted butyl rubber and varnished cambric as power-wire insulation. PVC materials are formulated to meet the requirements of the National Electric Code (NEC) as described below. PVC is used widely in many lesser categories of wire, such as appliance flexible cord. (See Pars. 9.3.7 and 9.3.8 for additional data on PE and PVC.)

9.3.1   Weatherproof line wire

Weatherproof line wire is used on pole lines and service connections. Sizes are 12 AWG through 2,000,000 circ. mils. It must be easy to string, and it should not stick on cross-arms. Its dense, homogeneous covering must withstand sunlight, weathering, fumes, acids, and alkalis. The wire must be of neat appearance when installed and not "festoon." Presently used insulation is black high-molecular weight PE or black cross-linked PE. The wire is not temperature-rated, but the annealing of its copper conductor above 80°C sets the limit. For service-drop cables, a high-molecular-weight PE with 75° C conductor temperature limit is used, or a cross-linked PE with a 90°C limit.

9.3.2   Building wire and cable

The National Electric Code and the Underwriters' "Standards for Safety" are the documents controlling the constructions and expected use of building wires and cables. The code lists the various approved insulations using letters or combinations of letters. For example, TW is a moisture-resistant thermoplastic. PVC meets the requirements.

TW type building wire has a single conductor, PVC insulated, for use up to 600 volts. Oil resistant, it is for general use in recognized raceways. Its maximum conductor temperature is 60°C in dry or wet locations. UL listings are No. 14 to No. 2 AWG. Copper conductors are solid or stranded. Insulation may be colored.

THW type building wire has a single conductor, PVC insulated, for use up to 600 volts. It is the general-purpose building wire for power and lighting as well as control wiring. It can be used in dry and wet locations and in the presence of oils. It is also used as switchboard wiring. Its maximum conductor temperature is 75°C dry or wet, 60°C as oil resistant, 90°C as appliance wiring. Its conductors are copper or aluminum. UL listings are No. 14 AWG to 1,000 M cm copper and No. 8 AWG to 1,000 M cm aluminum.

THHN or *THWN type building wire* can be used in the presence of gasoline and oils. Insulation is PVC with a nylon jacket. Maximum conductor temperatures are 75°C (for type THWN, in dry or wet locations, 90°C; for type THHN, in dry locations, as appliance wiring material, 105°C) UL listing is No. 14-8 AWG copper.

In addition the following wires are used in large quantities.

Nonmetallic-sheathed cable, PVC insulation. UL listing is No. 14-2 AWG copper, No. 12-8 AWG aluminum.

Underground feeder and branch circuit cable, PVC insulation and nonmetallic sheath.

Service entrance cable, three conductors flat, four conductors round. Insulation on conductors crosslinked is PE; jacket is PVC.

The code also makes provision for other insulations with assigned letters, such as R for rubber, MI for magnesium oxide, SA for silicone rubber, FEP for fluorinated ethylene propylene, V for varnished cambric, and A for asbestos. Paper and lead sheath carries no letter.

PVC has replaced rubber and a cotton braid as the insulation of most building wires. On line wires it has replaced cotton serving or braids saturated with asphalt.

9.3.3   Rubber and plastic power cable

This large group of wires and cables includes as its simplest member a one-conductor cable insulated with high-molecular-weight, low-density polyethylene intended for use as cathodic protection cable. It can be used in wet locations. Operation usually is direct current. A cable intended for use at 69,000 volts is insulated with thermoplastic or thermosetting polyethylene. Between these extremes are a large number of wires and cables. Many are insulated with PVC but others have different insulations, most of which we will mention in this section. Standards and specifications are issued by IPCEA-NEMA (*I*nsulated *P*ower *C*able *E*ngineers' *A*ssociation plus *N*ational *E*lectrical *M*anufacturers' *A*ssociation).

Single Conductor, Rubber Insulated 600-volt power cable carries UL listing types USE or RHW; its sizes are No. 12 AWG to 1000 M cm copper. Maximum conductor temperature is 75° for either type, in both wet and dry locations. It can be installed aerially, in conduit, or directly buried. It passes IEEE L260 Limiter Test as network cable.

Single-conductor power cable has chlorosulfonated polyethylene insulation. UL listings are RHW for dry or wet locations, RHH for dry locations, No. 14 AWG to 1000 M cm sizes. Maximum conductor temperatures are 75°C for RHW and 90°C for RHH. Both RHW and RHH types with a cross-linked PE insulation have a copper or aluminum conductor. Conductor temperature limits are unchanged.

Power cable with 5000-, 8000-, and 15,000-volt ratings is insulated with high-molecular-weight PE and has a PVC jacket. A resilient tape applied between the insulation and the copper shielding accommodates the large thermal expansion during load cycling. Tape is extruded simultaneously with the PE insulation.[3]

Single-conductor power cable for use up to 35,000 volts has ethylene-propylene insulations and an extruded strand shield. It is a flexible cable, easy to handle, and has superior ozone resistance. Maximum conductor temperature is 90°C. A similar cable has an insulation of cross-linked PE with a PVC jacket and is rated for 5 kV to 25 kV. Conductors are copper or aluminum. IPCEA-NEMA standard S66-524 governs. Largest 25 kV cable, shielded, is 1,000,000 cir. mils, with an approximate diameter of 2.01 in.

---

[3]   Anaconda Wire and Cable Co. patents 2,754,352 and 3,049,584.

9.3.4 Hook-up wire

An exact definition of "hook-up" has yet to be formulated by any recognized standards or specifications-issuing authority. However, a large number of wire constructions are covered by this term. Use is made of many insulating materials. In size, the range of diameters of such wire extends from 0.027 in., or No. 30 AWG (600 volts), to 0.494 in., or No. 0 AWG (2,500 volts). These values apply to MIL-W-81044 specification wire.

As the name implies, such wires are used in "hooking up" or connecting the components or parts of an electrical device. This may be done internally in a "black box" or electric device. It also applies to the connecting of two or more black boxes together. The wires may also be intended for aircraft and aerospace systems wiring.

The performance requirements for such wires can be voluminous and complex. Some needs may be met by a wire having an extruded covering of PVC. Colored wires are in demand for many types of wire.

The conductors are copper or high-strength copper alloy. Tin coating, silver plating, or nickel plating may be necessary to prevent chemical action between insulation and conductor.

There are seven military specifications which cover a very wide assortment of hook-up wires. The salient points covered by these specifications are given in the following resume of the specifications. The paragraphs are headed by the specification number.

MIL-W-5086A   Wire, electrical, 600 volts, copper (tinned), aircraft. Operating temperature, $-$ 55 to $+$ 105 °C. Although insulations are specifically named, requirements for Type I of this specification can be met with PVC, with nylon jacket. Type II has a glass braid over the PVC and a nylon jacket overall. Type III has PVC primary insulation and PVC jacket.

MIL-W-71398   Wire, electrical, polytetrafluoroethylene-insulated, 600 volts, copper. General-purpose aircraft wiring for high-temperature service, $-$ 55 to $+$ 200 °C. Insulation is extruded or tape-wrapped TFE on silver-plated copper. If nickel-plated copper is used, temperature range is $-$ 55 ° to $+$ 260 °C.

MIL-W-8777   Same as MIL-W-71398 except silicone primary insulation is rubber and primary cover is glass braid impregnated for high temperature. Jacket is polyester fiber braid impregnated for high temperature.

MIL-W-16878 (Navy)   Wire, electrical insulated, high temperature. Types are as follows: B, tinned copper, PVC, 600 volts, $-$ 54 ° to $+$ 105 °C; D, tinned copper, PVC, 3,000 volts, $-$ 54 ° to $+$ 105 °C; E, silver-coated, TFE, 600 volts, $-$ 65 ° to $+$ 200 °C; E, nickel-coated, TFE, 600 volts, $-$ 65 ° to $+$ 260 °C; F, tinned copper, silicone rubber, 600 volts, $-$ 54 ° to 200 °C; and K, silver-coated, FEP, 600 volts, $-$ 65 ° to 200 °C. The MIL-W-16878 is used for internal wiring in meters, panels, and electrical and electronic equipment generally.

MIL-W-81381   Wire, electric, polyimide (Kapton)[4]-insulated copper and

---

[4]   Kapton is a Du Pont tradename for polyimide insulations.

copper alloy, 600 volts, silver- or nickel-coated. Insulation is polyimide/FEP film, $-65°$ to $+200°C$. Used for general electronic wiring, internal wiring of meters, and the like. Applied in aerospace electric systems.

MIL-W-227598 (WEP) Wire, electrical, fluorocarbon-insulated copper. For silver-coated copper, with extruded TFE insulation containing abrasion-resistant mineral fillers, maximum temperature range is $-65°C$ to $+200°C$. For same insulation over nickel-plated copper, it is $-65°C$ to $+260°C$. TFE tape or TFE glass-reinforced tape may be used for insulation. A construction of TFE glass braid over primary insulation with extruded FEP jacket has $-65°C$ to $+200°C$ range. Specification applies to general-purpose aircraft wiring.

MIL-I-81044 Wire, electric, cross-linked polyalkene-insulated, copper (silver- or tin-plated copper), 600 to 2,500 volts. Maximum conductor temperature range is $-65°$ to $+150°C$. Primary insulation is cross-linked extruded polyalkene (polyethylene is chemically an alkene). Jacket is extruded cross-linked polyvinylidene fluoride.

Cross-linking is attained by irradiation of the extruded PE by high-energy electrons. The same process is followed *after* extrusion of the jacket. Irradiation mechanisms convert thermoplastics into cross-linked thermosets, thus increasing certain mechanical and other properties; for example, better resistance to stress cracking is achieved, as well as higher thermal limits. The wire specified here is used in high-density wiring and complex circuitry where, in addition to other requirements, workability characteristics must be outstanding. As might be expected, the wire is used, among other applications, in aerospace and aircraft work.

The seven military specifications on hook-up wire summarized above cover the following insulations: PVC, PE, polyvinylidene fluoride (Kynar), silicone rubber, TFE, FEP, polyimide/FEP film, nylon, and also glass-fiber and polyester-fiber materials. A large store of applicable data is contained in these specifications. They encompass the bulk of materials used as hook-up wire insulation.[5]

Other insulations may be suitable to meet some performance requirements. Just about any polymer that someone can find how to put on the wire will do. Since new polymers are introduced frequently, the engineer charged with design of hook-up wires has much to do with evaluating their possibilities. The entire wire and cable industry is served thereby. Among other materials are the urethanes, epoxies, neoprene, butyl rubber, fluorocarbon rubber, polyesters, polyamide-imides, polysulfone, epichlorohydrin rubbers, ethylene-propylene rubber, ionomers, and mineral insulation.

9.3.5 Communications wire and cable

In the communications field, polyethylene is the insulation most used. However, polypropylene (PP) may be able to displace PE from its present position. Recent formulations of PP appear to have overcome the previous objection to this material, the fact that it tended to contaminate copper (copper-poisoning) when used as insulation for copper conductors. This problem has been apparently solved by the

---

[5] Discussion on some of the polymers and films listed here has previously been given in Chaps. 6 and 8. The term "Kynar" is Pennwalt Corp. tradename for polyvinylidene fluoride.

addition of stabilizers that are permanent constituents of PP compounds used for insulation. The compound can be further improved by the addition of PE, which helps to increase overall stability. The performance characteristics of the improved PP compounds exhibit an improvement in comparison with those of PE. The brittle point of PP, however, is $-18\,°C$ compared to $-76\,°C$ for both the low- and high-density PE.

PE is used in jackets and sheaths as well as in primary insulation. Use of other insulations is mostly restricted to overall jackets where certain requirements can be met only by specific polymers or elastomers, for example, neoprene rubber.

Telephone cables   Two individual insulated wires are twisted together to form a pair that acts as the voice circuit. A selected number of twisted pairs are assembled together and stranded to form the active part of the cable. Over this assembly of twisted pairs a metal shield is applied to form electrostatic shielding and physical protection of the cable. The resistance and capacitance of the pairs between the wires and with the surrounding wires govern the telephone performance. Resistance is determined by cross-sectional area and the resistivity of the conductor. Capacitance is determined by the dielectric constant of the insulation and the amount of air present in the cable. The air tends to reduce the effective dielectric constant. Both polymeric and paper insulations are used for telephone cable.

Low-density PE is the polymeric insulation chosen both by the Bell and the General Telephone systems for insulation of telephone wires and cables. High-density PE is the polymer chosen by the Rural Electrification Authority in its loan program for rural telephone systems. This polymer has two disadvantages. Its insulation resistance in $75\,°C$ water and 600 volts applied tends to fall off rapidly. Stress caused by manufacturing operations is relaxed at higher temperatures, causing contraction of the polymer. Low-density PE resists the loss of insulation resistance for test times of 20 to 1 when compared with high-density PE.

Jackets for plastic telephone cables are made of black PE and copper or aluminum ribbons. Cables so made may be direct-buried or aerially suspended. It takes about a week for a moisture balance to be obtained between the outer air and the cable interior by diffusion through the insulation. Water that leaks into the cable will prevent its use, however. Water has a very high dielectric constant that is "seen" by the twisted pairs and causes their performance to change drastically.

Paper materials may also be used as wire insulation in the communications field. The paper is made from a mixture of rope and kraft pulps. The kraft pulp is of the sulfate type. Sulfite types are not used. (See Chap. 6 for detailed discussion of papers.)

Paper is applied by three methods: (1) Thin paper ribbon is helically applied; (2) paper is formed in place on the wire by means of a modified paper-making machine; or (3) a paper ribbon is applied as a longitudinal wrapping around the conductor. All three methods have been used to produce insulation for millions of feet of cable now in use. The first method, however, is of limited use in comparison with the other two.

Coaxial cables   Coaxial cables consist of a center conductor, the primary insulation, a shield, and an overall jacket. There are many cable designs that basically exhibit these parts. The primary insulation may be fluorocarbon polymers or solid or foamed polyethylene. The center conductor may be held in place by washers made of the above polymers. Foaming offers a method of control of the dielectric constant of the primary insulations.

The general design of coaxial cables is governed by the following formulas:[6]

$$Z_0 = (1.38 \ log \ 10 \ D \ / \ d) \ / \ \sqrt{e} \qquad (9.7)$$

$$C = 7.36 e \ / \ (log \ 10 D \ / \ d) \qquad (9.8)$$

in which:

$Z_0 =$ characteristic impedance, in ohms
$C =$ capacitance, in Pf/ft
$e =$ permittivity of primary insulation $= 1.0$ for air
$D =$ diameter over the primary insulation, in inches
$d =$ effective electrical diameter of center conductor, in inches

The permittivity of the primary insulation governs the cable diameter if the characteristic impedance and conductor size are specified.

9.3.6   High-voltage power cables

Where the load density is very high and the distances to be traversed are not great, it is economical to install high-voltage power cables to carry the load. The highest voltage presently in use in the USA is limited to 345,000 volts. The cable is a self-cooled system consisting of a pipe containing a 3-conductor cable immersed in oil. The pressure is 200 psi. The cable insulation is paper and oil. It is made by serving enough paper ribbons on the conductor until the desired thickness is attained. The conductors are assembled into a cable, which may be covered by a lead sheath and a PE jacket overall or the cable (no sheath) may be pulled in a pipe with the oil being put in place at the proper time. Shields and carbon-black paper tapes are a part of the paper-covered conductors. Oil-carrying tubes are incorporated in sheathed cables.

The largest market for high-voltage power cables is New York City and vicinity. The pavements cover more cable than has been installed by all other American users.

Crosslinked PE has been made for 138,000 volts. A more conservative practice is to limit power cables to 35,000 volts, using 69,000 volts only for special installations.

The industry is faced with the need for lower-loss cables that can carry large blocks of power over greater distances than those found in current practice. What is wanted is an insulation that is practicable for use in increasing the voltage at which losses are small as compared with those at present.

The reader is referred to an excellent report on present transmission systems using high voltages that appeared in the *IEEE Spectrum*, February 1972, p. 62, under the title, "Is Power to the People Going Underground?" The report (based on a conference devoted to this subject) describes the work being done to develop better cables and discusses various plans and methods for achieving improved transmission systems.

---

[6]   Source: Raychem Corp., Menlo Park, Calif. (Bulletin No. 7)

9.3.7   Summary of polyethylene properties

In view of the widespread use of PE in wire and cable insulations, further across-the-board data will be given here. PE has excellent electrical characteristics which it maintains over a wide range of frequencies and to temperatures near its melting point. The dielectric constant is low: 2.25 to 10^6 Hz. Dielectric losses are very low with a dissipation factor less than 0.0005 at 10^6 Hz. Dielectric strength is good: 500 volts/mil, step-by-step, for a 1/8-in. thickness. Water absorption is very low. Most solvents and weak acids and alkalis are resisted. Fabrication into wire and cable insulations has been well-mastered. Cross-linking by chemical or radiation converts the normal thermoplastic form into a thermoset. PE is compatible with PVC, which is often used as a jacket over it.

There are some disadvantages, however. The low-density material used in insulation is low in flexibility and tends to be stiff. Since it will burn, the maximum temperature for PE-insulated wire is low. If not protected against effects of sunlight, crazing occurs rapidly. Ethylene propylene-insulated wires are more flexible. Combinations of PE and EP polymers provide improved flexibility.

9.3.8   Summary of polyvinyl chloride properties

PVC has also become one of the most used insulations for wire and cable. When PVC was first introduced, the plasticizers needed to produce usable compounds were inadequate. In particular, there was difficulty with plasticizer stability in the compound. This problem has been long solved, and dependable PVC insulating compounds are now available that offer long-life reliability. PVC is soft and flexible but still strong enough to be used as jackets on cables of 2-in. diameter. The electrical properties, however, do not match those of PE. For a flexible PVC compound, the volume resistivity is 10^{11} to 10^{15} ohm-cm. Dielectric constant is 5.0 to 9.0 at 60 Hz and decreases as frequency increases. Dissipation factor is 0.09 to 0.14 at 60 Hz and also decreases as frequency increases. Dielectric strength is 300 volts/mil. PVC burns but slowly and may be self-extinguishing. It resists water, weak acids and alkalis, and most organic solvents. It is soluble and swells in esters and ketones and in aromatic hydrocarbons.

9.3.9   Summary of silicone rubber properties

Although the previous discussion on various types of wire and cable insulations has stressed PE and PVC materials, the unique properties of silicone rubber as a wire and cable insulation should not be overlooked. Silicone rubber is chemically unique among the insulations since it is an organo-inorganic polymer containing a silicone-oxygen backbone. There are no carbon linkages. This structure imparts to the polymer its outstanding combination of properties as insulation for many special-purpose wire and cable, such as nuclear cable, shipboard wire, and aircraft wire.

Silicone rubber-insulated wires resist extremes of temperature. Long life at 260°C is possible. And this can be exceeded by compounds containing selected silicone rubber gum. Brittle point is below − 65°C. Radiation resistance is good. Self-extinguishing compounds can be made to meet the vertical flame test of MIL-W-16878. Silicone rubber is fungus resistant, withstands weathering, and resists most solvents, water, acids and alkalis, oils, and hydraulic fluids. It is inert and usually odorless and tasteless. Loss of solids in vacuum is very low.

Silicone rubber is described as being easy to formulate. A suitable filler plus a vulcanizing agent enables the processor to produce a compound capable of providing required properties for a given application. There are no size limits to which silicone rubber insulation may be applied. It is usable on very small wires and larger ones up to 3,000,000 cir. mils. Postcures may be needed in some instances, but not usually.

Electrical characteristics are good over a wide temperature and frequency range. (Coaxial cables with silicon rubber insulation are used in Polaris missile work at frequencies up to 30 MHz.) Average values are as follows. The dielectric strength is 450 to 500 volts/mil at 150°C. The dielectric constant is 2.9 to 3.5, depending on the filler. The power factor is 0.00008 to 0.006, depending on the filler, with the value increasing as frequency rises. Silicone rubber is also remarkably resistant to corona effects, nearly as good as mica, and both ozone and sunlight effects are slight.

Flame-retardant grades of silicone rubber insulation are available. A unique characteristic of SR is that when it is burned, it leaves a *silica* ash that continues to maintain an insulating capability, particularly if the original compound had been reinforced with glass fibers. The silica ash may be able to provide insulating protection until the burned conductor is replaced. None of the other insulations leave a residue ash of comparable insulating value.

In construction of silicone rubber-insulated wire and cable, coated copper conductors should be used. For specifics on properties and construction, the wire and cable manufacturers should be consulted so that optimum types can be specified for given applications.

BIBLIOGRAPHY

A. Magnet Wire Insulation

Brancato, E. L., Johnson, L. M., and Walker, H. P., "Functional Evaluation of Motorette Insulation Systems," *Electrical Manufacturing*, Mar. 1959, p. 146.

Clark, F. M., *Insulating Materials for Design and Engineering Practice*, Wiley, 1962, pp. 766–787.

Goba, F. A., "Bibliography on Thermal Aging of Electrical Insulation," IEEE Cat. No. F87, 1969.

Lipsey, G. F., and Juneau, P. W., "Factors Affecting the Aging Characteristics of Various Wire Coating Materials in Transformer Oils," *AIEE Trans. Paper*, No. 86, 1960.

Payette, L. J., "Present Status of a Universal Magnet Wire," *Wire and Wire Products*, vol. 41, June 1966.

Pendleton, W. W., "Advanced Magnet Wire Systems," *Electro-Technology*, Oct. 1963, p. 95.

Pendleton, W. W., "Electrical Conductors for High-Temperature Operation," *Wire and Wire Products*, vol. 41, June 1966, p. 908.

Pendleton, W. W., and DeVries, D. H., "New High-Temperature Magnet Wires and their Application in Rotating Equipment," *Proceedings*, 5th IEEE-NEMA Electrical Insulation Conference, 1962.

Saums, H. L., and Pendleton, W. W., "Enameled Magnet Wire," *Electrical Manufacturing*, Oct. 1957, p. 130; *ibid*, "Evaluation and Application of Fibers-Insulated Magnet Wire," Feb. 1958, p. 93.

Saums, H. L., Pendleton, W. W., and Olson, E. H., "Compatability Tests of Enameled Wires in Closed Systems," *AIEE Conference Paper*, No. 560, 1956.

Standards and Specifications

NEMA: Spec. MW-1000*
ANSI: Stnd. ANSI-C9.100
MIL-W-583-C (revision W-583-D forthcoming)
IEEE: Thermal Stnd. IEEE #1 and Magnet Wire Test Procedure IEEE No. 57
ASTM: Magnet Wire Test'Procedure D-2307

B. Wire and Cable Insulation

"Chemicals and Coatings," staff review, *Wire and Wire Products,* April 1972. (Insulation materials are included.)
Couch, W. H., *et al,* "Modern High Voltage Insulation," *Power Apparatus and Systems,* AIEE Paper No. 693, Feb. 1956.
"Dielectric Constants of Rubber, Plastics, and Ceramics," a design guide, Electronic Properties Information Center, Hughes Aircraft Co., Culver City, Calif.
Elliott, D.K., "A Standard Procedure for Evaluating the Relative Thermal Life and Temperature Rating of Thin-Wall Airframe Wire Insulation," *IEEE Transactions on Electrical Insulation,* Mar. 1972, p. 16.
"Flat-Flexible Cable," a four-paper session in *Proceedings,* 10th IEEE-NEMA Electrical Insulation Conference, Sept. 1971, p. 99, p. 102, p. 309, and p. 331.
Lyons, C. J., and Leininger, R. I., "Radiation Effects on Insulation, Wire and Cable," *Insulation,* May 1959, p. 19; *ibid.,* June 1959, p. 38.
Milek, J. T., "Cable and Wire Insulation for Extreme Environments," *Interim Report No. 57,* June 14, 1967, Electronic Properties Information Center, Hughes Aircraft Co., Culver City, Calif. (Report contains 76 references plus summary reviews.)
Reed, R., "Corona Detection in Dielectric Materials," *Wire and Wire Products,* Dec. 1971.
Robertson, J. W., "New Flame Tests for Control and Power Cables," *Wire and Wire Products,* Feb. 1972.
Rubber Insulated Wire and Cable for Transmission and Distribution of Electrical Energy, NEMA Publication WC3-1959.
Spade. R., "Flammability Testing of Lead Wires," *Wire and Wire Products,* Dec. 1971.
"State-of-the-Art Review of Wire and Cable Insulations," a five-article series in *Insulation,* May 1966, June 1966, July 1966, Aug. 1966, and Feb. 1967.
Taylor, R. L., and Harriman, D. B., "Diffusions of Water Through Insulating Materials," *Ind. Eng. Chem.,* vol. 28, 1936, p. 1255.
Walder, V. T., "Synthetic Rubber for Wire Insulation," *Bell Laboratories Record,* vol. 23, 1945, p. 200.
Weber, H. M., "Synthetic Polymers for Cable Insulation," *Bull. SEV, Schweizer elektrotechnischer Verein,* vol. 50, 1959, p. 1109.
Western Electric Engineer, series of 18 articles on wire and cable manufacturing, issues of July through October, 1971. Western Electric Technical Information Dept., New York, N.Y.

C. Industry Sources

Note: Industry sources provide much useful properties and application data on insulation in the form of technical reports, specifications, and the like, Contact with most of the materials producers will make such information available. The brief selection below represents some industry data used as background in this chapter.

Anaconda Wire and Cable Co., *Specification Data Book.*
Budd Co., Polychem Div., *Materials for Multilayer Circuiting* (technical bulletin).

---

* See NEMA Magnet Wire Technical Committee report, *Nema MW-1000 Keeps Pace with The 70's,* covering presentation at Electrical Insulation Conference, Sept. 21, 1971, Chicago.

Dow Corning Corp., *Silicone Engineering Materials for Electrical and Electronic Equipment* (technical brochure).

E. I. du Pont de Nemours & Co., Technical Data, Kapton polyimide insulation.

General Electric Company, Silicone Div., *Silicone Rubber for Wire and Cable Application* (technical brochure).

Pennwalt Corp., "Evaluation of Kynar Vinylidene Fluoride Resin as an Insulator for Computer Applications" (technical paper).

Phelps Dodge Cable & Wire Co., *Underground Distribution Cable* (catalog).

Raychem Corp., "Insulation Materials Design to Achieve a High-Reliability Medium-Weight Wire" (technical report); also, "A New Extruded Alkene-Imide Wire."

CHAPTER
10

VARNISHES, FOAMS, MISCELLANEOUS COATINGS, ADHESIVES

This chapter deals primarily with insulating varnishes and impregnants, materials that have the dual purpose of providing electrical insulation as well as protection against the effects of moisture and contaminants and mechanical damage. Also discussed in this chapter is a miscellany of other materials that also combine a degree of insulating ability with some specialized function. These materials include foams, waxes, lacquers, paints, shellacs, and adhesives.

10.1   Insulating Varnishes

Varnishes used in electrical applications are liquid solutions of resins or are solventless liquid resins. They can be applied to a surface or into the interstices of windings. When applied to surfaces, the solvent is evaporated and the remaining resin cured to a smooth, glossy, attractive finish. In the interstices the solvent is removed, usually by baking or vacuum drying, and the remaining resin cured to form an impregnant of the winding. If a solvent is used, voids will be left since the remaining resin will be prevented from filling the interstices completely. If the varnish is solventless and the resin changes to a solid from a liquid without a loss of volume, complete filling of all voids will be attained.

Varnishes perform the following functions:

1. As a protective coating varnish provides a glossy, smooth surface that prevents corrosion and resists effects of moisture and other destructive agents. The coating also improves appearance.

2. As an impregnant varnish prevents the entrance of moisture, strengthens the winding, prevents movement of winding parts, increases insulating properties, and helps dissipate heat from the winding. The degree of attainment of these improvements depends on how well the voids are filled as well as on the characteristics of the varnish resin. Exclusion of air reduces oxidation and thermal degradation.

286

Fire retardation as well as inhibition of fungus growth can be obtained by use of specially compounded varnishes.

Impregnating varnishes are used to make varnished papers and cloths. Adhesive varnishes may be used as binders for mica, glass, and other flakes. They are also used in making laminates. (See Par. 10.6 on Adhesives.)

With so much to gain from the use of varnishes, the wide availability of basic polymers has naturally sparked the development of many diverse types. By now, these materials exhibit a broad range of performance characteristics. To facilitate description of varnishes and the selection procedures, it would be useful to follow several classification schemes.

10.1.1 Classification of insulating varnishes

The first approach to classification of varnishes is by type, as follows:

1. *Chemical type:* Asphalt (mineral); natural resin; synthetic resin.
2. *Engineering use:* Coating; impregnant; bonding agent (adhesive); protective function (fire-retardant, fungus inhibitor, and the like).
3. *Application methods:* Air drying; reactive cure; baking cure; solventless cure.

Regarding application methods it should be noted that in recent years there have been many developments resulting in new and better methods. Most varnishes are adaptable to the selected method of use. Some cooperation with the varnish manufacturer will help in fitting a varnish to a given process for optimum application results, or sometimes the reverse may be true—a certain process may be desired and a varnish is selected to match it.

Classification by thermal endurance is a *must.* Here the thermal ratings as given in publication No. 1 of the IEEE are used. Thus a varnish may be rated as suitable for use as Class 180 C, for example.

The ability of a varnish to meet the requirements of military specifications adds a method of classification. MIL-I-24092A and MIL-V-1137 require preproduction approval. The ASTM lists 26 specifications applicable to "varnishes." Of these ASTM D 2756, "Test for Weight Loss of Electrical Insulating Varnishes," and D 1830, "Thermal Stability of Coated Fabrics by Dielectric Breakdown," are noted. Most important is ASTM D 115, "Standard Methods of Testing Varnishes Used for Electrical Insulation." This test method lists eight classes of varnishes. Each class is designated by lower case letters starting with (a) for alcohol-soluble varnishes and ending with (h) for thermosetting laminating varnishes. ASTM D 1932-61T, "Thermal Endurance of Flexible Electrical Insulating Varnishes," gives a procedure for determining the relative thermal endurance of flexible insulating varnishes. ASTM D 2519-66T gives a test method for "Bond Strength of Electrical Varnishes by the Helical Coil Test." Also applicable is ASTM D 2307, which is the well-known "twisted-pair" test for enameled magnet wire but can also be employed for determining the relative thermal endurance of a varnish.

10.1.2 Varnish selection criteria

Selection factors in respect to varnishes should provide for information on the following major properties: Solids content; viscosity at $25\,°$ C ($77\,°$ F); specific gravity or weight per gallon; flash point; type of solvent. Additional information is desirable on the following to obtain more complete identification of the product: Color; thinners; shelf life; drying time (hours and temperature); curing time; dielectric strength (ASTM D 115); compatibility with magnet-wire and other wire and cable insulations; suitable application methods; outstanding features of the product, if any.

Varnish makers tend to supply the first five data items for most varnishes in general use. The remaining items are published where the data is applicable and needed for a better description of the varnish.

Table 10.1 has been compiled from technical data published by the makers of insulating varnishes. The identity numbers have been assigned to facilitate reference to the various varnishes in the following paragraphs, where a brief discussion on the characteristics of the varnishes and their indicated applications is supplied. Table 10.2 lists the identity numbers with corresponding applications and functional areas.

The first column of Table 10.1, "Types of Solids Content," provides a simple scheme for remembering the chemical classes of the varnishes. Notes on the other data tabulated follow:

Dielectric strength as determined by the method of ASTM D 115 is given as dry or wet: Dry after cooling in a desiccator over calcium chloride and wet after immersion in water for 24 hr at $23° \pm 1°$ C. Data is used mostly as an indication of quality. It is difficult to apply the data to actual service conditions.

Thermal endurance values are given as data for comparing performance and may or may not be realized in actual service because of the countless variables encountered in the makeup of electrical apparatus.

Suitable thinners must be known so varnish can be maintained in use. Note that Xylol tends to be the thinner for the high-temperature, high-strength varnishes. It is a most active solvent and care must be taken to prevent it from attacking the materials it comes in contact with. VM&P naphtha and mineral spirits are much less active and their use is preferred.

Solids $\pm 2\%$ Method D 115 is most important. The user may devise a related method for his own convenience in maintaining a varnish supply. The method affects the nonvolatile matter exhibited after a small quantity is baked at a chosen temperature and time as well as the apparatus and handling methods employed.

Specific gravity values are based on the ratio of the weight of a unit volume of sample as compared with the weight of the same volume of water at $23° \pm 1°$ C. Note that water density $= 1.000$ at $4°$ C. (This is maximum density of water.) One U.S. gallon contains 231 cu in., or 3,785.4 cu cm. One gallon of water at $4°$ C weighs 8.3452 lb and at $23°$ C, 8.333 lb. If the amount of solids is found to be 50 percent and the specific gravity is 0.95, the solids content in a gallon of varnish is 8.33 $\times$ 0.95 $\times$ 0.5 $=$ 3.95 lb.

Viscosity is given in centipoises (cps) at $25°$ C. The MacMichael viscosimeter is the instrument called for in ASTM D 115. In shop practice a convenient instrument is the Brookfield Viscosimeter, which is portable. A favorite method is by using a "cup" such as the Zahn Cup or the Ford Cup. These are cone-shaped cups of sheet metal with an orifice in the apex. Placed on a bath of varnish the cup floats apex down. The time for the cup to fill to a marked level is measured in seconds by a stop watch. A No. 4 Ford Cup will fill in 80 to 110 seconds at $25°$ C with varnish of 235–300 centipoises.

Table 10.1 Characteristics of Selected Insulating Varnishes*

Type of solids content	Identity No.	Dielectric strength Dry	Dielectric strength Wet	Thermal endurance, °C	Thinners	Solids ±2%	Specific gravity	Viscosity	Curing cycles hr	Curing cycles °F	Military Spec.
Modified polyester	31	4892	3030	180	Xylol	48	0.920–0.935	255–300	1 to 4	290–375	**
Modified polyester	21	4620	3110	180	Mineral spirits	50	0.895–0.905	235–285	2 to 6	350–400	**
Modified polyester	32	2650	1800	155	Xylol	50	0.925–0.975	130–210	2 to 6	290–325	
Clear thermoset polyester	51	3150	2250	200	Xylol	50	0.950–0.975	160–280	2 to 6	350–400	
100% solids polyester	772	—	—	180	None	100	1.070–1.100	675–750	8 to 15 min.	325	
Phenol formaldehyde	65	2000	1400	180	Isopropyl alcohol	55	1.010–1.040	180–250	2 to 3	275–300	
Oil-modified phenolic	150	2100	1200	180	Xylol	45	0.965–0.980	65–125	1 to 2	275–325	
Oil-modified phenolic	170	2000	850	130	VM & P naphtha	50	0.875–0.885	50–100	6 to 10	275–300	
Clear thermosetting	C150	2100	1200	180	Xylol	46	0.965–0.980	70–95	2 to 4	285–325	
Oil-modified	160	2000	1200	180	Xylol	50	0.940–0.960	125–200	3 to 8	275–325	
Red-pigmented	300	1000	350	155	Xylol	53	1.140–1.160	675–750	Air dry		
Black-oleoresinous	95	2500	1340	105	VM & P naphtha	40	0.840–0.850	630–830	Air dry	275–300	
Epoxy	59	3000	2800	130	3 of Xylol; 1 of cellosolve acetate	36 ±1	2 part 5	—	1 to 6		
Synthetic for printed circuits	64	2100	1850	130	Xylol	50	0.955–0.965	300–500	Air dry rapidly		
Silicone	155	1800	—	180	Xylene	50	8.3 lb/gal	100–200	Several bakes, end 300°F		**
Silicone	220	1800	—	200	N. butanol	60	8.6 lb/gal	125–200	2 to 6	300	**
Silicone	98	1800	—	200	Xylene	50	8.4 lb/gal	125–175	6	390	**
Silicone	88	1800	—	180	Xylene and petroleum spirits	10	6.79 lb/gal	1	Air dry (24 hr) or		
Silicone	77	1800	—	200	Dowanol EM	1.08	150	150	6 to 10 / 4 to 6	300 / 390	**

* Applicable units, standards, sources: Dielectric strength, volts/mil (ASTM D 115-55); Thermal endurance (makers' literature); thinners (makers' literature); solids, percent (ASTM D 115-55); specific gravity at 25°C; viscosity, centipoises at 25°C; curing cycles (makers' literature). Source: Sales literature as published by Schenectady Chemicals, General Electric, Dow Corning, and Sterling.

** Military Spec. MIL-I-24092-A, where applicable for items shown.

Table 10.2   Varnish Applications Versus Identity Number

Identity No.	Application area and/or function
31	For Class A (105) through Class H (200) motors
21	Most applications Class A through F. Easy to use mineral spirits thinner. Clear setting.
32	Many applications Class A through F. Check compatibility with magnet wire.
51	Organic varnish that can be used Class A through H. Check magnet wire. Does not decompose into components that will abrade brushes and commutators.
772	Solventless varnish for drip, roll coat, dip, and chain applications. Requires catalyst of 1-percent tertiary butyl perbenzoate. Cures very rapidly.
65	For high-speed rotating equipment. Cures to a very hard and strong mass.
150	Excellent chemical resistance.
170	Maximum penetration into compact windings.
C150	Suggested as not affected by refrigerants R 11, 12, 21, 22, 113, and 114.
160	Rotating apparatus, transformers, general purpose.
300	For finish coats, Air dry. Red.
95	For finish coats and many other uses. Air dry. Black.
59	Usual selections based on chemical resistance at Class 105 and 130
64	For coating printed circuits. Can be soldered through.
155	General purpose.
220	Used in vacuum impregnation of transformers.
98	Hard film. Strong bonding at high temperature.
88	After-coating for protection against moisture.

The curing cycle given in the table is that specified by the varnish manufacturer. In practice, however, wide variations are found.

10.1.3   *Selected polymers used in varnishes*

Epoxy resins   These resins find a wide use in insulating varnishes as they do in many electrical applications such as encapsulants, conformal coatings, and the like. For example, a clear epoxy ultrahigh temperature surface coating can be applied by dip, brush, or spray. It may be used for coating printed-circuit boards, or as varnish in coils or motor windings. Many epoxy compounds are intended for special-purpose varnish-type applications, for example: Fire-retardant dip; general-purpose pigmented; low-loss conformal coating and impregnant; high-rate thermal conductivity; flexibilized (resilient) clear coating; combination of high-temperature resistance (500° F) with high resistance to moisture and chemicals. Still another type is a conductive epoxy coating containing silver. Resistance is 0.3 ohms per sq ft.

Diallyl phthalate and diallyl isophthalates   Supplied as monomers and B-state prepolymers, these materials can be converted to coatings and sealants. As such they have no styrene odor, have excellent pot life, and exhibit the good electrical properties of the basic polymer. They may be applied by the vacuum process.

Polybutadiene liquid resins   Varnishes of these origins reflect the excellent electrical and other basic properties of the starting polymers. An asphalt/polybutadiene compound is an example of this group. It is in essence a solventless varnish for treating wires and cables to impart water-proofness. Electrical properties remain excellent over a wide range of temperatures. Cost is low.

Urethane varnishes   Several clean urethane varnishes are available to provide solderable thin films. Application of a hot solder iron to the film can be used to shrink the film and fold it back or to melt a spot on the film. Any damaged film can be replaced.

Miscellaneous varnishes   Several other types of varnish can be noted to indicate the wide range available:

Diphenyl oxide—Varnishes can be produced from these materials.

Polycarbonates—Varnishes that can be applied by brushing, dipping, and spray methods can be made from these polymers.

Alkyd—General-purpose clear alkyd coating that can be brushed, sprayed, and roller coated. They are used for weather coating outdoor or for indoor use. They are comprised of 43 percent solids and have a viscosity of 150–200 cps.

Rubber—Flexible, weatherproof, spray application.

Polystyrene—High Q, low loss, low dielectric constant, high insulation resistance.

10.2   Foams

Foams in electrical insulation start off with the advantage of low weight, an obviously desirable property in certain applications, such as aircraft systems. The foams also offer the advantage of adjustable dielectric constant, very low dielectric losses, and controllable density. The foams may be obtained in the form of boards or sheets and as powders, all in stated properties. In selected compounds, foams may be applied in place. Largest volume of foams are those of polystyrene. Foams of polyurethane, epoxy, silicone, and ceramics are among those used in foam-in-place compounds.

Open-cell foams have interconnected openings or voids. Closed-cell foams (unicellular) are those in which the cells are not connected. Syntactic foams are those in which the cells are separated through the introduction of minute hollow spheres (so-called "microspheres") into the polymer before or during the foaming process. In electrical compounds, the microspheres are either of glass or silica. Sizes range from 10 to 250 microns; wall thickness is 2 microns.

Some selected foam materials are briefly described below:

Foam-in-place liquid polyurethane resin   The liquid resin is mixed with a catalyst and poured immediately into a mold or cavity. Foaming begins almost at once and results in the formation of a rigid material. The process takes place at room temperature. Bulk density depends on the catalyst and may be 2, 8, or 14 lb/cu ft. For the 8-lb/cu ft density, dielectric constant is 1.12 over a frequency from 10^4 to 10^{10} Hz; dissipation factor at 10^{10} Hz is 0.002 and dielectric strength is 40 volts/mil.

Foam-in-place syntactic powder   This foam is a one-part epoxy system, used to fill all voids around a module or assembly and to provide a support. The powder cures at a temperature as low as 65 ° C. It can be extracted for repairs, if necessary. Dielectric constant is 1.38; dissipation factor, 0.006.

Foam-in-place silicone rubber   This foam is available in densities from 20.6 to 31.2 lb/cu ft. At 10^2 to 10^{10} Hz the dielectric constant is approximately 1.3 to 1.4 and the dissipation factor is below 0.01. Continuous use temperature is 200 ° C.

Polyurethane foam sheet   The closed-cell foam has densities of 7 lb/cu ft and 10 lb/cu ft. At 10^4 to 10^{10} Hz the dielectric constant is 1.18. At 10^{10} Hz, the dissipation factor is below 0.0004. In open-cell form, the polyurethane has a low dielectric constant of 1.05 and a dissipation factor of 0.0003, both at 10^4 to ^{10} Hz.

Silicone rubber foam sheet   The dielectric constant of this foam is 1.2 and the dissipation factor is 0.0007, both at 10^2 to 10^{10} Hz.

Polystyrene foam sheet, adjusted dielectric constant   This foam is available with dielectric constants of 1.02, 1.03, 1.07, and 1.1 to 1.9 in steps of 0.1, from 10^2 to 10^{10} Hz.

10.3   Waxes

Waxes find many uses in electrical insulation. They are used to coat papers, cloths, asbestos, and cordage such as yarns and cords. They are used as impregnants, potting compounds, sealers, and dipping baths. Lubrication can be obtained by wax coating the surfaces of wires and parts of assemblies. They can be added to compounds such as rubbers in order to improve some characteristic of the rubber in actual use. The extruded PVC on some wires may have some wax in the coating compound.

The great growth in polymer chemistry and the rapid introduction of new plastics has tended to reduce the use of waxes. But waxes are usually of low cost and are very easy to use.

Waxes are usually composed of high-molecular weight substances and can be grouped into five categories: (1) animal and insect sources (for example, beeswax, stearic acid); (2) vegetable (carnauba, bayberry, candelilla); (3) mineral (montan, ozocerite, ceresin); (4) petroleum (paraffin, microcrystalline, and other); and (5) synthetic (various polymer-derived waxes, such as those of polyethylene and chlorotrifluoroethylene as well as those of other chemical origin).

Fully refined paraffin waxes are composed mostly of hydrocarbons of the normal paraffinic, isoparaffinic, and naphthenic solid hydrocarbons. Cross sections of paraffin show large, well-formed crystals under the microscope. Microcrystalline wax, which is the wax in greatest use in electrical insulation, is nearly amorphous in structure since it is characterized by extremely fine crystals. The microcrystalline waxes consist of saturated aliphatic hydrocarbons.

The nonpetroleum waxes are esters of monohydric alcohols of the higher homologues. Of these, beeswax is used to coat yarns and cords to facilitate their use.

Carnauba (a hard, brittle, nontacky, and lustrous wax) finds an important use in the alteration of the properties of the microcrystalline waxes. The various types of waxes (as categorized above) can be combined, blended, or otherwise formulated to provide compounds of many desired properties. An extensive variety of such wax compounds is employed in the electrical insulating field for coating, potting, impregnating, sealing, and other functions.

The characteristics that identify a wax are melting point, flash point, specific gravity, acid number, saponification number, iodine number, refractive index, and percentage of unsaponifiable matter. There is a large literature on the use of waxes as electrical insulation. ASTM has issued a number of applicable specifications, of which ASTM D 1168, "Tests for Hydrocarbon Waxes for Electrical Insulation," deals with microcrystalline wax.

Electrical insulation data is often omitted from sales literature on waxes. This omission is probably due to the difficulty of measurement. Shrinkage, for example, can be the cause of performance variations. Testing of commercial compounds is therefore indicated to establish the electrical parameters, and specific requirements should be discussed with the suppliers to assure optimum properties. Exclusive dependence on catalogs would be inadvisable and might lead to a poor choice.

10.4   Lacquer, Enamel, Paint

This section reviews briefly the various liquid coatings that combine an electrical insulating function with that of surface protection and that may also be used to aid appearance.

Lacquer   A lacquer is a solution of a film-forming solid or solids in a volatile solvent. It is usually a clear solution and may be applied by any convenient method. The solvent escapes rapidly, leaving a smooth, glossy, and attractive coat. The cellulose polymers are in general use as the solid element. Lacquer has been used in coating the fiber jackets of ignition wires (automotive). Today the use of lacquer has nearly disappeared as a wire coating, but it finds many other uses.

Enamel   An enamel is a solution of film-forming solids in a solvent. Baking is required to evaporate the solvent and to set and cure the solid coating. Magnet wire enamels are used in large volume, as discussed in detail in Chap. 9.

Paint   One great difference between paints and lacquer and enamel is the large amount of pigment used in making the paint. The other ingredients are film-forming resins, solvents, or fluids for suspension of the solid particles. After application of the paint, the solvents and liquids evaporate. The film then usually sets or cures without baking. In some instances, however, baking is necessary.

Insulating paints are those that insulate as well as seal and finish electrical products. For example, an alkyd-base paint has been on the market for some 50 years and is still kept up-to-date in formulation. The widely known No. 1201 red alkyd provides a finish for rotating equipment, armatures, field coils, and basic machine and apparatus structures. A black paint is used for coils in rotating equipment. A high-temperature alkyd-base red paint is used at 125°C.

As might be expected, the proliferation in new polymers has stimulated the expansion of basic polymeric paint materials to replace the formerly used glyptals. The new paint materials include those based on epoxy resins, chlorinated rubber, acrylic resins, phenolics, silicones, and latex. Semiconducting paints are special types used to distribute voltage stress over the entire surface of the coated unit.

10.5   Adhesives

The process of adhesion is complex. In part it is the result of molecular bonding at the interface between the surfaces being joined (known as adherends or substrates); in part it may be the result of chemical reaction between these surfaces; and in part it may be the result of a "bridging" caused by the residual electrical forces or polarity surrounding the molecules at the interface. When materials of high polarity are being joined, this bridging action can be of appreciable strength, almost equal to that shown by chemical reactions.[1]

Adhesives used in electrical applications are largely of polymeric type and so can be put into two broad categories: thermoplastic and thermosetting. The method of application provides another classification scheme into four groups:

1. *Air-drying adhesives*—These adhesives (mostly thermoplastic) convert to a hardened state through solvent evaporation. Post-baking or air-drying may be used to remove the residual solvent. Pressure is applied for a suitable period to the closed joint.

2. *Fusible adhesives*—These materials (also thermoplastic) are converted to a finished form through melting, application to substrate, closing the joint, and cooling to hardness under pressure.

3. *Chemically reactive adhesives*—These adhesives, since they are thermosetting, are hardened by means of activation by catalysts and/or application of heat.

4. *Tape adhesives*—Pressure-sensitive-adhesives (mostly thermoplastic) are applied to the backing of p.s.a. tapes. As the term implies, application is by pressure alone. The thermosetting variety, however, does require heat cure for conversion.

The principal electrical insulating significance of adhesives lies in two areas: (1) as pressure-sensitive-adhesive electrical tapes (which are discussed in detail in Chapter 6), and (2) as adhesive compounds for joining or bonding various parts of electrical devices and apparatus where such form of construction is preferred to mechanical bonding means. In both areas the use of adhesive tapes or of adhesive compounds *per se* would assume that in addition to the mechanical function being performed, the adhesive contributes some degree of electrical insulating service. An excellent illustration is the use of adhesives in bonding the conductive copper foil to the plastic laminate in the fabrication of printed-circuit boards. As previously noted, insulating varnishes are made in adhesive formulations for use in various electrical/electronic applications, such as resin-bonded magnetic cores, fine coils, small motors, and other components and devices. The adhesive-type varnishes come in several forms, including putties, solventless types, and thixotropic.

There is a wide spectrum of possible adhesive compounds, both commercial and tailor-made. It is suggested that for specific requirements the manufacturers

---

[1] This discussion is drawn from "How to Use Adhesives for Solving Bonding Problems," I. Katz, *Electronic Design*, July 19, 1965, p. M14.

should be consulted. With a wide choice of available performance characteristics as well as of application methods (including induction bonding, ultrasonic bonding, and welding), an optimum combination of materials and processes can be developed. In fact, it may be advisable to consult with more than one potential supplier.

Many of the polymers discussed in Chap. 8 are used in the adhesive compounds. Included are the cellulosics, polyolefins, polyvinyls, polyamides, epoxies, silicones, phenolics, urethanes, acrylics, fluorocarbons, and also some of the elastomers and rubbers. Special adhesives are made for severe services such as those for cryogenic environments, ultrahigh vacuum of space, exposure to oxygen-rich atmospheres, to name a few. Other special types include conductive adhesives used in microelectronic circuit interconnections.

10.6   Shellacs

Lac is the exretion of an insect on branches of trees grown in India. It is gathered in crude form known as sticklac and is then subjected to several stages of purification to produce various grades used in commerce. (The term "sticklac" is based on the partly curved, stick-like pieces removed from the tree branches.) The grade used in the electrical industry comes in the shape of thin yellow or brown flakes which are put into solution with ethyl alcohol. The shellac flakes can be bleached, producing the "white" shellac known in industry. The electrical type of shellac can be applied in any convenient manner; it dries rapidly after application from solution.

Shellac is a natural thermoplastic with a specific gravity of 1.09 to 1.13, saponification value of 194 to 213, acid value of 48 to 64, and ester value of 137 to 163. Dry film breakdown voltage $\approx 1,000$ volts/mil; wet film breakdown voltage ≈ 500 volts/mil. Dielectric constant is 3.3 to 4.2 at 60 Hz to 10^5 Hz. Shellac resists action of transformer oil. Millions of pounds have been used in oil-immersed transformers. ASTM D784 and D411 are applicable to shellac.

BIBLIOGRAPHY

Note: *Industry sources provide much useful information on varnishes, coatings, and the like. Brochures, reports, and other published material from the following companies were used as reference and background in the preparation of this chapter: Acme Chemical & Insulation Co., New Haven, Conn.; Dow Corning Corp., Midland, Mich.; Emerson & Cuming, Inc., Canton Mass.; FMC Corp., Organic Chemicals Div., New York, N.Y.; General Electric Co., Schenectady, N.Y.; P. D. George Co., St. Louis, Mo.; Hull Corp., Hatboro, Pa.; Frank B. Ross Co., Inc., Jersey City, N.J.; Schenectady Chemicals, Inc., Schenectady, N.Y.; Sterling Division of Reichold Chemicals, Inc., Sewickley, Pa.; Westinghouse Electric Corp., Pittsburgh, Pa.; Zophar Mills, Brooklyn, N.Y.*

Anderson, O. E., "The Effect of Elevated Temperature on Silicone-Varnished Glass Fabric for Electrical Insulation," *ASTM Pub. 161*, 1954, p. 121.
de Bruyne, N. A., "The Physics of Adhesion," *J. Sci. Inst.*, vol. 24, 1947, p. 29.
Clark, F. M., *Insulating Materials for Design and Engineering Practice*, Wiley, 1962, pp. 854–912.
Haroldson, A. H., "Test Methods for Studying Thermal Stability and Heat Aging of Electrical Insulating Varnishes," *ASTM Pub. 161*, 1954, p. 41.

Herman C. J., and Viscusi, W., "Thermal Aging of Varnish," IEEE Conf. Paper No. 70CP238-PWR.

Hudrlik, R. E., and Hanson, W. M., "Thermal Classification of Pressure-Sensitive Adhesive Tapes," *Electrical Manufacturing,* Jan. 1961, p. 113.

Justus, D. A., "Epoxy Resin Adhesives," *Plastics Ind.,* vol. 17, 1959, p. 84.

Katz, I., *Adhesive Materials and Their Properties and Usage,* Foster Publishing, Long Beach, Calif., 1964.

Lever, A. E., and Rhys, J., *The Properties and Testing of Plastic Materials,* Chemical Publishing Co., New York, 1958.

Merriam, J. C., "Adhesive Bonding," *Materials in Design Eng.,* vol. 50, 1959, p. 111.

Parker, D. H., *Principles of Surface-Coating Technology,* Interscience, 1965.

"The Technology of Textiles", *International Science and Technology,* Sept. 1968.

Trivisonno, N. M., Lee, L. H., and Skinner, S. M., "Adhesion of Polyester Resin to Treated Glass Surfaces," *Ind. Eng. Chem.,* vol. 50, 1958, p. 912.

CHAPTER 11 | INORGANIC INSULATING MATERIALS

Inorganic materials such as asbestos, glasses, glass fibers, ceramics, micas, and others are used in many applications that require the ultimate safety factor of more or less indestructible spacing insulation. These materials appear in a variety of forms ranging from solid insulators and bushings to fibrous cloths and papers through built-up composites such as laminates. Organic binders may be present in composites.

11.1 Asbestos

Asbestos is a fibrous magnesium silicate mineral of the chrysotile family. It can be formed into roving, woven tape, wide cloth, papers, boards, and used as fillers in organic resins and laminates. Because of its very fine fibrous filament size, it offers a very large surface area for excellent reinforcement for resins. However, this same large area can hold a large amount of moisture, thus limiting asbestos to uses where moisture is not of prime concern.

Asbestos is very heat-resistant. It can be used up to 482° C, at which point it loses its water of crystallization. Above this temperature the structure is weakened and ceases to be useful, in most instances, over 500 ° C. The molecular structure continues to degrade up to 649 °C, at which point it disintegrates completely.

The uses for asbestos are confined to applications requiring chemical and structural stability at relatively high temperatures with a minimum of electrical demands. Uses in heating units such as power cords for ironing devices and toasters are frequent. Protective shielding and arc-quenching applications are also found. Often temporary taping, spacing, and isolation is obtained by means of asbestos because of its flexibility as well as its fire-proof properties.

Power equipment such as dry-type transformers has been the modern outlet

297

Table 11.1   Typical Properties of Chrysotile Asbestos

Property	Value
Density, gm/cc	2.38–2.59
Color	gray-white
Tensile strength, psi	130×10^3
Max. use temp., °C	315*
Commercial grade, °C	200
Specific heat, Btu/lb/°F	0.266
Thermal conductivity, Btu/ft²/hr/°F	1.90
Fusion temp., °C	1521
Weight loss at 300°C, percent	1.0
Dielectric strength, 20 mils, kV	
at room temperature	6.0
at 200°C	10.0
Moisture gain (1 hr at 65 percent RH), percent	1.15
Dielectric constant	
dry	6.0
16 hr at 91 percent RH	30.0
Dissipation factor	
dry	0.10
16 hr at 91 percent RH	0.95
Resistivity, ohm-cm $\times 10^{13}$	
dry	2.0
48 hr at 100 percent RH	0.003

* Aluminum phosphate-bonded asbestos can be used in rigid structures up to 315°C but is vulnerable to effects of moisture (as is untreated asbestos).

for asbestos in treated form. With silicone varnish, the combination can perform at temperatures comparable to glass fiber plus silicone.

Table 11.1 lists some of the properties of chrysotile asbestos in fiber form.

11.2   Ceramics

Unlike asbestos and glass fibers, ceramic bodies are nonflexible, having been preformed before firing to the rigid state. Machining is possible under some difficulty. Many insulator types are formed in two steps: firing of body and glazing (a glass-like added coating). Glazing permits the insulator to be more easily cleaned, adds mechanical strength, and prevents moisture pickup.

Ceramic insulators appear as standoff insulators, feed-through bushings, lightning-arrester fixtures, arc chutes, lamp bases, electronic separators, resistor and high temperature coil forms, heater forms, among other forms. They have excellent stability to heat, moisture, chemical, and electrical degradation as well as outstanding mechanical properties.

The raw materials in most ceramics are oxides of magnesium, calcium, aluminum, beryllium, silicon, zirconium, and barium. The alumina-silica (mullite) combination in clay is the basis for porcelain and its derivatives. Steatite is formed from a combination of magnesia-silica (forsterite) and magnesia-alumina-silica (cordierite) with talc, a clinoenstatite form of magnesia-silica.

A complete listing of ceramic types with composition, limiting usage temperatures, resistivity, dissipation factor, dielectric constant, dielectric strength, thermal conductivity, and expansion and tensile properties is shown in Table 11.2.

The Alsimag[1] ceramics are combinations of alumina, silica and magnesia in forms similar to cordierite. They are characterized by good to fair low-frequency losses and good to excellent high-frequency properties.

Ultra-steatite, not shown in Table 11.2, has a higher level of loss at power frequency but shows an extremely low loss at 10 MHz (0.000003 tan at 25°C). This makes it an excellent choice for substrates in forming integrated circuits. The most used substrate for IC's, however, is alumina, since it has a higher thermal conductivity and a lower expansion coefficient than ultra-steatite, and has a low dissipation factor over a wide range of frequency. Also used for substrates are beryllia, forsterite, glasses, glass-ceramics (such as Corning's Pyroceram), and glass-bonded mica. Since beryllia has outstanding thermal conductivity (over ten times that of alumina), it is particularly useful in applications where electrical insulation properties have to be combined with heat-dissipation capability.

Many glass coatings are becoming popular for IC substrates because of their smoothness. Glasses can be made in smooth coatings of 50 Å compared to that of 200 Å for carefully ground ceramics. For alumina, in an "as fired" condition, 500 Å are required, which is equivalent to approximately 2 microinches. (Paragraph 11.3 discusses solid and fibrous glass insulations.)

The titanates—barium titanate in particular—are used extensively in electronic capacitors in which a sizable capacitance is to be achieved in units of very small dimensions.

11.3 Glasses and Glass Fibers

Glass is considered a frozen liquid in that no ordered structure exists and the solid phase is readily returned to the liquid by raising the temperature through the transition region. Glass can be made to have an ordered structure by the process called "nucleation." This process depends on seeding the molten glass with microscopic grains of a metal oxide or ceramic that act as nuclei for the crystal growth process. The end product is a ceramic formed from glass.

Glasses fall into major categories such as soda lime, borosilicate, lead glass, silica, and quartz. Generally, the best electrical glasses are low in alkali content. The best radiation resistance is obtained by low boron content. Low melting points are obtained by the use of lead-oxide components.

Table 11.3 lists the common glass types in various physical and electrical property categories.

Two forms of glass fibers made from alumino-boro-silicate glasses are used—staple (short fibers) and continuous filament. The low alkali content gives fiberglass constructions excellent electrical properties up to high-use temperatures (200°–220°C). The inherent high tensile strength of glass fiber is one of its advantages over natural fiber materials. Values as high as 2×10^6 psi have been measured, but commercial fibers are at the 200,000 psi level.

Heat dissipation with glass fibers is more efficient than with other fibers and, if impregnation is complete, can be better than that with resinous materials without glass fibers.

[1] American Lava division of 3M Company.

Table 11.2   Ceramic Insulation Properties

Material	Composition	"Te" max. temp., °C*	Resistivity, at 25°C, ohm-cm	Dissipation factor at 25°C		
				1 kHz	1 MHz	10 MHz
Porcelain	$3Al_2O_3 \cdot 2SiO_2 + SiO_2$	370	5×10^{13}	.0140	.0076	.0075
Steatite	$3MgO \cdot 4SiO_2 \cdot H_2O$	840	10^{17}	.0063	.0045	.0041
Forsterite	$2MgO \cdot SiO_2$	1040	10^{17}	.0015	.0002	.0001
Alumina	$Al_2O_3 + SiO_2$	1070	10^{16}	.0010	.0006	.0007
Zircon	$ZrO_2 \cdot SiO_2$	870	10^{15}	.0140	.0010	.0005
Zirconia	ZrO_2	—	10^{8}	—	.010	—
Cordierite	$2MgO \cdot 2Al_2O_3 \cdot 5SiO_2$	780	10^{16}	.0100	.0030	.0030
Ceramic foam	ZrP_2O_7	—	—	—	.0010	—
Titania (rutile)	TiO_2	520	10^{13}	—	.0003	—
Magnesia	$MgO + MgO \cdot SiO_2$	850	10^{17}	.0002	.0002	.0002
K-6	$BaO \cdot MgO \cdot 4Al_2O_3 \cdot 12SiO_2$	—	—	.056	.0018	—
Barium titanate	$2BaO \cdot 3TiO_2$	—	10^{13}	—	.0105	—
P-2	$PbO \cdot Al_2O_3 \cdot SiO_2$	—	—	—	.0010	—
Spodumene	$Li_2O \cdot Al_2O_3 \cdot 4SiO_2$	—	—	—	.0040	—
Lithium alumino-silicate	$Li_2O \cdot 2Al_2O_3 \cdot 7SiO_2$	—	10^{12}	—	.0040	—
Boron nitride	BN	—	10^{14}	.0010	.0010	—
Silicon nitride	SiN	—	—	—	—	—
Thoria	ThO_2	—	10^{10}	—	.0003	—
Pyroceram	$SiO_2 + PbO$	800	—	—	.0017	—
Quartz + mica	$SiO_2 + Mica$	800	10^{14}	—	.0020	—
Mica	$K_2O \cdot 3Al_2O_3 \cdot 6SiO_2 \cdot 2H_2O$	—	10^{8}	—	—	—
Hafnia	HfO_2	—	10^{9}	—	.0100	—
Ceria	CeO_2	—	10^{12}	—	.0007	—
Spinel	$MgO \cdot Al_2O_3$	900	10^{16}	—	.0004	—
Mullite	$3Al_2O_3 \cdot 2SiO_2$	—	—	—	.0040	—
Wollastonite	$CaO \cdot SiO_2$	—	—	—	.0004	—
Beryllia	BeO	1700	10^{17}	.0084	.0010	.0005

* The temperature at which a cubic centimeter of ceramic possesses a resistance of 1 MΩ.

Table 11.2 Ceramic Insulation Properties (Cont'd)

Material	Dielectric constant			Dielectric strength at 25°C, vpm	Thermal properties		
	1 kHz	1 MHz	10 MHz		Conductivity, Btu/hr/ft²/°F/ft	Expansion, in./in./ °C × 10⁻⁶	Tensile strength, psi
Porcelain	6.8	6.2	6.0	150	1.5	6.0	6,000
Steatite	5.5	6.0	5.5	250	2.0	9.0	10,000
Forsterite	6.2	6.2	5.8	250	2.0	10.6	9,000
Alumina	9.0	10.0	8.5	350	10.0	8.0	25,000
Zircon	9.0	9.5	8.4	350	4.0	5.0	13,000
Zirconia	—	12.0	—	—	14.3	5.0	18,000
Cordierite	—	5.0	5.0	180	1.8	2.3	5,000
Ceramic foam	—	1.5	—	—	0.05	6.3	—
Titania (rutile)	100	100	100	—	2.0	8.0	—
Magnesia	—	8.2	—	—	23.0	12.8	12,000
K-6	—	5.5	—	—	—	2.4	—
Barium titanate	9000	—	—	—	—	—	—
P-2	—	5.8	—	—	—	5.1	—
Spodumene	—	6.4	—	—	—	2.0	—
Lithium alumino-silicate	—	5.3	—	—	1.0	0.5	5,000
Boron nitride	4.2	4.2	—	1000	16.6	4.3	3,500
Silicon nitride	—	—	—	—	—	—	18,000
Thoria	—	13.5	—	—	8.0	8.0	18,000
Pyroceram	—	6.0	—	250	2.1	3.0	—
Quartz + mica	—	7.2	—	400	2.0	12.0	9,000
Mica	—	6.0	—	1500	0.4	20.0	—
Hafnia	—	12.0	—	—	1.0	6.5	—
Ceria	—	15.0	—	—	7.0	10.0	—
Spinel	—	7.5	—	300	4.4	6.6	—
Mullite	—	6.6	—	—	1.5	5.0	4,000
Wollastonite	—	6.6	6.3	—	1.5	7.0	7,000
Beryllia	4.5	5.8	4.2	300	125	6.0	14,000

Table 11.3   Quality of Glasses

Property	Fused quartz	Alumino-silicate	96 percent silica	Borosilicate	Lead	Soda lime
Dissipation factor 60 Hz at 25°C	0.0002	0.001	0.0003	0.015	0.0020	0.040
Dielectric constant	3.7	—	3.8	4.5	8.0	7.5
Coefficient of expansion per °F $\times 10^{-6}$	16	—	25	125	280	250
Modulus of elasticity, psi $\times 10^6$	—	—	9.7	10	7.5	9.8
ASTM percentage of Na_2O extraction	0	—	0.002	0.005	0.15	0.04
Resistance to thermal shock, °C	—	—	1170	150	160	70
Resistivity at 250°C, ohm-cm $\times 10^6$	10^7	10^8	10^6	10^2	10^4	1.0
Dissipation factor at 1 MHz, 20°C	—	0.0018	0.0005	0.0046	0.0015	0.010
Dielectric constant at 1 MHz, 20°C	3.8	6.5	3.8	5.2	6.8	7.4
Max. use temp., Te, °C	900	850	550	350	450	250
Dielectric strength on 2 mm at 250°C, kV	—	—	10.0	2.0	10.0	0.4
Dissipation factor at 8.6 kMHz, 250°C	0.0001	0.0060	—	—	—	—
Dielectric constant at 8.6 kMHz, 250°C	3.7	6.3	—	—	—	—

Woven glass fibers depend for dielectric strength on the resin impregnant and its ability to wet and fill all interstices. Glass with silicone rubber can be an excellent combination for radiation and corona resistance. Woven glass fibers are used extensively in laminated sheets and boards, both flexible and rigid, for barrier construction, switchboards, mechanical support, and the like.

Glass papers made from filaments and from flake have been available for use in special applications.

11.4   Mica

Much of the success of the electrical power industry has been built on the use of mica in electrical insulation positions where long-time resistance to high voltage phenomena is needed. Mica has outstanding resistance to the attack of corona and extremely good dielectric strength and dielectric loss. Mica is flexible in thin sections and can withstand the expansion forces in the heat cycling of apparatus. Finally, mica has superior thermal properties such as thermal conductivity and useful temperature range.

Muscovite mica, a potassium alumino-silicate, is superior to the phlogopite type, a magnesium-modified mica, although both are used for insulation purposes. As supplied, mica is split into platelets less than 1 mil in thickness. Quality is based on useful area and freedom from blemishes. NEMA Standard No. MEI-1965 shows a method of grading mica and lists many forms of mica in tapes and sheets. Although mica is found in the natural state in many countries, the largest size splittings come from India, which supplies about 80 percent of the electrical-grade mica requirements of the world. Table 11.4 provides the general properties of muscovite mica.

The development of mica papers made by paper-making techniques that are essentially those used for cellulosic papers has brought to the mica family an important new group of members. These papers are made from imported natural mica that is not otherwise suitable for standard applications and also from domestic micas which are not produced in standard sizes and grades. The mica papers are utilized in the fabrication of various rigid, semirigid, and flexible materials that parallel those made from natural micas. In some respects they offer competitive features. NEMA standards for various fabricated forms now include mica paper products.

Muscovite mica enjoys many new uses (besides the time-honored electrical power insulation uses). In electronics, mica is used as a microwave window for low electrical loss and excellent mechanical features. Cryogenic applications for mica have been developed. High-heat uses that involve vacuum are also on the application list. Thin-film techniques are also used.

Mica has also had a long history in both power and electronic capacitors. Recently, silver electrodes made by silk-screen methods have replaced the foil electrodes in these units.

Mica finds an excellent use in transistor-mounting washers. The mounting aids dissipation of heat in addition to providing high-quality insulation.

Class H electronic transformers now use a combination of mica splitting plus silicone-bonded mica dust for intercoil insulation that provides an easy winding surface and an excellent but thin insulation for miniaturization.

Mica can be made synthetically. Fluorophlogopite synthetic mica has no hydroxyl group to be converted to water as does natural mica. As a consequence, synthetic mica can be used at temperatures exceeding natural micas by approximately 400 °C (900 °C). It is used extensively in composites such as glass-bonded mica for tight dimensional control of molded parts and in some cases for machinability. Use temperatures run to 1,100 °F for composites.

Table 11.4   Typical Properties of Muscovite Mica

Property	Value
Specific gravity, gm/cc	3.0
Specific heat, gm cal/cc	2,500
Thermal conductivity, Btu/ft^2/hr/ °F/ft	0.31
Dielectric strength, 1 mil, vpm	5,000*
Dielectric constant at 25°C	6.5
Dissipation factor at 25°C	0.0001
Resistivity, ohm-cm at 25°C	10^{16}
Moh hardness	3.0
Modulus of elasticity, psi	25×10^6
Tensile strength, psi	10,000
Compressive strength, psi	32,000
Shear stress, kg/mm	10
Linear coefficient of expansion, per °C	0.032
Max. operating temperature, °C	500
Max. operating gradient at 60 Hz	300

* 30,000 at 0.01 mils

Table 11.5   Properties of Glass-Bonded Mica

Property	Units	Test results
Specific gravity	gm/cc	3.8
Specific heat	cal/gm/$^\circ$C	0.2
Coefficient of expansion	per $^\circ$C $\times$ 10^{-6}	10.5
Heat conductivity	cal/cm/sec/cm^2/$^\circ$C $\times$ 10^{-5}	107
Heat distortion	$^\circ$C	375
Tensile strength	psi $\times$ 10^3	7.5
Modulus of elasticity	psi $\times$ 10^6	10.0
Compression strength	psi $\times$ 10^3	20.0
Impact strength Izod	ft-lb/in. of notch	1.80
Brinell hardness	—	100
Dielectric strength	vpm	400
Dielectric constant	60 Hz	8.0
	1 kHz	7.5
	1 MHz	7.7
Dissipation factor at 25°C, $\times$ 10^{-4}	60 Hz	5.3
	1 kHz	28
	1 MHz	14
Resistivity at 25°C	ohm-cm $\times$ 10^{16}	5
Arc resistance*	sec.	300

* ASTM D495.

11.5   Mica Composites

Low-temperature composites (105°–180°C) of mica with resinous binders and paper or other fiber backing have been used in high-voltage machine insulation for many years. Here, mica splittings are anchored in place in a tape form for wrapping on the coil windings for ground insulation.

Mica in laminated composites with low temperature resins (105°–180°C) are used as high-voltage barrier members in dry-type transformers and switchgear.

Glass-bonded mica is a rigid form of composite insulation. The mica is ground into flakes and added to melted glass. The combination is molded into rigid plates and other shapes for extremely useful structures. The combination of low loss and high dielectric strength, low moisture absorption, and good chemical resistance in high-temperature applications has gained for glass-bonded mica a unique position among composite insulations. Table 11.5 lists some of the more important properties of this material. As noted in Par. 11, 4, synthetic mica is also used in glass-bonded composites.

Mica paper, mentioned in Par. 11.4, is used extensively in composite insulations with other papers, films, and the like. (A detailed discussion of flexible insulations, discrete as well as composite, is found in Chap. 6.)

BIBLIOGRAPHY

Note: Information on Ceramics is also available from the following sources: Alumina Ceramic Manufacturers Assoc., New York, N.Y.; American Ceramic Society, Inc., Columbus, O.; Mica Industry Assoc., Newtown, Conn.; National Institute of Ceramic Engineers, Columbus, O.; Steatite Manufacturers Assoc., New York, N.Y.

Clark, F. M., *Insulating Materials for Design and Engineering Practice,* Wiley, 1962, pp. 975–1179.

DuBois, J. H., "Glass-Bonded Mica," *SPE Journal,* vol. 9, 1953, p. 20.

Duncan, G. I., and Felger, M. M., "The Properties of Electrical Insulation at Ultrahigh Temperatures," AIEE Trans. Paper 58–115, 1958.

Lemmon, J. W., "Ceramics," *Ind. Eng. Chem.,* vol. 50, 1958, p. 1433.

Littleton, J. T., and Morey, G. N., *Electrical Properties of Glass,* Wiley, 1933.

Mathes, K. N., and Stewart, H. J., "Asbestos and Glass-Fiber Magnet Wire Insulation," *AIEE Transactions,* vol. 58, 1939, p. 291.

NASA Contributions to the Technology of Inorganic Coatings, J. D. Plunkett, ed., NASA SP-5014, 1964.

NEMA, *Mica Standard,* 1–18–1956.

Properties of Selected Commercial Glasses, B-83, Corning Glass Works, 1957.

Shand, E. B., *Glass-Engineering Handbook,* McGraw-Hill, 1958.

Shaver, W. W., and Stookey, D. D., "Pyroceram," *SAE Journal,* vol. 66, 1958, p. 34.

Shell, H. R., Comeford, J. E., and Eitel, W., *Synthetic Asbestos Investigations,* U.S. Bureau of Mines Report 5417.

Sinclair, W. E., *Asbestos,* Mining Publications, Ltd., London, 1955.

Spooner, L. W., "High-Temperature Inorganic Insulation," *Electrical Manufacturing,* Apr. 1957, p. 130.

Thurnauer, H., "Ceramic Insulation—A State-of-the-Art Review," *Electro-Technology,* Feb. 1962, p. 102.

Vondracek, C. H., and Croop, E. J., "New Inorganic Insulation for 500°C Electrical Equipment," *Insulation,* Jan. 1958, p. 35.

CHAPTER 12

LIQUID AND GASEOUS INSULANTS

Liquid and gaseous insulants and coolants have been used for many years because of a number of advantages not offered by solid insulation. Among them may be mentioned self-healing after discharge; convective and conductive heat transfer; intimate contact with electrodes and structural parts for heat dissipation; corona control; expansion control; oxidation control; high and uniform dielectric strength; uniform dielectric constant; and chemical and thermal stability. In addition, pure materials can be obtained and, with some materials, low cost.

The use of liquids has far outstripped that of gases. With the discovery of the electronegative gases and their ability to control corona, however, gases are now being applied more frequently.

A condensed discussion of a few of the more important liquid and gaseous insulants is given in the following paragraphs. The Bibliography at the end of the chapter provides sources of more detailed information.

12.1 Mineral Oils

Mineral oil is used in a great many power applications, principally power and distribution transformers, cables, capacitors, circuit breakers, and switchgear. The continued high insulating quality of mineral oil is threatened by deterioration caused by: (1) oxidation, (2) discharge decomposition, and (3) partial discharge as in corona conditions.

Oxidation in mineral oils causes the formation of conducting byproducts such as peroxides, acids, and soap. The conductive molecules reduce the dielectric strength of the oil, a condition that significantly increases its conductivity. If not checked, this increase will cause ultimate failure.

To the oxidative decomposition is added the deterioration caused by corona and internal discharge. The effect of this deterioration is to produce carbonization and further increase the conductivity of the system. The oxidative effects are accompanied by the formation of water and sludge, the latter a resinous solid that tends to plug cooling tubes and hinder the convective flow of the oil and its function as a coolant.

As the ultimate molecular deterioration occurs, corona is enhanced by the formation of gases. Thus the various degradative mechanisms are interrelated and finally completely destructive.

Various means are used to slow down and mitigate the aging processes. Inhibitors and retarders are used in oils to diminish the oxidation rate. Largely used as an inhibitor is DBPC (ditertiary-butyl-paracresol). Inert gases are used above the liquid level to eliminate oxygen from the system. Fluorocarbon gases are employed to reduce corona and other discharge effects. Forced cooling by fins and radiators and internal construction is used to prevent excessive temperatures from developing and thereby prolonging the life of the apparatus. Hermetic sealing with expansion chambers removes exposure to air and prevents oxidation. In cables, hydrostatic filling eliminates all gases and limits degradation from corona and oxidation. Metal deactivators reduce the catalytic activity of conductors. Acid neutralizers, corrosion inhibitors, detergents, and the like, all prolong the life of the oil. Even the papers used in oil-filled devices are treated with nitrogenous compounds to prolong their flexibility and the life of the oil.

Mineral oils may be reconditioned by (a) centrifuging, (b) filtering, (c) clay treatment, (d) refining with acid and clay, and (e) addition of aromatic hydrocarbons, inhibitors, and other stabilizing ingredients. In continuous percolation methods, oil is used and reconditioned constantly. Clay and/or alumina beds provide continuous means for removal of acids, sludge, water, carbon, and similar materials, as the oil is circulated in the system. Approximately 50 lb of alumina are required to treat 700 gal of oil in such a system.

Quality control on oil in transformers is maintained by sampling tests. The ASTM D-877 dielectric test cup with 1-in. diameter, cylindrical, sharp-edge electrodes spaced 0.100 in. apart is used. The test oil is broken down with end points greater than the 22 kV required for acceptable levels. Fresh oils will often show 30 kV values or higher. The ASTM color range of 2–4 must also be maintained for the oil sample.

Acidity must be no higher than that requiring 1 mg of KOH per gram of oil to neutralize. Finally, the sludge content must not exceed 0.15 percent by weight of the oil. All apparatus used for control tests must be clean and dry. The worst condition of the oil, taken at bottom-of-tank locations after adequate settling time, should be sought so that no chance of misinterpretation of condition is possible. A list of ASTM tests and typical values for transformer mineral oils is given in Table 12.1.

Some active metals—such as iron, copper, copper alloys, and lead—have a deleterious effect on insulating oils. The passive metals are aluminum, tin, magnesium, zinc, cadmium, nickel, stainless steel, and their oxides. Copper and its alloys may be plated by passive metals to eliminate the threat of oxidative action. Passivation of active metals by suitable surface treatment is also utilized. Parkerization, a proprietary acid phosphate pretreatment, is common.

Some nonmetals should also be avoided in oil-filled devices. These are fiber board, asphalt, wax, rubber (both synthetic and natural), tars, chloride flux, greases, rosin, acrylic plastics, saran, polyvinyl chloride, and silicone resins. All soldered

Table 12.1   Mineral Oil Tests

ASTM	Description	Typical transformer oil values
D-1250	Specific gravity	0.885
D-92	Flash point, °C	135
D-92	Burn point, °C	148
D-877	Dielectric strength, kV	30
D-155	Color	1
D-974	Acidity, mg KOH/gm	0.02
D-446	Viscosity, 37.8°C, S.U.S.*	58
D-97	Pour point, °C	−45
—	Boiling point, °C (1 atm.)	> 150
D-924	Dissipation factor at 60 Hz, 100°C, percent	0.10
D-924	Dielectric constant at 60 Hz, 100°C	2.25
—	Evaporation (8 hr at −100°C), percent	8
D-1250	Refractive index at 25°C	1.48
D-971	Interfacial tension, dynes/cm	30.0
D-1275	Free sulfur, percent	nil
D-1275	Total sulfur, percent	0.1
D-1250	Specific heat, cal/gm/°C	0.425
D-1169	Resistivity at 100°C, ohm-cm	35×10^{12}
D-1250	Coefficient of expansion, cc/cc/°C	0.00070
ASME	Thermal conductivity, CGS**	327×10^{-6}
—	Aniline point, °C	76
—	Specific optical dispersion†	115

* S.U.S. = Saybolt Universal Seconds
** (gm-cal/sec) (cm^2) (°C/cm)
† A value of 135 for S.O.D. is for a highly aromatic oil, whereas a 103 S.O.D. indicates over-refining and low aromaticity.

joints in oil-filled apparatus should be carefully cleaned of flux before the oil is pumped in.

Other nonmetallic structural and functional components used in oil-filled apparatus are not considered harmful to oil in its oxidation resistance. In this class are the enamels for magnet wire such as Formvar, polyurethane, nylon, epoxy, terephthalate polyesters, polyesterimides, polyamide-imides, polyesteramide-imides, and polyimides. Other nonmetals in this class include silicone rubber, melamine resin, cellulose esters, glass, aluminum oxide, paper, cotton, cork, masonite, pressboard, wood, shellac, silica gel, and alkyd resins.

The modern trend in transformers is to replace copper with aluminum conductors. This move, prompted primarily by economics, has very decided benefits in reducing the deterioration effects on the oil. The aluminum, being passive, has little catalytic effect in the oxidative process.

12.2   Askarels

Askarels overcome the flammability of mineral oils and also possess a higher dielectric constant (approximately double). Thus they are used in transformers in hazardous positions and in capacitors where, for equal capacitance, units can be half the size of mineral-oil-filled types.

Askarels are composed of chlorinated diphenyls. If, for example, they are

Table 12.2   Typical Askarel Tests

ASTM	Description of test	Test results
D-1250	Specific gravity at 25°C	1.56
D-92	Flash point, °C	None
D-877	Dielectric strength, kV	35
D-446	Viscosity at 37.8°C, S.U.S.	54
D-974	Acidity, mg KOH/gm	0.008
D-1250	Coefficient of expansion, cc/cc/°C	0.00067
D-97	Pour point, °C	−35
—	Boiling point at 1 atm., °C	302
D-924	Dissipation factor at 60 HZ, 100°C, percent	1.5
D-924	Dielectric constant at 60 Hz, 100°C	3.6
D-1169	Specific resistivity at 90°C, ohm-cm	0.1×10^{12}
ASME	Thermal conductivity, CGS	298×10^{-6}
D-1250	Refractive index, 25°C	1.614
APH	Color	150
D-1250	Specific heat, cal/gm/°C	0.264

decomposed by the effects of arcing, they give off HCl gas, which, although non-flammable, has corrosive effects on structural members of a system and leads to degradation of insulating materials.

Askarels have equal or better dielectric strength than that shown by mineral oils, but poorer dielectric loss characteristics. At elevated temperatures they can exert solvent action on some wire enamels and on other resinous insulations. As a result, the list of materials that should not be used in the presence of askarels is longer than a similar list for mineral oils. The materials best suited for use in askarels are: cellulose paper, press boards, wood, silicones, polytetrafluoroethylene, phenolics, asbestos, glass, mica, ceramics, epoxy resins, urethane resins, and silicone-glass laminates.

As with mineral oil, askarels may be reconditioned by use of activated alumina beds. The contaminants are effectively removed, restoring high dielectric strength and lower dielectric loss. Tests for askarels are described in Table 12.2.

12.3   Silicone Oils

For excellent high temperature stability and low loss at high frequency as well as a favorable (flat) temperature-viscosity slope, the dimethyl and phenylmethyl silicone fluids have found wide usage in electrical devices. This use is in small capacitors and transformers and in a variety of high-frequency applications.

Temperatures as high as 200 °C can be maintained with little decomposition of the silicone fluid, but gradual polymerization does occur and viscosity approaches that of a gel in long exposure to heated air. The fire hazard with silicone fluids is minimal, with silicon, dioxide, water, and carbon dioxide the main decomposition products. Silicone fluids are vulnerable to arcing since carbon and other conductive products formed reduce dielectric strength and increase the dissipation factor. Insulations of all types are largely unaffected by exposure to silicone oils and very few contaminate the oils.

Silicone oils also exhibit very low pour points (−50°C and lower). Typical properties are shown in Table 12.3.

By replacing some of the methyl groups in dimethyl silicone with fluorocar-

Table 12.3   Typical Properties of Silicone Oil

ASTM	Description of test	Test results
D-1250	Specific gravity, 25°C	0.963
D-92	Flash point, °C	318
D-877	Dielectric strength, kV	35
D-446	Viscosity at 37.8°C, S.U.S.	189
D-974	Acidity, mg KOH/gm	0.04
D-1250	Coefficient of expansion, cc/cc/°C	0.00098
D-97	Pour point, °C	−54
—	Boiling point, °C	> 350
D-924	Dissipation factor at 60 Hz, 100°C, percent	0.04
D-924	Dielectric constant at 60 Hz, 100°C	2.48
D-1169	Specific resistivity at 100°C, ohm-cm	2000×10^{12}
ASME	Thermal conductivity, CGS	360×10^{-6}
D-971	Surface tension, dynes/cm	21
D-1250	Specific heat, cal/gm/°C	0.35
D-877	Dielectric strength after 200 megarads (gamma), kV	20
D-1169	Resistivity at 100°C after 200 megarads (gamma), ohm-cm	2×10^{12}

bon groups, fluoro-silicone oils are formed with very high dielectric constants. These oils are useful in electronic applications.

Silicone oils withstand mild gamma radiation about as well as the organic liquid insulants but must also be shielded from extreme dosages. Gelation, loss of electrical properties, and evolution of exposive gases are the results of exposure to large dosages.

12.4   Fluorocarbon Oils

The fluorinated hydrocarbon liquids, as a group, are available in great variety, but the fluorinated aliphatic hydrocarbons are not only the least volatile but the most practical for use in electronic transformers, which provide the widest application area for these liquids to date. High voltages in a minimum of space are possible. Research and development has been expended in work on long-chain fluorocarbons with increased dielectric strength, higher boiling points, reduced coefficient of expansion, but little or no change in dielectric loss characteristics. Structure of these fluorocarbons is as follows:

Properties of this fluorocarbon liquid and two others are shown in Table 12.4.

The fluorocarbon liquids are excellent in cooling electrical apparatus, but because of their low boiling point and relatively high cost they are not a serious com-

Table 12.4 Typical Properties of Fluorocarbon Liquids

ASTM	Test description	$(C_4F_9)_3N$	$(C_3F_6O)_5C_2HF_5$	$C_8F_{16}O$*
D-1250	Specific gravity, 25°C	1.87	1.79	1.77
D-92	Flashpoint	None	None	None
D-877	Dielectric strength	40	50	37
D-446	Viscosity at 37.8°C, S.U.S.	32	35	27
D-1250	Coefficient of expansion, cc/cc/°C	0.0012	0.00036	0.0016
D-97	Pour point, °C	−50	−84	−100
—	Boiling point, °C	177.8	224.2	102.2
D-924	Dissipation factor at 60 Hz, percent	< .05	< .006	< .05
D-924	Dielectric constant at 60 Hz	1.90	2.45	1.86
D-1169	Specific resistivity, ohm-cm	3×10^{14}	$> 4 \times 10^4$	6×10^{14}
ASME	Thermal conductivity, CGS	210×10^{-6}	210×10^{-6}	347×10^{-6}
D-971	Surface tension, dynes/cm	16.1	—	15.2
D-1250	Specific heat, cal/gr/°C	0.27	—	0.25
D-1250	Refractive index at 25°C	1.291	1.464	1.277

*FC-75 trade designation

petitor of the chlorinated askarels. For ecological reasons, however, some of the exotic liquids may find increased use in the future.

12.5 Organic Esters

In their purest form (free from moisture, polar contaminants, and acids) the organic esters have been used in power cables and in high-frequency capacitors. Castor oil (largely a triglyceride of ricinoleic acid) has excellent resistance to corona, as well as low volatility, nontoxicity, and high flash and fire points. It is recommended for d-c applications.

Dibutyl sebacate (D.B.S.) has been used extensively in high-frequency power capacitors. Hermetic sealing is necessary with these liquids to prevent oxidation and hydrolysis. Dibutyl sebacate shows a better dielectric strength at 100°C and 60 Hz than fresh transformer mineral oil even though its dielectric loss is many times that of the oil. Much less carbon is formed after arcing than in mineral oil. The minimum dielectric loss for D.B.S. occurs at 1 MHz and 100°C. Table 12.5 shows typical properties of the organic ester oils.

Silicate-ester-base fluids are characterized by nontoxicity, high-voltage stabil-

Table 12.5 Typical Properties of Organic Esters

Property	Castor oil	Dibutyl sebacate	Butyl stearate	Silicate ester
Pour point, °C	−23	−10	21	−60
Flash point, °C	291	175	167	188
Viscosity at 37.8°C, S.U.S.	—	46	48	12.2
Dielectric strength at 100°C, kV	—	38	—	27
Dielectric constant at 60 Hz	3.74	4.4	3.3	2.65
Dissipation factor, percent				
at 60 Hz	6.0	1.0	1.0	—
at 1 MHz	—	0.015	0.013	0.42
Resistivity, ohm-cm	3×10^{10}	5×10^{12}	5×10^{12}	9×10^{10}
Thermal conductivity	0.000422	—	—	0.000327

ity, low losses at high-frequency, fire resistance, and a wide temperature range ($-54°$ to $+204°$C). These fluids must be hermetically sealed to remove the effects of moisture. They have good cooling properties and are compatible with many common insulations (but not silicones).

Finally, ethylene glycol, which has a dielectric constant of 38.8 at $20°$C, is used in capacitors.

12.6   Refrigerants

In the same class of halogen-type compounds as the askarels (Par. 12.2) and the fluorocarbons (Par. 12.4) are the volatile low-boiling liquid fluorochlorohydrocarbons used in air-conditioning and refrigeration units. This class of refrigerant compounds is also discussed in Par. 12.8.

The boiling point determines whether a given compound in this group is useful as a liquid coolant, a refrigerant, or a gaseous dielectric. Those that are useful as refrigerants are maintained under pressure in hermetic systems, at first partly as liquids, then, after expansion takes place, as vapors. Hermetic motor insulation encounters both phases of the refrigerant. The overall effect is to cool the motor. However, under abnormal conditions, the coolant is prevented from cycling or is "lost." The motor insulation may then be subjected to over-runs in temperature, followed possibly by flooding with refrigerant liquid. Thus cold shock, heat evaporation, solvent action, and the need for starting under load are all factors that must be considered in the design of an insulation system for hermetic motors.

Table 12.6 lists some of the more common refrigerants, with their formulas, boiling points, and specific heats at constant pressure. Of the numbered refrigerants, R-22 has the greatest specific heat—0.330 calories per gram per $°$C at constant pressure. This property is partly responsible for its high efficiency. It is used in increasing amounts in modern hermetic systems for air conditioning where space requirements are limited. However, R-22, more than any other refrigerant, has a softening effect on wire enamels and other insulation components. R-22 is absorbed by the wire enamel and may cause blistering of the coating if the winding is subjected to a rapid rise in temperature (as may occur with loss of refrigerant).

Table 12.6   Some Common Refrigerants

Refrigerant No.	Name	Formula	Boiling point, $°$C	Specific heat, $cal/gm/°C$
R-11	Trichloromonofluoromethane	CCl_3F	$+$ 24.1	0.209
R-12	Dichlorodifluoromethane	CCl_2F_2	$-$ 30.0	0.243
R-13	Monochlorotrifluoromethane	$CClF_3$	$-$ 80.0	0.250
R-14	Tetrafluoromethane	CF_4	-128.0	0.270
R-21	Dichloromonofluoromethane	$CHCl_2F$	$+$ 8.9	0.250
R-22	Monochlorodifluoromethane	$CHClF_2$	$-$ 40.8	0.330
R-113	Trichlorotrifluoroethane	$C_2Cl_3F_3$	$+$ 45.8	0.220
R-114	Dichlorotetrafluoroethane	$C_2Cl_2F_4$	$+$ 3.8	0.230
R-115	Monochloropentafluoroethane	C_2ClF_5	$-$ 38.0	0.290
—	Tetrafluoroethylene	C_2F_4	$-$ 78.4	—
—	Hexafluoroethane	C_2F_6	$-$ 78.2	—
—	Sulfur dioxide	SO_2	$-$ 10.0	0.152

This blistering has not caused a large reduction in dielectric strength (turn-to-turn) in the windings and, aside from appearance, is not a serious threat to motor life. Despite this viewpoint, industry *is* moving toward development of magnet wire enamels that will show a minimum of blistering. (See previous discussion in Chap. 5, Par. 5.9.1.)

12.7   Dielectric Gases

Discussed here are air, hydrogen, helium, carbon dioxide, and nitrogen at various pressure levels, all of which can be used as gas dielectrics and cooling media. To complete the coverage of gases, the electronegative gases will be discussed in Par. 12.8.

Compressed air at 100 psig has a dielectric strength comparable to mineral oil. However, air promotes oxidation of other components of the system and is readily ionized, forming corona at high voltages. *Compressed nitrogen* has less dielectric strength (75 percent) than air but does not promote oxidation.

Hydrogen has been used for many years as a gaseous dielectric and coolant for high-speed turbo generators. Its use has an added advantage in reducing windage loss. The thermal conductivity of the hydrogen is approximately seven times that of air whereas its viscosity is only half that of air. The specific heat of hydrogen is more than fourteen times that of air; as a result, it is a very efficient coolant.

Helium as a coolant gas is also efficient because of attributes similar to those of hydrogen but it is not used because of its scarcity. Unlike hydrogen, helium has a relatively high viscosity, a factor which makes it less desirable in rotating machinery.

Carbon dioxide has no outstanding advantage that would dictate its preference as a dielectric over air or nitrogen. This statement also applies to a number of hydrocarbon gases such as methane, ethane, and butane. It is only when these gases are chlorinated or fluorinated (or both) that they take on engineering value as electronegative gases.

12.8   Electronegative Gases

Since the breakdown of gases such as air, nitrogen, and carbon dioxide is caused by the presence of free electrons during the process of ionization (corona), any gas that *captures* electrons would, in comparison, show superior dielectric strength. Such gases (known as electronegative gases) are formed by halogenation of hydrocarbons and sulfur. They *do* capture free electrons and exhibit both high corona-voltage levels and high dielectric strengths at all pressures. These gases include those known as the freons and also sulfur hexafluoride (SF_6).

Although SF_6 is not a fluorocarbon, it shows electron capture capability to a remarkable degree, thus ranking as an electronegative gas. Its dielectric strength is better than twice that of compressed air at up to 20 atmospheres pressure. SF_6 is widely used in insulating applications.

Sulfur dioxide (SO_2) is somewhat of a misfit in this group, but it has been used as a refrigerant in the past. A number of the materials listed are fluorinated or chlorinated hydrocarbons and show potential as insulants.

Table 12.7 compares the various gases with air on a relative dielectric strength basis, that of air at one atmosphere being considered unity. In a summary

Table 12.7   Relative Dielectric Strength of Gases

Gas	Relative dielectric strength
Helium	0.15
Hydrogen	0.65
Carbon dioxide	0.80
Nitrogen	0.90
Air (1 atm)	1.00
CH_2ClF	1.03
$CF_4(F14)$	1.10
$CHCl_2F(F21)$	1.33
CF_3Br	1.35
$CHClF_2(F22)$	1.40
$CClF_3(F13)$	1.43
C_2F_6	1.88
C_3F_8	2.00
C_2ClF_5	2.3
SF_6	2.35
$C_4F_8(C318)$	2.40
$CCl_2F_2(F12)$	2.42
C_4F_{10}	2.50
$C_2Cl_2F_4$	2.52
ClO_3F	2.73
$CCl_3F(F11)$	3.50
Oil (Mineral)	3.70
$CHCl_3$	4.24
Air (20 atm)	6.30
CCl_4	6.33
SF_6 (20 atm)	12.5

of halogenated liquids and gases, Table 12.8 lists the compounds in the order of increasing boiling points and also by their application areas. Although some freons (F14, F23, F13) have very low boiling points, most freons and other refrigerants are in the middle group, between the electronegative gases and the liquid coolants.

The reader is also referred to an earlier discussion on electronegative gases in Chap. 2, Par. 2.12.)

BIBLIOGRAPHY

Baker, W. P., "The Nondestructive Testing of Electric Strength of Liquids," *IEE Proceedings* (London), vol. 103, Part A, 1956, p. 337.

Balsbaugh, J. C., Assaf, A. G., and Pendleton, W. W., "Mineral Oil Deterioration," *Ind. Eng. Chem.,* vol. 33, 1941, p. 1321.

Camilli, G., Gordon, G. S., and Plump, R. E., "Gaseous Insulation for High-Voltage Transformers," *AIEE Transactions,* vol. 71, 1952, p. 348.

Camilli, G., "An Experimental Gas-Insulated 138-kV Current Transformer," *AIEE Transactions,* vol. 74, Part III, 1955, p. 100.

Church, H. F., "Factors Affecting the Life of Impregnated Paper Capacitors," *IEE Proceedings,* vol. 98, Part 3, 1951, p. 113.

Table 12.8 Summary of Halogenated Insulants

Boiling point, °C	Electronegative gases	Refrigerants	Liquid coolants
−128	CF_4(F14)	—	—
− 82.2	CHF_3(F23)	—	—
− 81.0	$CClF_3$(F13)	—	—
− 78.6	CH_3F	—	—
− 78.4	C_2F_4	—	—
− 78.2	C_2F_6	—	—
− 70.0	$C_2H_2F_2$	—	—
− 63.8	SF_6	—	—
− 51.6	CH_2F_2	—	—
− 46.8	$C_2H_3F_3$	—	—
− 40.8	$CHClF_2$	$CHClF_2$(F22)	—
− 38.7	C_2ClF_5	C_2ClF_5(F115)	—
− 37.7	C_2H_5F	C_2H_5F	—
− 36.7	C_3F_8	C_3F_8	—
− 29.8	CCl_2F_2	CCl_2F_2(F12)	—
− 24.7	$C_2H_3ClF_2$(G100)	$C_2H_3ClF_2$(G100)	—
− 24.7	$C_2H_4F_2$	$C_2H_4F_2$	—
− 23.0	CH_3Cl	CH_3Cl	—
− 10.1	C_3H_7F	C_3H_7F	—
− 10.0	—	SO_2	—
− 10.0	C_3H_5F	C_3H_5F	—
− 5.8	C_4F_8(C318)	C_4F_8(C318)	—
− 2.0	C_4F_{10}	C_4F_{10}	—
− 0.6	$C_3H_6F_2$	$C_3H_6F_2$	—
+ 3.8	$C_2Cl_2F_4$	$C_2Cl_2F_4$(F114)	—
+ 8.9	$CHCl_2F$	$CHCl_2F$(F21)	—
+ 10.0	$C_2H_4F_2$	$C_2H_4F_2$(G101)	—
+ 12.8	C_2H_5Cl	C_2H_5Cl	—
+ 16.0	C_4H_9F	C_4H_9F	—
+ 24.1	CCl_3F	CCl_3F(F11)	—
+ 30.0	C_8H_7F	C_8H_7F	—
+ 32.0	C_4H_9F	C_4H_9F	—
+ 40.1	CH_2Cl_2	CH_2Cl_2	—
+ 45.8	—	$C_2Cl_3F_3$(F113)	$C_2Cl_3F_3$
+ 56.0	—	—	C_2H_3F
+ 61.3	—	—	$CHCl_3$
+ 62.8	—	—	$C_5H_{11}F$
+ 76.8	—	—	CCl_4
+ 84.9	—	—	C_6H_5F
+ 91.5	—	$C_2Cl_4F_2$(F112)	$C_2Cl_4F_2$(F112)
+102.0	—	—	$C_8F_{16}O$(FC75)
+102.4	—	—	$C_7H_5F_3$
+114.0	—	—	C_7H_7F
+130.0	—	—	C_6H_4ClF
+132.0	—	—	$C_7H_6F_2$
+138.0	—	—	C_2Cl_5F
+140.0	—	—	C_7H_7F
+142.6	—	—	$C_7H_5ClF_2$
+175.0	—	—	C_6H_6NF
+177.8	—	—	$C_{12}F_{27}N$ (HCF TBA)
+178.0	—	—	$C_7H_5Cl_2F$
+201.0	—	—	$C_7H_4F_3NO_2$
+212.0	—	—	$C_{10}H_7F$
+224.0	—	—	$C_{17}HF_{35}O_5$

Clark, F. M., *Insulating Materials for Design and Engineering Practice,* Wiley, 1962, pp. 58–250, pp. 278–356.

Clark, F. M., *et al,* "The Use of Nonflammable Liquid Impregnants in Electrical Capacitors and Transformers," Paper 119, *Conference Internationale des Grands Reseaux Electriques Haute Tension,* 1958.

Cohen, E. H., "The Electrical Strength of Highly Compressed Gases," *IEE Proceedings,* vol. 103, Part A, 1956, p. 57.

Cooper, R., "Experiments on the Electric Strength of Air at Centimeter Wavelengths", *IEE Proceedings* (London), vol. 94, 1947, p. 315.

Currin, C. G., and Dexter, J. F., "Dielectric Properties of Silicone Fluids," Conference on Electrical Insulation (National Research Council), Oct. 1956.

Dakin, T. W., "Electrical Insulation Deterioration Treated as a Chemical Rate Phenomenon", *AIEE Transactions,* vol. 67, Part I, 1948, p. 113.

Dakin, T. W., and Works, C. N., "Impulse Strength of Liquid-Impregnated Pressboard," AIEE Paper No. 52-228, 1952.

Devins, J. C., and Sharbaugh, A. H., "The Fundamental Nature of Electrical Breakdown" (an examination of breakdown phenomena in gases), *Electro-Technology,* Feb. 1961, p. 103.

Ganster, W. F., *et al,* "Bibliography on Gaseous Dielectric Phenomena," AIEE S-97, 1957.

Gemant, A., *Liquid Dielectrics,* Wiley, 1933.

Hall, H. C., and Skipper, D. J., "The Impulse Strength of Lapped Impregnated Paper Dielectric," *Proceedings of the IEE* (London), Paper No. 2025S, 1956, pp. 571 and 589.

Hill, C. F., "Temperature Limits Set by Oil and Cellulose Insulation," *AIEE Transactions,* vol. 58, 1939, p. 484.

Jones, H. F., "Impulse Voltage Strength of Oil-Immersed Insulation," *Electrical Energy,* Dec. 1956, p. 102.

Kilham, L. F., Jr., *et al,* "Fluorochemical Vapor-Cooling Techniques in Electronic Equipment", *Electrical Manufacturing,* Aug. 1959, p. 88.

Kilham, L. F., Jr., "Transformer Miniaturization using Fluorochemical Liquids and Conduction Techniques," *Proceedings of the IRE,* vol. 44, 1956, p. 155.

Loeb, L. B., *Fundamental Processes of Electrical Discharge in Gases,* Wiley, 1939.

Olyphant, M., Jr., and Brice, T. J., "Dielectric and Coolant Studies of Inert Fluorochemical Liquids," Special AIEE Paper, 1957.

Millek, J. T., *Fluorochemical Gases—Data Sheets,* Electronic Properties Information Center, Hughes Aircraft Co., Culver City, Calif., Report #DS-142, Nov. 1964. Also, *Sulfur Hexafluoride—Data Sheets,* Report #DS-140, Oct. 1964.

Peek, F. W., *Dielectric Phenomena in High-Voltage Engineering,* McGraw-Hill, 1929.

Petley, J. E., *Evaluation of Silicone-Fluid-Impregnated Capacitors,* General Electric Co., TIS Report 58 CT 117, May 1958.

Sommerman, C.M.L., *et al,* "Impulse Ionization and Breakdown in Liquid Dielectrics," *AIEE Communications and Electronics,* May 1954, p. 147.

Warner, A. J., "A New Chlorinated Liquid Dielectric," *Electrical Engineering,* vol. 72, Feb. 1953, p. 68.

APPENDIX A

Design Aids and Guidelines

Potting and Encapsulating Applications

1. Many environmental, electrical, and mechanical problems are solved by the use of encapsulants and potting compounds. Whenever an unusually tough situation presents itself, and there are no other less expensive ways to solve the problem, these compounds may provide a very effective solution. A typical example is the use of encapsulated or potted construction for protection against all combinations of very severe environmental exposure such as radiation and extreme temperature.

2. Certain resin compounds use catatalysts and curing agents. These materials include the epoxies, polyesters, polysulfides, silicone rubbers, furanes, isocyanates, phenol formaldehydes, vinyls, polysyrenes, and the plasticols. On the other hand, melts of asphalt, waxes, polyamides, polyethylene, and polyisobutylene can be used without curing agents and can be injection molded.

3. Some compounds can be foamed in place. In this group can be included the phenol formaldehydes, isocyanates, silicones, epoxies, polyesters, and polystyrenes. Foams are useful where flexibility is needed, where light weight may be a factor, or where some special protective function is involved, such as the exclusion of vacuum of outer space from given devices. Foams, however, do not offer the best dielectric strength.

4. Encapsulants should be chosen by reference to the use. Devices in outdoor locations need nonhydrolizable resins; those for high frequency use need the low-loss polyethylene or polystyrene while power-frequency devices can be potted in epoxy or polyester resins.

5. Temperatures of use are important. In the low to 80 °C region asphalts, waxes, and the like, may be acceptable. At higher temperatures the thermosetting phenol formaldehyde resin may be useful. And for very high temperatures, silicone rubber might be the only type available.

6. Resiliency at high temperatures is difficult to obtain unless silicone rubber is used.

7. For good heat conductivity with low thermal expansion, fillers are employed. This approach is needed where internal heating is necessary and where metallic leads protrude through the encapsulant.

8. Special environmental or operating conditions involve the selection of a potting or encapsulating compound with the capability of resisting the effects of such conditions. For example, where mechanical stress is encountered, tough resins such as polyethylene or phenol formaldehyde may be indicated. Where arc and tract resistance under given weather conditions have to be considered, an effective filler such as aluminum hydrate is recommended for the compound. Operating conditions where solvent action or corrosive atmospheres may be expected call for potting or encapsulating materials with a high enough resistance to such environments.

9. Encapsulation offers an inexpensive method for obtaining a hermetic seal and shock resistant medium. This method, however, demands complete compatibil-

317

ity of components including the curing agent. Some amine curing agents are deleterious to magnet wire enamels, particularly to the hydrolizable group.

10. Repair of cured resin units is not feasible whereas hot-melt potted units may be repaired and repotted.

11. Large blocks of resin must be cured slowly, and it is often desirable to use glass cloth filler for reducing the thickness of resin at any one point.

12. A wide range of heat-distortion points is available for encapsulating and potting compounds. Typical values include 85°C for polystyrene, 165° to 270°C for polyesters, 158° to 260°C for epoxies, and 150°C for silicones. Actual values are a function of compound constituents, such as hardeners and fillers, as well as the type of basic resin.

Rotating Machinery and Transformers

1. The design engineer must be well aware of the various methods of securing windings within a motor or transformer. Such simple means as tie cords and slot wedges in motors are augmented by several dips and bakes of a good bond strength varnish. In rotors, special banding techniques have been devised.

2. Wedging and blocking of closely packed pancake coils assure stability in transformers.

3. If added stability is needed, the complete potting of motor stators or rotors and smaller transformers has produced remarkable mechanical strength. En-capsulation techniques (in which less resin is used than in potting to cover the windings) have also been applied with success.

4. A process described as "trickle varnishing" has been developed by some manu-facturers as an inexpensive method of securing windings in rotors or stators. In essense, this method calls for the application of a 100-percent solids resin (in unpolymerized liquid form) in a drip action.

5. Although the impregnation of windings with varnish is an important method, it is often poorly accomplished.

6. Wet-winding techniques (application of impregnating varnish during winding) can achieve better "fill" than post-winding impregnation.

7. Vacuum impregnation of 100-percent solids resin is sometimes employed for completely potting windings.

8. It is always desirable to dry out windings prior to treatment since trapped moisture at operating temperatures might cause serious ruptures in the insula-tion.

9. Magnet wire with thermoplastic cement outer coating can be used to design various coils that can be cemented into place by application of heat.

10. Insulation failure often begins with mechanical failure. The evidence, however, is often destroyed after a power failure. However, sufficient data are in hand to show that a mechanically sound system will outlast a poorly constructed one, insulation itself being of equal quality. Vibration of windings in itself is often sufficient to abrade wire insulation to the point of failure. Initial weak spots put in by hammer blow during fabrication may eventually cause failure. High-speed winding machines with worn parts may impart scrapes and elongated sections that become focal points for further deterioration. Short-circuit condi-tions in transformers cause pressure on wire insulation simultaneously with

flash heating, with the possibility of mechanical cut-through in the insulation. In high-speed armatures centrifugal force may cause failure through loss of bond strength or by cut-through. IEEE No. 57 test procedure is set up to detect mechanical cracking in adjacent wires by means of an overvoltage procedure. End of life is recorded when two cracks line up sufficiently to produce air breakdown at 1,000 volts.

11. Areas of mechanical weakness occur not only because of external causes or operating conditions but may also be inherent in the weak structure of the insulation material, such as occurs, for example, from internal voids. Weak sites of this nature may lead to internal corona and shortened life.

APPENDIX B
Graphical Analysis of Statistical Data

Since much data-taking in regard to properties of insulations shows a spread of values, it is of great interest to express the variation in terms of recognized statistical quantities.[1]

Statisticians regard the average value or mean value as the center of a cluster of values extending outward positively and negatively in discrete zones based on the standard deviation and assuming a more or less uniform pattern of distribution. The larger the "population," the closer to a dependable distribution curve one obtains.

If the distribution is a normal frequency one, a graphical analysis can be made in a very simple procedure.

Table B.1 shows the spread of 25 breakdown values on film-coated magnet wire as an example. These data plotted on a frequency distribution curve (Fig. B.1) show a close fit to a normal distribution. In a different way, the cumulative frequency curve results in an ogive type (Fig. B.2).

If the outer frequency ends are stretched as shown in Fig. B.3 (on probability paper), the data plot appears as nearly a straight line. All statistical data can be learned from this plot.

The mean value is at 50 percent, and the standard deviation, σ, is at 84.13 percent minus the mean value, which can be expressed as a percentage for the coefficient of dispersion. The range, R, is the value at three times σ (99.87 percent) minus the mean value. Both σ and R are expressed as $\pm$ values as shown in Fig. B.3. N represents the number of observations.

The procedure is as follows:

1. Arrange the data in ascending order.

2. Divide 100 percent by N to obtain the incremental percentage increase per datum.

3. Divide the increment by 2 for the first percentage assigned to the lowest datum. Example: For $N = 25$, the increment is 4 percent and the starting value is 2 percent, as shown.

4. Assign ascending incremental percentages to the data in sequence.

5. Plot the data versus percentage values on probability graph paper, leaving room for 3σ values at 99.87 percent or at 0.13 percent.

6. Determine the mean, $\overline{X}$, at 50 percent, using best straight line.

7. Determine the standard deviation at 84.13 percent (σ).

8. Determine the coefficient of dispersion by dividing the value for σ by the mean $\overline{X}$.

9. Determine the range at 99.87 percent $= 3\sigma$.

---

[1] See Moses, G. L., *Electrical Insulation*, McGraw-Hill, 1951, p. 11, and Worthing, A. G. and Geffner, J., *Treatment of Experimental Data*, Wiley, 1948.

Table B.1   Breakdown Data for Magnet Wire

kV	Percent	kV	Percent
4.2	2	8.2	54
5.0	6	8.2	58
5.8	10	8.2	62
6.1	14	8.8	66
6.5	18	9.0	70
6.9	22	9.0	74
6.9	26	9.0	78
6.9	30	9.7	82
7.6	34	9.7	86
7.6	38	10.1	90
7.6	42	10.6	94
7.6	46	11.3	98
8.2	50		

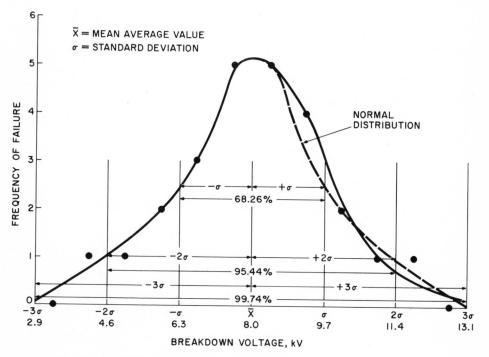

Fig. B-1   Frequency-distribution curve for breakdown data for film-coated magnet wire.

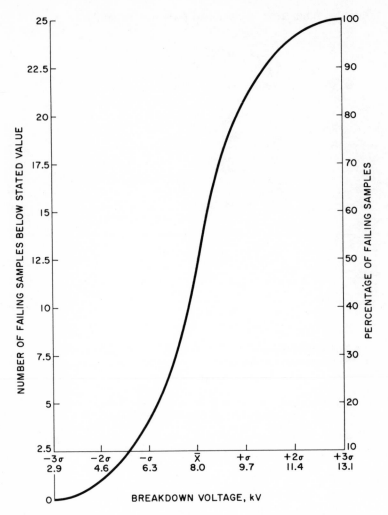

Fig. B-2   Galton's ogive curve for distribution of breakdown data for film-coated magnet wire.

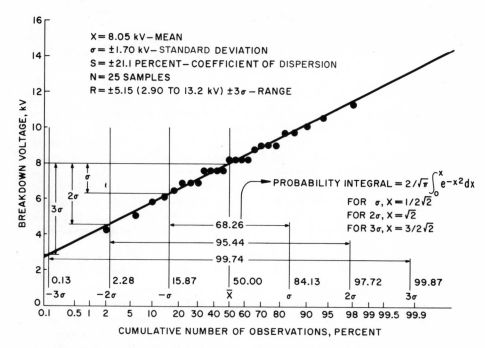

Fig. B-3   Cumulative probability of dielectric failure for film-coated magnet wire.

APPENDIX C

Test Procedures, Standards, and Classification Systems

Influence of the IEEE

The work of T. W. Dakin in 1948 led to a general acceptance of the theory that the aging process in insulating materials can be treated as a chemical rate reaction.[1] Under stimulus of this theory, many members of the American Institute of Electrical Engineers (AIEE), which later became one of the parent organizations of the present Institute of Electrical and Electronics Engineers (IEEE), formed working groups to develop the details of test procedures for thermal evaluation of insulating materials and systems. Without doubt, the role of the AIEE and now IEEE has been a major factor in keeping the field of insulation investigation alive and in bringing *engineering judgment* to bear on what was formerly no more than an "art."

Among the test procedures developed were AIEE No. 57 (now IEEE No. 57, and also ASTM 2307) for magnet wire, AIEE No. 510 (now IEEE No. 117) for random-wound motorette evaluation, IEEE No. 259 for specialty transformers, IEEE No. 275 for formwound statorette evaluation, and IEEE No. 501 for d-c machines.[2] IEEE continues to work on new or modified test codes for varnishes, solid insulations, gases, dry-type transformers, small motors, sealed tubes, hermetic systems, encapsulated coils, and motor armatures. As noted later, the IEEE works closely with the ASTM in many test-procedure projects.

Insulation systems that combine inorganic as well as organic insulations have been difficult to study because of indecision on test end points. Progress of such systems projects has therefore been relatively slow-moving. Also, it has been difficult to evaluate and set up standards for materials such as insulating varnishes that are never used as discrete materials but invariably go through aging in combination with other materials.

In general, there are many pitfalls in evaluating discrete materials that are parts of systems. It is found that the same material used in different systems and for different applications may not always yield the same test results for life expectancy. Such behavior may be caused by an environmental influence, the effect of some incompatibility, or some electrochemical or electromechanical imbalance or variance.

Therefore, material classification is made for comparison purposes only, with the precaution that design engineers must be aware of the deleterious effects of special parameters that may exist in some apparatus but may be entirely absent in others. Usually the ratings mean "for use under optimum conditions" where heat may be the only aging factor present. Such factors as thermal conductivity, vibration, moisture, radiation, corona contamination, overload, and mechanical forces are assumed to be controlled. If not, the design engineer would be wise to derate the materials accordingly.

---

[1] Dakin, T. W., "Electrical Insulation Deterioration Treated as a Chemical Rate Phenomenon," *AIEE Transactions*, vol. 67, Part I, 1948, p. 113.

[2] Goba, F. A., "Bibliography on Thermal Aging of Electrical Insulation," *IEEE Transactions on Electrical Insulation*, June 1969, vol. EI-4, No. 2, p. 31.

Systems tests are somewhat different in that these are designed to cover all the constituents of the system at the functional level of expected performances. It is true that, realistically, not all the expected environmental effects may be present and other effects may be abnormally prominent. Such factors should be considered. For example, test methods have been devised that employ hermets (hermetic motorettes) to bring within the compass of systems tests the effects of the refrigerant since the latter may appreciably alter life of the device.

Classification systems, of course, also aid the supplier of materials and components to assign thermal ratings to his products and use such information for technical promotion and sales.

For the design engineer the classification systems point the way to selection of materials and components that should yield desired design objectives. Systems tests should be run, as a regular procedure, to prove the functional usability of each selected component.

On the above-described basis, the IEEE has set up the following temperature-limited classifications for materials systems, which are used as well for classifying discrete materials:

Class O	90° C	Unimpregnated cellulose
Class A	105° C	Formvar, urethane, impregnated cellulose
Class B	130° C	Epoxy, polyester resin-glass systems, phenolic glass
Class F	155° C	Terephthalate polyester, modified silicone
Class H	180° C	Silicone, silicone glass, esterimide, THEIC polyester
Class K*	200° C	Amide-imide overcoated esterimide, amide-imide overcoated polyester
Class M*	220° C	Amide-imide, polyimide, high-temperature polyamide fibers (Nomex)
Class C**	250° C up	Polytetrafluoroethylene, glass, and ceramics

* Not in the original IEEE classification: Class K added by U.S. Navy, Class M by industry.

** See discussion in latter section of the significance of Class C.

Influence of ASTM

The American Society for Testing and Materials (ASTM) has been a continuing influence on general developments in electrical insulation test procedures. Many of the ASTM workers in this field are also active in the IEEE and, consequently, full cooperation exists between the two organizations.

As noted above, ASTM has issued test procedure D2307 (based on IEEE No. 57) which has been accepted nationally as the latest procedure to be followed in testing enameled magnet wire for thermal endurance. ASTM has also produced, among other standards, D1830 for coated insulating fabrics and D1932 for flexible insulating varnishes.

Influence of Government Agencies

The insulation of electrical apparatus and electronic devices has long been of major concern to the armed services. The problem of thermal stability, in particular, seems to have been pursued most consistently by the naval branch. The Naval Ship Engineering Center (NAVSEC), its predecessor, the Bureau of Ships, and the two Navy laboratories (Naval Research Laboratory and Marine Engineering Laboratory) have continued to give valuable aid to the development of thermal endurance studies.

Many of the personnel in NAVSEC, NRL, and MEL have served and are serving on committees that are developing testing procedures in this vital subject. These individuals and others in the Navy have presented many papers concerning thermal life of both systems and materials. The deep involvement of the Navy in insulation problems is motivated by the fact that shipboard electric systems installations are more demanding than many land-based systems. Therefore, the Navy has favored the "worst condition" tests as the best approach to evaluate systems. An illustration of this approach is the motorette test in which test units are voltage checked while dripping wet with condensed dew.

The other service branches are, of course, also concerned with problems of insulation, and a considerable amount of work is being done by the U.S. Army Electronics Command, the Air Force Materials Laboratory, and the cognizant NASA facilities, with a good deal of emphasis, as may be expected, on electronics systems and space problems. Of the civilian agencies, the National Bureau of Standards has made significant contributions to research in dielectrics.

Of great importance in the advancement of research in dielectrics is the Conference on Electrical Insulation and Dielectric Phenomena sponsored annually by the National Academy of Sciences–National Research Council (NAS–NRC). The *Annual Reports* of the Conference provide abstracts of the papers presented. "Work in progress" is stressed. *The Annual Digest of Literature on Dielectrics* is another outstanding publication of the Conference. It constitutes an invaluable and authoritative review of the field with comprehensive bibliographies.

Because of the great need for mechanical integrity under the worst possible combat conditions, the armed services have devised extremely severe mechanical tests to prove out electrical–electronics systems. Such tests include those for high shock, acceleration, and vibration. The development of such test requirements by the armed services has influenced test procedures by industry. Many manufacturers put their products through these tests not only to meet military specifications, but also to satisfy civilian users who want the ultimate in mechanical strength and stability.

Influence of NEMA and IEC

The National Electrical Manufacturers Association (NEMA) has several groups within its organization concerned with electrical insulation in its various forms—laminates, sleevings and tubing, magnet wire, and so forth. An outstanding example is the work of the Magnet Wire Technical Committee. Among other endeavors, this Committee is interested in thermal stability of the enamelled wire types. NEMA has established two points in the procedure based on log life versus reciprocal absolute temperature. The 20,000-hr mark must be at or above the proposed rating temperature, and the 5,000-hr point must be equal to or greater than 20 degrees above the operating temperature. In this way a fixed limiting slope is demanded of life lines for candidates for each classification level.

NEMA interests are not only national but also international. It currently sponsors participation in the projects of the International Electrotechnical Commission (IEC), with the objective of either reaching agreement with IEC procedures or helping to modify them to meet USA needs. The activities of both NEMA and IEC has a sustaining influence on the activities of IEEE, thus encouraging continuing long-range projects. Such national and international activities help to keep major areas of research active with a continuing influx of new engineers and members of related disciplines, such as chemists, as older workers in the field retire. Most of the work of the IEC in the field of insulation is centered in its Technical Committee 15. Some of its proposals are included in footnote (2).

Influence of Underwriters' Laboratories

Underwriters' Laboratories' (UL) prime concern is with the safety of electrically operated appliances and equipment of all categories, whether for home, commercial, or industrial use. Naturally, the reliable performance of electrical insulation is a major factor in their tests and ratings. By use of methods established by IEEE, UL compares systems and estimates the relative thermal endurance of new products proposed by manufacturers. The use of UL testing methods and ratings provides a broad base of uniform evaluation throughout the country and offers the ultimate consumer a reliable guide to safe equipment. The UL has also been active in work aimed at compatibility of components in given systems. The UL leans on the Navy as well as the IEEE for interpretation of test results.

In the area of magnet wire UL has set up a system of magnet wire substitution groupings for various classification temperatures. Users of magnet wire may contact UL to determine which wires are available from suppliers for specific applications. Suppliers must submit aging data and infrared curves to the UL to establish rating and identification. Samples must be submitted semiannually for continuous monitoring of the products.

Influence of Other Organization

Many other organization than those discussed in the preceding paragraphs have interest in insulation. The American Ceramic Society and the Electrochemical Society, for example, represent professional societies with groups working actively in areas of insulation of particular interest to them. Trade groups such as the Society of the Plastics Industry and the Institute of Printed Circuits have similar specialized interests. A selected list of organizations is given at the conclusion of this appendix.

The Significance of Class C Thermal Classification

Thermal endurance studies based on the Arrhenius equation (see discussion in Chap. 4) apply strictly to discrete organic insulation and to organic constituents of composite insulation. More recently, an entirely new field of *all*-inorganic insulation has opened up as a result of studies of materials for service under conditions of intense radiation, high heat, or both in combination.

Such insulation systems do not show degradation attributable to heat alone. Other influences such as vibration, melting, neutron capture, dielectric breakdown, and corona bombardment may cause failure in inorganic systems, but these influences are only indirectly linked with the thermal life.

The class of materials and materials systems that exists from 250°C up to and above red heat (600°C) is known as Class C. Mica, glass, ceramics, carbides, borides, oxides, asbestos, minerals, and inorganic polymers may be included in this class.

Polytetrafluoroethylene (TFE), although usually placed in Class C, is in fact an organic insulation because of its carbon molecular "backbone." In some respects TFE has Class C thermal stability, but it lacks comparable mechanical stability and radiation resistance.

Not all Class C materials, however, are radiation resistant to high levels. Borosilicate glasses and other boron compounds are only mildly radiation-resistant. The low resistance in the boron atom is due to its susceptibility to capturing neutrons and overheating. Cracking and mechanical powdering can result from a high neutron flux.

Some inorganics become semiconductors at lower temperatures than others. These materials are those with ionic impurities found in poorly formed alumina, in some natural minerals, and in metallic ion inclusions such as iron, copper, and silicon. Alkali metal ions, such as sodium, potassium, lithium, rubidium, and cesium, impart conduction to glass at high temperatures. However, all Class C insulations finally become semiconductors as the temperature is increased to very high levels.

As noted earlier in this book, except for a few materials, there is a correlation between thermal endurance and radiation resistance. Beyond the range of organic materials, however, it is difficult to determine thermal maxima, and only the extremely stable oxides such as the alumina and magnesia show evidence of resisting *both* the extremes of heat and radiation.

Inorganics, such as mica, contain water of crystallization that is released at high temperatures. The use of mica, therefore, is limited to temperatures below red heat.

APPENDIX D

Selected List of Organizations Involved in Insulation

American Ceramic Society, 4055 N. High St., Columbus, Ohio 43214.

American Chemical Society (ACS), 1155 Sixteenth St., N.W., Washington, D.C. 20036.

American National Standards Institute (ANSI), 1430 Broadway, New York, N.Y. 10018.

American Society for Testing and Materials (ASTM), 1916 Race St., Philadelphia, Penna. 19103

CIGRE (International Conference on Large Electric Systems), 112 Boulevard Haussman, Paris, France.

Conference on Electrical Insulation and Dielectric Phenomena, National Academy of Sciences-National Research Council (NAS-NRC), 2101 Constitution Ave., N.W., Washington, D.C. 20037

Electrical/Electonics Insulation Conference (co-sponsored by IEEE and NEMA), 155 East 44th St., New York, N.Y. 10017

Electrochemical Society, Inc., 30 E. 42 St., New York, N.Y. 10017

Institute of Electrical and Electronics Engineers (IEEE), United Engineering Center, 345 E. 47 St., New York, N.Y. 10017

Institute of Printed Circuits (IPC), 3525 W. Peterson Road, Chicago, Ill. 60645.

Insulated Power Cable Engineers Association (IPCEA), Montclair, N.J. 07042.

International Electrotechnical Commission (IEC), Geneva, Switzerland.

National Electrical Manufacturers Association (NEMA), 155 E. 44 St., New York, N.Y. 10017.

Society of Plastics Engineers (SPE), 656 W. Putnum Ave., Greenwich, Conn., 06830.

Society of the Plastics Industry (SPI), 250 Park Ave., New York 10017.

Underwriters' Laboratories (for systems insulation aging and magnet wire testing) 333 Pfingsten Rd., Northbrook, Ill. 60062.

Note: National professional society, trade, and government bodies, such as those of England, France, Germany, Japan, and Switzerland, are active in the field of electrical insulation. Names and addresses of cognizant organizations are available from respective consular offices, directories, and the like, and also from American National Standards Institute, listed above.

INDEX

INDEX